CHAPMAN & HALL/CRC
APPLIED ENVIRONMENTAL STATISTICS

SAMPLING TECHNIQUES FOR FOREST INVENTORIES

CHAPMAN & HALL/CRC
APPLIED ENVIRONMENTAL STATISTICS

Series Editor
Richard Smith, Ph.D.

Published Titles

Timothy G. Gregoire and Harry T. Valentine, Sampling Strategies for Natural Resources and the Environment

Steven P. Millard and Nagaraj K. Neerchal, Environmental Statistics with S Plus

Michael E. Ginevan and Douglas E. Splitstone, Statistical Tools for Environmental Quality

Daniel Mandallaz, Sampling Techniques for Forest Inventory

Forthcoming Titles

Thomas C. Edwards and Richard R. Cutler, Analysis of Ecological Data Using R

Bryan F. J. Manly, Statistics for Environmental Science and Management, 2nd Edition

Song S. Qian, Environmental and Ecological Statistics with R

CHAPMAN & HALL/CRC
APPLIED ENVIRONMENTAL STATISTICS

SAMPLING TECHNIQUES FOR FOREST INVENTORIES

DANIEL MANDALLAZ

ETH ZURICH, DEPARTMENT OF ENVIRONMENTAL SCIENCES

CRC Press
Taylor & Francis Group
Boca Raton London New York

CRC Press is an imprint of the
Taylor & Francis Group, an **informa** business

A CHAPMAN & HALL BOOK

CRC Press
Taylor & Francis Group
6000 Broken Sound Parkway NW, Suite 300
Boca Raton, FL 33487-2742

First issued in paperback 2019

ISBN-13: 978-1-58488-976-2 (hbk)
ISBN-13: 978-0-367-38843-0 (pbk)

Library of Congress Cataloging-in-Publication Data

Mandallaz, Daniel.
    Sampling techniques for forest inventories / Daniel Mandallaz.
        p. cm. -- (Applied environmental statistics ; 4)
    Includes bibliographical references and index.
    ISBN 978-1-58488-976-2 (alk. paper)
    1. Forest surveys. I. Title. II. Series.

SD387.S86M36 2007
634.9'285--dc22   2007028682

Visit the Taylor & Francis Web site at
http://www.taylorandfrancis.com

and the CRC Press Web site at
http://www.crcpress.com

CV 08 14 2020 1408

A Léa Marine

Le hasard fait bien les choses

French adage and statistical paradigm

Contents

Preface

Inventories are the bases for forest management planning, with the goal being the optimal utilization of resources under given constraints. To accomplish this, managers must collect, summarize, and interpret information – that is perform statistical work. The development and improvement of forest management practices, which began toward the end of the Middle Ages, have strongly depended on the parallel evolution of inventory techniques and statistical methodology, in particular sampling schemes. Without these, current forest inventories would be impossible to conduct.

Over the last 80 years, the number of techniques, the demand for more and better information, and finally the mere complexity of their incumbent investigations seem to have grown exponentially. Furthermore, the increased importance of related problems in landscape research and ecology (keywords e.g. biomass, carbon sequestration, and bio-diversity) as well as their interactions with the sociological and economic environment have required specialized procedures for data collection and statistical inference. However, their accompanying economic constraints have necessitated cost-efficient approaches in performing all of these tasks.

The objective of this textbook is to provide graduate students and professionals with the up-to-date statistical concepts and tools needed to conduct a modern forest inventory. **This exposition is as general and concise as possible.** Emphasis has been placed deliberately on the mathematical-statistical features of forest sampling to assess classical dendrometrical quantities. It is assumed that the reader has a sufficient understanding of elementary probability theory, statistics, and linear algebra. More precisely, one must be able to calculate unconditional and conditional probabilities and understand the concepts of random variables, distributions, expectations, variances (including their conditional versions as derived and summarized in Appendix B), central limit theorem and confidence intervals, as well as utilize the least-squares estimation technique in linear models (using matrix notation). The standard notation of naive set theory (e.g. $A \cup B$, $A \cap B$, $A \setminus B$, $A \subset B$, $A \supset B$, $x \in A \ni x$, $x \notin A$) is presented throughout. Likewise, the reader will ideally have some prior knowledge of the general economic-political background of forest inventories and aspects of mensuration (e.g. the handling of instruments), plus skills in remote sensing and geographical information systems (GIS). MSc and PhD students in Forestry, and particularly in Forest Management, will almost

be useful to experienced forest biometricians who wish to become rapidly acquainted with a modern approach to sampling theory for inventories, as well as some recent developments not yet available in book form.

The fundamental concepts and techniques, as used primarily in sociological and economics studies, are presented in chapter 2 and can be summarized as **design-based survey sampling and inference for finite populations** e.g. geographical areas, enterprises, households, farms, employees or students), usually so large that a full survey (census) is neither feasible nor even meaningful. **Inclusion probabilities** and the **Horvitz-Thompson estimator** form the cornerstone of this chapter and are also essential to a forest inventory. More advanced topics are addressed in chapter 3. Excellent classical works at the intermediate mathematical level include those by Cochran (1977) and Särndal et al. (2003), and in French by Gourieroux (1981) and Tillé (2001). Likewise, Cassel et al. (1977), Chaudhuri and Stenger (1992), Chaudhuri and Vos (1988), and Tillé (2006) describe more complicated mathematical and statistical themes.

Key references (in English) for sampling theory in forest inventories are from de Vries (1986) and Schreuder et al. (1993). Those compiled by Kangas and Maltamo (Eds, 2006) and Köhl et al. (2006), the latter containing an extensive bibliography, give broad and up-to-date introductions to this subject, but without proofs of the exhibited statistical techniques. Johnson (2000) provides an elementary and encyclopedic (900 pp!) review of standard procedures, while the writing of Gregoire and Valentine (2007) is an excellent introduction to modern concepts in sampling strategies with interesting chapters on some specific problems. Pardé and Bouchon (1988) and Rondeux (1993), both writing in French, as well as Zöhrer (1980), in German, present basic overviews with emphases on practical work. Unfortunately, none of these authors, except Gregoire and Valentine (2007) in some instances, utilize the so-called **infinite population or Monte Carlo approach** that is much better suited to forest inventories and, in many ways easier to understand. Therefore, this formalism for inventories, within a **design-based** framework, is developed here in chapter 4 (foundations and one-phase sampling schemes) and in chapter 5 (two-phase sampling schemes). It rests upon the concept of **local density** which is essentially an adaptation of the Horvitz-Thompson estimator. These two chapters give a full treatment of **one-phase** and **two-phase** sampling schemes at the point (plot) level, under both **simple random sampling** and **cluster random sampling**, with either **one-stage** or **two-stage** selection procedures at the tree level. These techniques usually suffice for most routine inventories or serve as building blocks for more complex ones. The treatment of cluster-sampling differs markedly from the classical setup, being simpler and easier from both a theoretical and a practical point of view. Simulations performed on a small real forest with full census illustrate the techniques discussed in chapters 4 and 5. Those results are then displayed and critiqued in

its interplay with **model-assisted** techniques (g-weights), as well as **small area estimations** and **analytical studies**, are dealt with in chapter 6. **Geostatistics** and the associated **Kriging** procedures are presented in chapter 7. Using a case study, chapter 8 describes various estimation procedures. Chapter 9 tackles the difficult problem of **optimal design** for forest inventories from a modern point of view relying on the concept of anticipated variance. The resulting optimal schemes are illustrated in chapter 10 with data from the Swiss National Forest Inventory. Chapter 11 outlines the essential facts pertaining to the estimation of growth and change. Finally, chapter 12 provides a short introduction to **transect-sampling** based on the stereological approach. A small number of exercises are also proposed in selected chapters.

It is worth mentioning that the formalism developed in chapters 4, 5 and 6 can be used to estimate the integral of a function over a spatial domain – a key problem in such fields as soil physics, mining or petrology. This is a simpler alternative to the geostatistical techniques developed in chapter 7, which are usually more efficient, particularly for local estimations.

This book is based partly on the writings of C.E. Särndal, as adapted to the context of a forest inventory. In addition, references are made to research, both recent and older, by outstanding forest inventorists, including B. Mat´ and T.G. Gregoire, as well as to the author's own work and lectures at ETH Zurich. Whenever feasible, proofs are given, in contrast to most books on the subject. These occasionally rely on heuristic arguments to minimize the amount of mathematics to a reasonable level of sophistication and spacing. It cannot be overemphasized that readers should not only have a good command of definitions and concepts but also have at least a sufficient understanding of the proofs for the main results.

The scope of this book is restricted when compared to the seemingly unlimited field of applications for sampling techniques within environmental and sociological-economic realms. Nevertheless, the average reader will need time and endurance to master all of the topics covered. Many sections are therefore intended for either further reading or specific applications on an as-needed basis, or they will facilitate one's access to more specialized references. Readers who desire to familiarize themselves quickly with the key aspects of a forest inventory can in a first perusal focus their attention on the following topics: chapters 1, 2 (sections 2.1 to 2.6), 4 and 5, plus a brief glance at the case study in chapter 8. This should suffice for tackling standard estimation problems (without the planning aspects). Courageous readers who persevere through this entire tome should be able to consult all of the current literature on forest sampling (and partly on general survey sampling) and, why not, eventually contribute their own solutions to existing and oncoming challenges?

Acknowledgments

I would like to express my thanks to Professor em. P. Bachmann, former chair of Forest Inventory and Planning at ETH, and to Professor H.R. Heinimann, chair of Land Use Engineering at ETH, for their continuous support as well as for the working environment they have succeeded in creating. Thanks are also due to Professor H.R. Künsch, Department of Mathematics at ETH, for scrutinizing some mathematical aspects as well as to Dr. A. Lanz and E. Kaufmann from the FSL Institute in Birmensdorf, to Dr. R. Ye from the Chinese Academy of Forest Sciences in Beijing and to H.P. Caprez at ETH, all for their technical support, and to Priscilla Licht for editing the manuscript. Last but not least, I thank Professor T.G. Gregoire, Yale University, for encouraging me to write this book, and the publisher for his assistance and patience.

CHAPTER 1

Introduction and terminology

We now proceed to define the terminology and notation that will be used throughout this work. A particular population $\mathcal{P}$ of N individuals (sometimes also called elements or units) $\{u_1, u_2, \ldots u_n\}$ are identified by their labels $i = 1, 2, \ldots N$. $\mathcal{P}$ may consist of all the students at ETH, of all the trees of Switzerland (where, in this case, N is unknown), of all the employees older than 18 years on August 1st 2007 in Switzerland. In the set theoretical sense **it must be clear whether something belongs to the population or not**. Surprisingly, this seemingly simple requirement can be the source of great problems in applications (what is a tree, an unemployed person, etc. ?). Defining the population under study is a key task at the planning stage, often requiring intensive discussions and frustrating compromises, a matter we shall not discuss any further in this book. For each individual i in $\mathcal{P}$ one is interested in p response variables with numerical values $Y_i^{(m)}, m = 1, \ldots p, i = 1 \ldots N$ which can be measured at a given time point in an error-free manner. Note that any qualitative variable can always be coded numerically with a set of 0 indicator variables. Whenever ambiguity is excluded we shall drop the upper index that identifies the response variable. An error-free assumption can be problematic even when dealing with physical quantities (e.g. the volume of a tree) and can also be a source of great difficulties in the case of non-response during interviews. Usually the quantities of primary interest are population totals, means and variances. That is

$$Y^{(m)} = \sum_{i=1}^{N} Y_i^{(m)} \tag{1.1}$$

$$\bar{Y}^{(m)} = \frac{Y^{(m)}}{N} \tag{1.2}$$

$$S^2_{Y^{(m)}} = \frac{\sum_{i=1}^{N}(Y_i^{(m)} - \bar{Y}^{(m)})^2}{N-1} \tag{1.3}$$

Sometimes, more complicated statistical characteristics of the population are needed, such as ratios, covariances, or correlations

1

$$R_{l,m} = \frac{Y^{(m)}}{Y^{(l)}} \qquad (1.4)$$

$$C_{l,m} = \frac{\sum_{i=1}^{N}(Y_i^{(l)} - \bar{Y}^{(l)})(Y_i^{(m)} - \bar{Y}^{(m)})}{N-1} \qquad (1.5)$$

$$\rho_{l,m} = \frac{C_{l,m}}{\sqrt{S_{Y^{(l)}}^2 \, S_{Y^{(m)}}^2}} \qquad (1.6)$$

In any case, the estimation of totals will be a key issue. In pursuing a forest inventory the spatial mean of additive quantities is frequently more important than the population total. Suppose that a forested area F with a surface area $\lambda(F)$ in ha contains a well-defined population of N trees. Moreover, say that all trees have at least a $12cm$ diameter at $1.3m$ above the ground (diameter at breast height, or DBH) and that the response variables of interest are $Y_i^{(1)} \equiv$ and $Y_i^{(2)} =$volume in m^3. Then the spatial mean $\bar{Y}_s^{(m)} = \frac{Y^{(m)}}{\lambda(F)}$ represents the number of stems per ha ($m = 1$) and the volume per ha ($m = 2$) respectively. Note that N is usually unknown and will have to be estimated via the variable $Y_i^{(1)}$. Likewise, the mean volume per tree can be obtained by estimating the ratio $R_{2,1}$.

In practice N can be very large, making a complete evaluation of the entire population impossible (and usually not even meaningful, not to mention the illusion of almost unlimited resources). For this reason, one must restrict one's investigation to a subset s of the population $\mathcal{P}$, also known as a "sample". An element $u \in s$ is then called a sampling unit. The problem is to draw conclusions for the entire population based solely on the sample s. The next question is how to choose that sample. Essentially two ways are possible: by expert judgement (purposive sampling) or by some random mechanism. The criterion used here is that the sample should be representative of the population (this is of course rather vague because if one considers a sample to be representative, one presumably knows roughly what the population looks like already). It is now widely (but not universally!) accepted that representativeness can be insured only by introducing at least partial random selection. Therefore, in the next chapter we shall define and analyze some of the most important sampling schemes (i.e. procedures for sample selection) as well as the estimation techniques that allow us to make inferences from the sample at hand to the entire population.

CHAPTER 2

Sampling finite populations: the essentials

2.1 Sampling schemes and inclusion probabilities

Here we consider a population $\mathcal{P}$ of N individuals and their associated response variables $Y_i^{(m)}$. A sampling scheme is a procedure that involves one or more random mechanisms to select a subset $s \subset \mathcal{P}$ of the population, i.e. the sample. The set of all possible samples s is denoted by $\mathcal{S}$, which is a subset of the set of all subsets (the power set) of $\mathcal{P}$. A well-known example might be a lottery machine that may choose 6 balls out of 45. In that case the set consists of the $\binom{45}{6}$ potential outcomes, which are all equally possible with a probability $\binom{45}{6}^{-1}$. In a survey one usually needs a **sampling frame**, i.e. a list of all individuals in the population, which are identified by a key in the data base (e.g. the social security number of Swiss residents). In this book the identifying key is an integer number called the **label** and is simply denoted by $i = 1, 2 \ldots N$. Again, a forest inventory is peculiar in that no such list can exist, but this difficulty can be circumvented as we shall see. Using pseudo-random numbers (e.g. generated by a computer program and not by a physical mechanism of some kind) one can draw, in most instances sequentially, the individuals forming the sample. At this point it is not necessary to describe the practical implementation of such schemes; these will be discussed later.

We introduce the **indicator variables** I_i which for each individual informs us whether it belongs to the sample or not:

$$I_i = \begin{cases} 1 & \text{if } i \in s \\ 0 & \text{otherwise} \end{cases} \tag{2.1}$$

The probability that the sample s will be selected is denoted by $p(s)$. We emphasize the fact that in this setup the same individual may be drawn many times (i.e. sampling with replacement) and the number of distinct individuals in that sample also is generally a random variable. The order in which individuals are drawn can be important. Therefore, the set theoretical interpretation of the sample s is not quite appropriate in the general case, see Cassel et al. (1977) for details. The probability that a given individual will be included in

$$\pi_i = \mathbb{P}(I_i = 1) = \mathbb{E}(I_i) = \sum_{s \ni i} p(s) \tag{2.2}$$

The symbols $\mathbb{P}$ and $\mathbb{E}$ denote probability and expectation with respect to the sampling schemes. Note that the inclusion probabilities are not assumed to be constant over all individuals and that the most efficient procedures precisely rest upon unequal π_i. The number $n_s = \sum_{i=1}^{N} I_i$ of **distinct** elements in the sample satisfies the relationships given in the next two theorems

Theorem 2.1.1. *The effective sample size $n = \mathbb{E}n_s$ is given as*

$$\mathbb{E}(n_s) = \mathbb{E}\left(\sum_{i=1}^{N} I_i\right) = \sum_{i=1}^{N} \pi_i = n$$

Calculating the variance requires knowledge of the so-called pair-wise inclusion probabilities defined according to

$$\pi_{ij} = \mathbb{P}(I_i = 1, I_j = 1) = \mathbb{E}(I_i I_j) \tag{2.3}$$

Note that $\pi_{ii} = \pi_i$. The variances and covariances of indicator variables are

$$\mathbb{V}(I_i) = \pi_i(1 - \pi_i) = \Delta_{ii}, \quad \mathbb{COV}(I_i, I_j) = \pi_{ij} - \pi_i \pi_j = \Delta_{ij} \tag{2.4}$$

The following properties are important:

Theorem 2.1.2. *The pair-wise inclusion probabilities satisfy the relationships*

$$\mathbb{E}\left(n_s(n_s - 1)\right) = \sum_{i,j \in \mathcal{P}, i \neq j} \pi_{ij}$$

and

$$\sum_{j \in \mathcal{P}, j \neq i} \pi_{ij} = \pi_i \left(\mathbb{E}(n_s \mid I_i = 1) - 1\right)$$

In particular for a fixed sample size $n_s \equiv n$ one has

$$\sum_{j \in \mathcal{P}, j \neq i} \pi_{ij} = \pi_i(n - 1)$$

The first equality follows by calculating $\mathbb{E}n_s^2$ and using Theorem 2.1.1, the second equality by noting that

$$\mathbb{E}(n_s \mid I_i = 1) = 1 + \sum_{j, j \neq i} \mathbb{E}(I_j \mid I_i = 1) = 1 + \sum_{j, j \neq i} \frac{\pi_{ij}}{\pi_i}$$

2.2 The Horvitz-Thompson estimator

We can now define what is probably the most important estimator used in sampling theory, introduced in 1952 by D.G. Horvitz and D.J Thompson, now simply called the Horvitz-Thompson (HT) estimator or also the

$$\hat{Y}_\pi^{(m)} = \sum_{i \in s} \frac{Y_i^{(m)}}{\pi_i} = \sum_{i \in \mathcal{P}} \frac{I_i Y_i^{(m)}}{\pi_i} \tag{2.5}$$

An estimator $\hat{T}(s)$ is considered **unbiased** for a population quantity θ (mean, total, variance, etc.) if its expected value under the random mechanism that generates the samples s is equal to the true value θ, that is if $\mathbb{E}_s T(s$ $\sum_s p(s)T(s) = \theta$. The HT estimator then yields an unbiased point estimate of the population total as long as $\pi_i > 0$ for all i

$$\mathbb{E}(\hat{Y}_\pi^{(m)}) = Y^{(m)} \tag{2.6}$$

The proof is immediate because according to Eq. 2.2 the $\mathbb{E}(I_i)$ and the cancel each other. Note that $\hat{Y}_\pi^{(m)}$ is a random variable because the indicator variables I_i are random. In this model the response variables $Y_i^{(m)}$ are fixed. This is the so-called **design-based approach**. Under hypothetical repeated sampling we know that the point estimates will be distributed around the true unknown value of the population total in such a way that the expected value of the point estimates is precisely the quantity we want to predict. We can say that, in some sense, the randomization procedure allows us to draw conclusions from the observed values in the available sample and apply them to the unobserved values of the remaining individuals of the population under study. Again, we can drop the upper index (m). To calculate the variance we recall the simple fact that for random variables X_i and real numbers a_i has

$$\mathbb{V}(\sum_i a_i X_i) = \sum_i a_i^2 \mathbb{V}(X_i) + \sum_{i \neq j} a_i a_j \mathbb{COV}(X_i, X_j)$$

Using 2.4 we obtain for the theoretical variance:

Theorem 2.2.1.

$$\mathbb{V}(\hat{Y}_\pi) = \sum_{i=1}^N \frac{Y_i^2(1 - \pi_i)}{\pi_i} + \sum_{i=1,j=1,i\neq j}^N \frac{Y_i Y_j(\pi_{ij} - \pi_i \pi_j)}{\pi_i \pi_j}$$

In practice one also needs an estimate of the variance. To do so let us note that the first sum in 2.2.1 can be predicted by considering the new variable $\frac{Y_i^2(1-\pi_i)}{\pi_i}$, and then estimating it by HT with the π_i^{-1} weights. Likewise, for the second sum, we estimate by HT over the population of all pairs i with the weights π_{ij}^{-1}. Hence, the following is an unbiased point estimate of the theoretical variance, provided that $\pi_{ij} > 0$ for all $i, j \in \mathcal{P}$

Theorem 2.2.2.

$$\hat{\mathbb{V}}(\hat{Y}_\pi) = \sum_{i=1}^N \frac{I_i Y_i^2(1 - \pi_i)}{\pi_i^2} + \sum_{i=1,j=1,i\neq j}^N \frac{I_i I_j Y_i Y_j(\pi_{ij} - \pi_i \pi_j)}{\pi_i \pi_j \pi_{ij}}$$

The condition $\pi_{ij} > 0 \; \forall \; i, j \in \mathcal{P}$ is crucial. Of course $\pi_{ij} > 0 \; \forall \; i, j$

Theorem 2.2.2 can be totally misleading.

We introduce the following notation and terms which occur repeatedly in formulae: $\check{Y}_i = \frac{Y_i}{\pi_i}$ which is called the expanded value. Likewise, we define $\check{\Delta}_{ij} = \frac{\Delta_{ij}}{\pi_{ij}}$. Then one can rewrite

$$\hat{Y}_\pi = \sum_{i \in s} \check{Y}_i = \sum_{i \in \mathcal{P}} I_i \check{Y}_i \tag{2.7}$$

$$\mathbb{V}(\hat{Y}_\pi) = \sum_{i,j \in \mathcal{P}} \check{Y}_i \check{Y}_j \Delta_{ij} \tag{2.8}$$

$$\hat{\mathbb{V}}(\hat{Y}_\pi) = \sum_{i,j \in s} \check{Y}_i \check{Y}_j \check{\Delta}_{ij} \tag{2.9}$$

Although the above general formulae are useful for theoretical considerations, calculating those double sums can be prohibitive in practice. Instead, one usually obtains, as we shall see, computationally simpler expressions for specific sampling schemes.

We say that an estimator $\hat{T}(s)$ of the population parameter θ is **consistent** its expected mean square error $\mathbb{E}_s(\hat{T}(s) - \theta)^2$ tends to zero as the sample size increases. Let us note that the mere concept of asymptotic is rather difficult to define in finite populations. We must consider an increasing sequence of population and samples, i.e. N, $n_s \to \infty$ (for a short introduction see Särndal et al. (2003)). In practice, this means large samples in much greater populations.

To obtain a $1 - \alpha$ confidence interval one can rely for large samples on the central limit theorem:

$$CI_{1-\alpha}(\hat{Y}_\pi) = \left[\hat{Y}_\pi - z_{1-\alpha} \sqrt{\hat{\mathbb{V}}(\hat{Y}_\pi)} , \ \hat{Y}_\pi + z_{1-\alpha} \sqrt{\hat{\mathbb{V}}(\hat{Y}_\pi)} \right] \tag{2.10}$$

where $z_{1-\alpha}$ is the two-sided $1 - \alpha$ quantile of the standard normal distribution. Recall that, for example, $\alpha = 0.05$, i.e. 95 percent confidence intervals, $z_{1-\alpha}$ $1.96 \approx 2$.

Under hypothetical repeated sampling 95 percent of these random intervals will contain the true unknown total. Note that **this does not mean** that the true value has a 95% chance of lying within the confidence interval calculated with the survey data, because the true value is either in or out. Although we do not know which alternative is correct, we have a statistical certainty. Out of the thousands of surveys conducted each year, roughly 95% of them will give a confidence interval containing the true unknown total (if the job has been done properly!) but we will not know for which surveys this holds. This is the classical frequentist interpretation. Of course, other philosophical approaches exist, such as the Bayesian school; which, very roughly speaking, contends that prior to the survey the true value could be anywhere and that the a posteriori probability (given the data) for the true value to be in $CI_{1-\alpha}(Y$ is approximatively $1 - \alpha$.

ulation, one arrives at the following result:

$$\mathbb{V}(\hat{Y}_\pi) = -\frac{1}{2} \sum_{i,j\in\mathcal{P}} \Delta_{ij}(\check{Y}_i - \check{Y}_j)^2 + \sum_{i\in\mathcal{P}} \frac{Y_i^2}{\pi_i}\left(\mathbb{E}(n_s \mid I_i = 1) - \mathbb{E}(n_s)\right) \quad (2.11)$$

A similar but equivalent form of Eq. 2.11 has been described by Ramakrishnan (1975b). Under a fixed sample size the second term vanishes and one obtains the so-called Yates-Grundy formula:

$$\mathbb{V}(\hat{Y}_\pi) = -\frac{1}{2} \sum_{i,j\in\mathcal{P}} \Delta_{ij}(\check{Y}_i - \check{Y}_j)^2$$

$$\hat{\mathbb{V}}(\hat{Y}_\pi) = -\frac{1}{2} \sum_{i,j\in s} \check{\Delta}_{ij}(\check{Y}_i - \check{Y}_j)^2 \quad (2.12)$$

It is worth noting that the above theoretical variance is, under that fixed sample size, the same as in Theorem 2.2.1, even though the point estimates of variances from Theorem 2.2.2 and Eq. 2.12 will generally differ. The Yates-Grundy formula tells us that if the inclusion probabilities π_i are proportional to the response variables Y_i then $\check{Y}_i$ is constant and therefore the variance is zero, thereby making this the ideal sampling scheme! Nevertheless, this is no longer true with random sample sizes. To implement such a sampling scheme would usually require us to know the Y_i for the entire population, which, of course, defeats the point. However, if prior auxiliary information is available in the form of a response variable X_i that is known for all individuals (from, say, a previous census) and if one can expect a strong correlation between the X_i and the Y_i then one should sample with a probability proportional to the X_i. Such an approach is called **Probability Proportional to Size (PPS** sampling. This is intuitively obvious: suppose that you have to estimate the total weight of a population consisting of 5 elephants and 10'000 mice, then you evaluate the elephants and consider the mice less so! We shall later see that this technique is fundamental in optimizing sample surveys. In practice one usually must investigate many response variables with the same survey. It is clear that a sampling scheme efficient for one variable may be inefficient for another one. In other words, one has to choose a design based on priorities while respecting the objectives.

Estimating the population mean is straightforward. If N is known, one simply sets

$$\hat{\bar{Y}} = \frac{\hat{Y}_\pi}{N} \quad (2.13)$$

$$\mathbb{V}(\hat{\bar{Y}}_\pi) = \frac{\mathbb{V}(\hat{Y}_\pi)}{N^2} \quad (2.14)$$

$$\hat{\mathbb{V}}(\hat{\bar{Y}}_\pi) = \frac{\hat{\mathbb{V}}(\hat{Y}_\pi)}{N^2} \quad (2.15)$$

is known (which is almost never the case in a forest inventory), it is rather surprising that one can construct estimates than can be better than that from Eq. 2.13 in some circumstances. The so-called weighted sample mean is such an example, defined as

$$\tilde{Y}_s = \frac{\hat{Y}_\pi}{\hat{N}}$$

where

$$\hat{N} = \sum_{i=1}^{N} \frac{1}{\pi_i}$$

is an estimate of the population size. We shall revisit this point in the section on the estimation of ratios. For now we will consider the most important schemes used in applications and we will examine the previous general results in these particular situations.

2.3 Simple random sampling without replacement

This scheme has constant inclusion probabilities with a fixed sample size One example would be the common lottery machines. This type of sampling is frequently used as a building block for more complicated schemes. Because of Theorem 2.1.1 this implies that $\pi_i \equiv \frac{n}{N}$. All samples s have the same probability of being chosen, i.e.,

$$p(s) = \frac{1}{\binom{N}{n}} = \frac{(N-n)!n!}{N!}$$

Also

$$\pi_i = \sum_{s \ni i} p(s) = \frac{\binom{N-1}{n-1}}{\binom{N}{n}} = \frac{n}{N}$$

This combinatorial argument follows from the fact that if a particular individual i is included in the sample we can choose the remaining $n-1$ in the sample only out of the remaining $N-1$ in the population. The rest is simple algebra. Likewise, one can obtain with $\pi_{ij} = \mathbb{P}(I_i = 1 \mid I_j = 1)\mathbb{P}(I_j = 1)$ the pair-wise inclusion probabilities according to

$$\pi_{ij} = \frac{n(n-1)}{N(N-1)}$$

Tedious but elementary algebra can be applied to the general results from the previous section for this particular case (Note that it is a good exercise to write down the proofs). One then obtains

$$
\hat{Y}_\pi = N\frac{1}{n}\sum_{i\in s}Y_i = N\bar{Y}_s
$$

$$
\mathbb{V}(\hat{Y}_\pi) = N^2(1-\frac{n}{N})\frac{1}{n}\frac{\sum_{i=1}^{N}(Y_i-\bar{Y})^2}{(N-1)}
$$

$$
\hat{\mathbb{V}}(\hat{Y}_\pi) = N^2(1-\frac{n}{N})\frac{1}{n}\frac{\sum_{i\in s}(Y_i-\bar{Y}_s)^2}{(n-1)} \tag{2.16}
$$

Remarks:

- $\bar{Y}_s$ is the ordinary mean of the observations in the sample. It is obviously equal to the unbiased estimate $\hat{Y}_\pi$ from Eq. 2.13. Its theoretical and estimated variances can be obtained from the above equations by dropping N^2.

- With unequal probability sampling the ordinary sample mean does not generally estimate the mean of the population and can, therefore, be totally misleading. This is also true for haphazard sampling where the inclusion probabilities are unknown (e.g. interviews carried out with students sampled in the cafeteria, where heavy coffee drinkers will have a much higher probability of being sampled).

- In elementary textbooks the notation $\frac{1}{n}\sum_{i=1}^{n}Y_i = \bar{Y}_s$ is occasionally used. This is misleading because the labels for individuals in the sample are almost never precisely those of the first n individuals in the population.

- $\frac{n}{N}$ is the so-called sampling fraction. For a census it is 1, in which case we logically obtain zero for the theoretical and estimated variance.

- $S_Y^2 = \frac{\sum_{i=1}^{N}(Y_i-\bar{Y})^2}{(N-1)}$ is the population variance and can be estimated without bias by the sample variance $\frac{\sum_{i\in s}(Y_i-\bar{Y}_s)^2}{(n-1)} = s_Y^2$.

To implement this scheme with a sampling frame such as with a list of the individuals in a file, one can proceed as follows:

- Step 1: Generate for each individual a random variable U_i that is uniformly distributed on the interval $[0,1]$. Most statistical packages provide this facility.

- Step 2: Rank in increasing order the individual according to their U_i values to obtain the sequence $U_{(1)}, U_{(2)}, \dots U_{(N)}$. This gives a random permutation of the initial ordering of the individuals in the list.

- Step 3: Select the first n individuals from that permutated list. These form the sample s.

The above algorithm is easy to perform. In contrast, the implementation of sampling schemes can be difficult with unequal inclusion probabilities and with

fixed sample sizes (e.g.   ), such that calculation of the $_{ij}$
see (Särndal et al., 2003) for details and further references. Therefore, the next
two sections will present simple procedures for conducting unequal probability
sampling with random sample sizes. As we shall see, in a forest inventory,
unequal probability sampling is not too difficult to implement, even though
the number of trees selected will almost always be random during practical
applications.

2.4 Poisson sampling

Given a set of inclusion probabilities π_i the **Poisson sampling** design has a
simple list-sequential implementation. Let $\epsilon_i, \ldots \epsilon_N$ be N independent random
variables distributed uniformly on the interval $[0, 1]$. If $\epsilon_i < \pi_i$ the individual
is selected, otherwise not, which by definition occurs with the required probability π_i. Poisson sampling is a scheme without replacement, that is a selected
individual occurs only once in the sample. The sample size n_s is obviously
random with mean $\mathbb{E}(n_s) = \sum_{i=1}^{N} \pi_i$ and variance $\mathbb{V}(n_s) = \sum_{i=1}^{N} \pi_i (1 - \pi$
Because of the independence of the ϵ_i our pair-wise inclusion probabilities
satisfy $\pi_{ij} = \pi_i \pi_j > 0$ and consequently $\Delta_{ij} = \check{\Delta}_{ij} = 0$. In this special case,
the general results provide the following formulae:

$$\hat{Y}_\pi = \sum_{i \in s} \check{Y}_i \tag{2.17}$$

$$\mathbb{V}(\hat{Y}_\pi) = \sum_{i=1}^{N} \pi_i (1 - \pi_i) \check{Y}_i^2 \tag{2.18}$$

$$\hat{\mathbb{V}}(\hat{Y}_\pi) = \sum_{i \in s} (1 - \pi_i) \check{Y}_i^2 \tag{2.19}$$

The variances $\mathbb{V}(\hat{Y}_\pi)$ and $\hat{\mathbb{V}}(\hat{Y}_\pi)$ can be unduly large because of variability in
the sample sizes. A better, but slightly biased, estimator can be obtained with
model-assisted techniques:

$$\hat{Y}_{po} = N\tilde{Y}_s \tag{2.20}$$

$$\mathbb{V}(\hat{Y}_{po}) \approx \sum_{i \in \mathcal{P}} \frac{(Y_i - \bar{Y})^2}{\pi_i} - NS_Y^2 \tag{2.21}$$

$$\hat{\mathbb{V}}(\hat{Y}_{po}) \approx \left(\frac{N}{\hat{N}}\right)^2 \sum_{i \in s} \frac{(1 - \pi_i)}{\pi_i^2} (Y_i - \tilde{Y}_s)^2 \tag{2.22}$$

where $\tilde{Y}_s = \frac{\hat{Y}_\pi}{\hat{N}}$ with $\hat{N} = \sum_{i \in s} \frac{1}{\pi_i}$. This estimate of the true population mean
$\bar{Y}$ should be used even if N is known.

To implement Poisson sampling with **PPS**, $\pi_i \propto X_i$, and the expected sample
size $n = \mathbb{E}(n_s)$, it suffices to take

$$\pi_i = \frac{nX_i}{\sum_{k \in \mathcal{P}} X_k}$$

It is theoretically possible that for some individuals $\pi_i \geq 1$. All these units, say, $\{i_1, i_2, \ldots i_k\}$ will have to be included in the sample. Then one considers the reduced population $\mathcal{P}^* = \mathcal{P} \setminus \{i_1, i_2 \ldots i_k\}$ and iterates, if necessary, the procedure.

The special case of $\pi_i \equiv \pi = \frac{n}{N}$ (i.e. $X_i \equiv 1$) is called **Bernoulli sampling** There, we would see that $\hat{Y}_{po} = \frac{n}{n_s}\hat{Y}_\pi$ and that $\mathbb{V}(\hat{Y}_{po})$ is very nearly the same as would be found for simple random sampling with a fixed size In contrast, the variance of the unmodified HT estimator is usually much larger. These examples demonstrate that a good strategy must consider both sampling schemes and estimators.

The next section presents a sampling scheme and an estimator that attempt to combine **PPS** and simplicity.

2.5 Unequal probability sampling with replacement

We now consider a population $\mathcal{P}$ with response variables Y_i and an auxiliary variable X_i known for all $i \in \{1, 2 \ldots N\}$. The sampling frame is the list of all individuals ordered, without loss of generality, according to their labels i's. We can then define the cumulative sums $S_0 = 0, S_1 = X_1, S_k = S_{k-1} + X_k, k$ $2, 3 \ldots N$. Note that $S_N = \sum_{k=1}^{N} X_k$.

The sampling procedure consists of n, fixed, consecutive identically but independently distributed draws of points Z_l, $l = 1, 2 \ldots n$ that are uniformly distributed on the interval $[0, S_N]$ (i.e. $Z_l \sim S_N \times U[0, 1]$, with $U[0, 1]$ being a uniformly distributed random variable on the interval $[0, 1]$). The individual labeled i is selected at the l-th draw if $S_{i-1} \leq Z_l < S_i$. This obviously occurs with probability $p_i = \frac{X_i}{S_N}$. Note that by construction $\sum_{i=1}^{N} p_i = 1$. The number of times $T_i \in \{0, 1, \ldots n\}$ a given individual i is included in the sample follows therefore a binomial distribution with parameter n (number of draws) and p_i (probability of success). This is a sampling procedure with replacement because the same individual can be selected more than once (maximum of times). The following facts are well known from elementary probability theory (although it is always a good exercise to prove them from scratch).

- The random vector $T_1, T_2, \ldots T_N$ follows a multinomial distribution with parameter $p_1, p_2, \ldots p_N$.
- $\mathbb{E}(T_i) = np_i$ and $\mathbb{V}(T_i) = np_i(1 - p_i)$
- Given $T_i = t_i$, T_j follows a binomial distribution with parameter $n - t_i$ $\frac{p_j}{1 - p_i}$
- $\mathbb{E}(T_i T_j) = \mathbb{E}(T_i \mathbb{E}(T_j \mid T_i)) = n(n-1)p_i p_j$

$$(\quad _j) = \qquad _j$$

The so-called pwr estimator (p-expanded with replacement, due to Hansen and Hurwitz, 1943) is defined as

$$\hat{Y}_{pwr} = \frac{1}{n} \sum_{i=1}^{N} \frac{T_i Y_i}{p_i} = \frac{1}{n} \sum_{i \in s} \frac{T_i Y_i}{p_i} \tag{2.23}$$

where the sample is described as $s = \{i \in \mathcal{P} \mid T_i \neq 0\}$. Using the afore-mentioned facts it is easy to show that the pwr estimator yields an unbiased estimate of the population total and that its variance is given by

$$\mathbb{V}(\hat{Y}_{pwr}) = \frac{1}{n} \left(\sum_{i=1}^{N} \frac{Y_i^2}{p_i} - \sum_{i=1}^{N} Y_i^2 \right) = \frac{1}{n} \sum_{i=1}^{N} \left(\frac{Y_i}{p_i} - Y \right)^2 p_i \tag{2.24}$$

With a little more algebraic work one can obtain the following unbiased estimate of the variance

$$\hat{\mathbb{V}}(\hat{Y}_{pwr}) = \frac{1}{n(n-1)} \sum_{i=1}^{N} T_i \left(\frac{Y_i}{p_i} - \hat{Y}_{pwr} \right)^2 \tag{2.25}$$

The exact implementation of unequal probability sampling schemes without replacement is rather difficult (see Särndal et al. (2003) and Tillé (2006)). A simple approximation can be gained by considering the previous scheme within a Horvitz-Thompson framework when the p_i are all much smaller than 1. Looking at the complementary events $i \notin s$ we arrive at $\pi_i = 1 - (1-p_i)^n \approx np$ by using the first-order Taylor development of $(1-x)^n$ for small x. Recalling that $\mathbb{P}(A) = \mathbb{P}(A) + \mathbb{P}(B) - \mathbb{P}(A \cap B)$ we have $\pi_{ij} = 1 - \mathbb{P}(i \notin s \cup j \notin s) = 1 - (1-p_i)^n - (1-p_j)^n + (1-p_i-p_j)^n$. Then, using a second order Taylor approximation, we learn after performing some algebra that $\pi_{ij} \approx n(n-1)p_i$ Furthermore, the expected number of distinct elements n_s in the sample has an expected value $\mathbb{E}(n_s) = \sum_{i=1}^{N} \pi_i \approx \sum_{i=1}^{N} np_i = n \sum_{i=1}^{N} p_i = n$. Because $n_s \leq n$ we have $n_s \approx n$. Therefore, when all the p_i are small, we have an almost exact **PPS** and Y_{pwr} is very close to the Horvitz-Thompson estimate Y_π. This is the reason for the popularity of the pwr procedure.

2.6 Estimation of ratios

In many applications one must estimate the ratios of two population totals (or, equivalently, of their means). In other words, we need a quantity of the form $R_{l,m} = \frac{Y^{(l)}}{Y^{(m)}}$, usually for many pairs of such variables. For example one may want to estimate the proportion of a spruce infected with a given fungi in a certain area. Neither the total number of spruce trees, nor the number that are infected is known. We shall assume, which is nearly always the case, that the same sampling scheme can be performed for variables involved in the

$$\hat{R}_{l,m} = \frac{\hat{Y}_{\pi}^{(l)}}{\hat{Y}_{\pi}^{(m)}} = \frac{\hat{\bar{Y}}_{\pi}^{(l)}}{\hat{\bar{Y}}_{\pi}^{(m)}} \qquad (2.26)$$

The theoretical and estimated variances of the estimated means are usually of order n^{-1} (n being the expected sample size) so that the expected value of the ratio is equal to the ratio of the expected values also up to an term. Hence, $\hat{R}_{l,m}$ is asymptotically (i.e. for large samples) unbiased. It can be shown (see e.g. Särndal et al., 2003) that the bias ratio satisfies

$$\frac{\mathbb{E}(\hat{R}_{l,m} - R_{l,m})^2}{\mathbb{V}(\hat{R}_{l,m})} \leq \frac{\mathbb{V}(\hat{Y}_{\pi}^{(m)})}{(Y_{\pi}^{(m)})^2}$$

The variance follows from these arguments:

$$\mathbb{V}(\hat{R}_{l,m}) \approx \mathbb{E}\left(\frac{\hat{\bar{Y}}_{\pi}^{(l)}}{\hat{\bar{Y}}_{\pi}^{()}} - \hat{R}_{l,m}\right)^2 \approx \frac{1}{(\bar{Y}^{(m)})^2}\mathbb{E}\left(\hat{\bar{Y}}_{\pi}^{(l)} - R_{l,m}\hat{\bar{Y}}_{\pi}^{(m)}\right)^2$$

Furthermore, $\mathbb{E}\left(\hat{\bar{Y}}_{\pi}^{(l)} - R_{l,m}\hat{\bar{Y}}_{\pi}^{(m)}\right) \approx 0$ as $\hat{R}_{l,m} \approx R_{l,m}$. Hence, one has up to the order of n^{-2}

$$\mathbb{E}\left(\hat{\bar{Y}}_{\pi}^{(l)} - \hat{R}_{l,m}\hat{\bar{Y}}_{\pi}^{(l)}\right)^2 \approx \mathbb{V}\left(\hat{\bar{Y}}_{\pi}^{(l)} - \hat{R}_{l,m}\hat{\bar{Y}}_{\pi}^{(l)}\right)$$

Let us now define the **residual response variable**

$$Z_i = Y_i^{(l)} - R_{l,m}Y_i^{(m)} \approx Y_i^{(l)} - \hat{R}_{l,m}Y_i^{(m)}$$

Note that we need an approximation because Z_i is per se unobservable as we do not know the true ratio. However, we do know that the total and the mean of this new variable are approximately zero. Nevertheless, the trick is to use the Horvitz-Thompson technique to estimate the variance of its predicted zero mean. Putting the pieces together we have the following important result for the estimation of ratios:

$$\hat{R}_{l,m} = \frac{\hat{\bar{Y}}_{\pi}^{(l)}}{\hat{\bar{Y}}_{\pi}^{(m)}} \qquad (2.27)$$

$$\hat{\mathbb{V}}(\hat{R}_{l,m}) = \frac{1}{(\hat{\bar{Y}}^{(m)})^2}\hat{\mathbb{V}}(\hat{\bar{Z}}_{\pi}) \qquad (2.28)$$

Although the above heuristic arguments can be made rigorous under some regularity conditions, the main point is that $\hat{\mathbb{V}}(\hat{R}_{l,m})$ tends to zero as and is correct up to the order n^{-2}. This also shows that the absolute bias $\mathbb{E}(\hat{R}_{l,m} - R_{l,m})$ will generally tend to zero as n^{-1}. The residual technique can obviously be applied to estimators other than HT.

To illustrate the technique let us consider an opinion poll carried out with

population $\mathcal{P}$ consists of all persons who voted at the last election and who have the right to participate in the next voting. We, of course, assume that they all will tell the truth if interviewed! The subpopulation $A \subset \mathcal{P}$ consists of the N_A electors who voted for the political party A at the last election and $B \subset \mathcal{P}$ of the N_B electors who intend to vote for a legislative act B at the next election (we disregard their party affiliation). The variable $Y_i^{(1)}$ is 1 if $i \in A \cap B$, otherwise 0 and the variable $Y_i^{(2)}$ is 1 if $i \in A$ and 0 otherwise. Let us denote by N_A, $N_{A \cap B}$ the corresponding unknown number of electors in and, likewise, by n_A, $n_{A \cap B}$ the number in the sample. $R_{1,2} = \frac{N_{A \cap B}}{N_A}$ is the proportion of electors intending to select B among those who voted for A. In this case, one obtains the intuitive result $\hat{R}_{1,2} = \frac{n_{A \cap B}}{n_A}$. The residual variable $Z_i = Y_i^{(1)} - \hat{R}_{1,2} Y_i^{(2)}$ is then

$$
Z_i = \begin{cases}
1 - \left(\frac{n_{A \cap B}}{n_A} \right) 1 & \text{if } i \in A \cap B \\
0 - \left(\frac{n_{A \cap B}}{n_A} \right) 1 & \text{if } i \in A \text{ and } i \notin B \\
0 - \left(\frac{n_{A \cap B}}{n_A} \right) 0 & \text{if } i \notin A
\end{cases}
$$

There are $n_{A \cap B}$, $n_A - n_{A \cap B}$ and $n - n_A$ individuals in the sample corresponding to the three possible values for Z_i (they form a partition of $\mathcal{P}$). It is easily seen that the sample mean of the Z_i is 0. Applying some algebra will then yield

$$
\hat{V}(\hat{\bar{Z}}) = \frac{n_{A \cap B} \left(1 - \frac{n_{A \cap B}}{n_A} \right)}{n(n-1)}
$$

One finally obtains

$$
\hat{V}(\hat{R}_{1,2}) = \frac{n^2}{n(n-1)} \frac{1}{n_A} \frac{n_{A \cap B}}{n_A} \left(1 - \frac{n_{A \cap B}}{n_A} \right) \approx \frac{1}{n_A} \frac{n_{A \cap B}}{n_A} \left(1 - \frac{n_{A \cap B}}{n_A} \right)
$$

for n large. The second expression is a well-known formula for the variance of the binomial case. However, note that in this example the denominator n_A also random.

Here, we also show that **indicator variables are extremely useful when one wants to estimate a total or mean over sub-populations, also called domains**. For instance, let $\mathcal{P}$ represent the population of trees above $12cm$ DBH and consider a partition of this population according to diameter classes $C_1, C_2, \ldots C_k$, $C_i \subset \mathcal{P}$. When optimizing inventories one has to estimate the average squared volume of the trees in each class. Let Z_i be the square of the volume and define the indicator variables $Y_i^{(k)} = 1$ if $i \in C_k$ and 0 otherwise. Next, set $Z_i^{(k)} = Z_i Y_i^{(k)}$. The average squared volumes in each class are then defined by the ratios $R^{(k)} = \frac{Z^{(k)}}{Y^{(k)}}$.

In a particular case (i.e. $Y_i^{(m)} \equiv 1$) let us consider the estimation of the

$$\tilde{Y}_s = \frac{\sum_{i \in s} \frac{Y_i}{\pi_i}}{\sum_{i \in s} \frac{1}{\pi_i}} = \frac{\hat{Y}_\pi}{\hat{N}} \qquad (2.29)$$

A variance estimator is then

$$\hat{V}(\tilde{Y}_s) = \frac{\sum_{i,j \in s} \check{\Delta}_{ij} \left(\frac{Y_i - \tilde{Y}_s}{\pi_j}\right) \left(\frac{Y_j - \tilde{Y}_s}{\pi_j}\right)}{\hat{N}^2} = \frac{\hat{V}(\hat{Z}_\pi)}{\hat{N}^2} \qquad (2.30)$$

where $Z_i = Y_i - \tilde{Y}_s$.

For some designs, the estimators $\hat{\tilde{Y}}_\pi$ from Eq. 2.13 and $\tilde{Y}_s$ are identical (this is the case for simple random sampling and stratified simple random sampling, which will be defined in the next section). When N is unknown (as in a forest inventory), there is no choice between $\hat{\tilde{Y}}_\pi$ and $\tilde{Y}_s$; only the latter can be used. However, if N is known and the two estimators differ, a choice must be made, despite the estimation of an a priori known quantity N. Although it is hard to pinpoint the exact conditions under which $\tilde{Y}_s$ is preferred, here are some arguments in favor of $\tilde{Y}_s$.

1. If $Y_i \equiv c$ (i.e. the individual values are concentrated around the population mean) then $\tilde{Y}_s = c$ gives the correct result for each sample s compared with $\hat{\tilde{Y}}_\pi = c\frac{\hat{N}}{N}$. Hence, $\tilde{Y}_s$ is a better estimator under random sample size. If the realized sample size n_s happens to be greater than average, both the numerator sum and denominator sum of $\tilde{Y}_s$ will have more terms, and vice versa if n_s is small. The ratio thereby retains a certain stability. By contrast $\hat{\tilde{Y}}_\pi$, whose denominator remains fixed, lacks this stability.

2. Let again be $Y_i \equiv c$ and implement Bernoulli sampling with $\pi_i \equiv \pi$. Then $\tilde{Y}_s = c \ \forall s$, whereas $\hat{\tilde{Y}}_\pi = c\frac{n_s}{N\pi}$ can have a large variability due to the random sample size n_s.

3. $\tilde{Y}_s$ is usually better when the π_i are poorly correlated with the Y_i.

Finally, another alternative estimator of the mean has good properties if N known and the sample size is random. That is

$$\hat{\tilde{Y}}_{alt} = \frac{n}{N n_s} \sum_{i \in s} \frac{Y_i}{\pi_i} = \frac{n}{n_s} \hat{\tilde{Y}}_\pi \qquad (2.31)$$

where n_s is the random sample size with an expected value of $n = \mathbb{E}(n_s$ $\sum_{i=1}^{N} \pi_i$. If the sample size is fixed, then $\hat{\tilde{Y}}_{alt} = \hat{\tilde{Y}}_\pi$ and if all the π_i are equal $(= \frac{n}{N})$, then according to Eq. 2.29, $\hat{\tilde{Y}}_{alt} = \tilde{Y}_s$. The approximately unbiased estimate from 2.31 may have substantially smaller variance than both $\hat{\tilde{Y}}_\pi$ $\tilde{Y}_s$ when 1) the sample size is random, 2) there is considerable variation in the π_i and 3) $\frac{Y_i}{\pi_i}$ is roughly constant (**PPS**). To calculate the variance of $\hat{\tilde{Y}}_{alt}$ note that one can rewrite it as $\hat{\tilde{Y}}_{alt} = \frac{n}{N} \hat{R}_{1,2}$, where $\hat{R}_{1,2} = \frac{\hat{\tilde{Y}}_\pi^{(1)}}{\hat{\tilde{Y}}_\pi^{(2)}}$ with $Y_i^{(1)} = Y_i$

i (,) = $_n$. We can use Eq. 2.27 to finally obtain with $Z_i = Y_i - \hat{R}_{1,2}\pi_i$ the approximate variance estimate:

$$\hat{\mathbb{V}}(\hat{\bar{Y}}_{alt}) = \left(\frac{n}{n_s}\right)^2 \hat{\mathbb{V}}(\hat{\bar{Z}}_\pi) \tag{2.32}$$

2.7 Stratification and post-stratification

In stratified sampling, the population is divided into non-overlapping sub-populations called strata. A probability sample is selected in each stratum. Selections from within the different strata are independent and can be performed through various techniques. Stratified sampling is widely used as a powerful and flexible method. Let us examine a few reasons for its popularity.

- Suppose that estimates of specified precision are wanted for certain sub-populations (domains of study), e.g. coniferous and broadleaves species, or mice and elephants. Each domain of study can be treated as a separate stratum if the domain membership is specified in the frame.

- Practical aspects related to the response, measurement, and auxiliary information may differ considerably from one sub-population to another, so that wish to treat each sub-population as a separate stratum.

- For practical or administrative reasons, the survey organization may have divided its entire territory into several geographic districts. Here, it is natural to let each district be a stratum.

The choice of the stratification variable, the resulting cut-off points, and the number of strata are important aspects (Särndal et al., 2003). In principle a stratification variable should be strongly correlated with the response variable of primary interest. Each stratum should then be as homogeneous as possible and differ markedly from the others, such that variance among strata is larger than the variance within strata (again think of mice and elephants). Often, but not always, the selection of a sampling design as well as the choice of the associated estimator is made uniformly for all strata. Nevertheless, the variety of demands often makes it impossible to arrive at a global optimum for the stratified sampling design.

Let us introduce some notation and definitions. By a **stratification** of a finite population $\mathcal{P}$ with N individuals we mean a partition of $\mathcal{P}$ into sub-populations, called strata, which are denoted by $\mathcal{P}_1, \mathcal{P}_2, \ldots \mathcal{P}_H$ with individuals, $(h = 1, 2 \ldots H)$. We assume that the N_h are known. By **stratified sampling** we mean that a probability sample s_h is chosen from $\mathcal{P}_h$ according to a design $p_h(\cdot)$ $(h = 1 \ldots H)$ and that the selection in one stratum is independent of the selections in all others. The resulting total sample is s $s_1 \cup s_2 \cup \cdots \cup s_h$. Because of this independence $p(s) = p_1(s_1)p_2(s_2) \cdots p_H(s_H$ The totals, means, and variance per strata are then defined as for the entire population:

$$Y_h^{(m)} = \sum_{i \in \mathcal{P}_h} Y_i^{(m)} \tag{2.33}$$

$$\bar{Y}_h^{(m)} = \frac{Y_h^{(m)}}{N_h} \tag{2.34}$$

$$S_{Y^{(m)},h}^2 = \frac{\sum_{i \in \mathcal{P}_h}^{N} (Y_i^{(m)} - \bar{Y}_h^{(m)})^2}{N_h - 1} \tag{2.35}$$

Obviously one has

$$Y^{(m)} = \sum_{h=1}^{H} Y_h^{(m)} = \sum_{h=1}^{H} N_h \bar{Y}_h^{(m)} \tag{2.36}$$

From the above definitions it is apparent that one can combine the estimates in the different strata to obtain an overall unbiased estimate for population. As usual we drop the upper index in the following:

$$\hat{Y}_\pi = \sum_{i=1}^{H} \hat{Y}_{\pi,h} \tag{2.37}$$

$$\mathbb{V}(\hat{Y}_\pi) = \sum_{i=1}^{H} \mathbb{V}(\hat{Y}_{\pi,h}) \tag{2.38}$$

$$\hat{\mathbb{V}}(\hat{Y}_\pi) = \sum_{i=1}^{H} \hat{\mathbb{V}}(\hat{Y}_{\pi,h}) \tag{2.39}$$

Again, to obtain the corresponding formulae for the mean of the population one must divide the above expressions by N and N^2 respectively.

For easier reference we state the result for simple random sampling without replacement of n_h out of N_h individuals in each stratum

$$\hat{Y} = \sum_{h=1}^{H} N_h \bar{Y}_{s,h} \tag{2.40}$$

$$\mathbb{V}(\hat{Y}) = \sum_{h=1}^{H} N_h^2 \frac{1 - f_h}{n_h} S_{Y,h}^2 \tag{2.41}$$

$$\hat{\mathbb{V}}(\hat{Y}) = \sum_{h=1}^{H} N_h^2 \frac{1 - f_h}{n_h} s_{Y,h}^2 \tag{2.42}$$

in which the $f_h = \frac{n_h}{N_h}$ are the sampling fractions and

$$\bar{Y}_{s,h} = \frac{\sum_{i \in s_h} Y_i}{n_h}$$

$$s_{Y,h} \qquad n_h - 1 \sum_{i \in S_h}(Y_i \quad Y_{s,h})$$

are the ordinary sample means and the sample variances in the h-th stratum.

For some problems one does not know the strata membership from the sampling frame, a detail that can be determined only after an individual is selected. This is called post-stratification. The general case is technically rather complex and we shall consider the following setup, which is the most widely used in practice. We consider a population $\mathcal{P}$ of N individuals partitioned into strata $\mathcal{P}_h$ of sizes N_h and apply simple random sampling without replacement of n out of N individuals, such that the overall sampling fraction is $f = \frac{n}{N}$

Once the sample is drawn the strata membership can be determined. Here we would find that n_h individuals belong to the h-th stratum, $h = 1, 2 \ldots H$ We assume therefore that n is large enough to ensure that $n_h > 0 \ \forall h$, i.e. that $(1 - \frac{N_h}{N})^n \approx 0$. Although the strata membership is known only after sampling is completed, **we assume that the relative sizes of the strata** $W_h = \frac{N_h}{N}$ **are known prior to sampling**, at least approximately (which is often the case). If not, one must use a more complicated two-phase procedure for stratification (which will be dealt with in the context of a forest inventory). The **post-stratified** estimate of the population mean is defined as

$$\hat{\bar{Y}}_{post} = \sum_{h=1}^{H} \frac{N_h}{N} \bar{Y}_{s,h} = \sum_{h=1}^{H} W_h \bar{Y}_{s,h} \qquad (2.43)$$

where $\bar{Y}_{s,h}$ is defined as in 2.40. Thus the post-stratified estimate is simply the weighted mean of the estimated strata means. The only but important difference to stratification is that the strata sample sizes n_h are now random variables, though the total $\sum_{h=1}^{H} n_h = n$ is fixed.

Researchers such as Särndal et al. (2003) have previously shown that the variance and estimated variance of the post-stratified estimates can be given by

$$\mathbb{V}(\hat{\bar{Y}}_{post}) = \frac{1-f}{n} \sum_{h=1}^{H} W_h S_{Y,h}^2 + \frac{1-f}{n^2} \sum_{h=1}^{H}(1 - W_h)S_{Y,h}^2 \qquad (2.44)$$

$$\hat{\mathbb{V}}(\hat{\bar{Y}}_{post}) = \frac{1-f}{n} \sum_{h=1}^{H} W_h s_{Y,h}^2 + \frac{1-f}{n^2} \sum_{h=1}^{H}(1 - W_h)s_{Y,h}^2 \qquad (2.45)$$

The first term is the variance of the stratified estimate under proportional allocation (i.e. $\frac{n_h}{N_h} \equiv \frac{n}{N}$). The second term, due to post-stratification, is of order n^{-2}. Hence, in large samples, post-stratification is nearly as good as stratification.

If the survey costs differ markedly among strata one can optimize the allocation of the n_h. If the total survey costs in each stratum is a simple linear

costs for a given required variance of the stratified estimate or conversely. This solution is called the Neymann allocation and is expressed as

$$\frac{n_h}{N_h} \propto \sqrt{\frac{S_h^2}{c_h}} \tag{2.46}$$

Its proof is simple if one knows the Lagrange multiplier rule for maximizing the functions of several variables under given constraints (see e.g. Särndal et al. (2003); Cochran (1977)).

2.8 Two-stage sampling

In many applications it is difficult or even impossible to establish a sampling frame that lists all the individuals, while the population can be naturally partitioned into so-called **primary sampling units, PSU**, for which a sampling frame either exists or can be established at low costs. After a sample of PSU has been drawn (the first-stage sample), one can establish a sampling frame for each PSU in this sample. Then, in each selected PSU, one draws a sample (the second-stage sample) of so-called **secondary sampling units, SSU** which are the elements of the population. For instance, if one wants to assess the performance of high-school students in Zürich, the PSU could be the schools (one take a sample of the schools, for which a list certainly exists) and the SSU would be the classes within each school (each school will have a list of its own classes). Note that the population elements are classes and not students. Likewise, one can divide Switzerland into districts (such a list exists), take a sample of those, and within each chosen district, sample separate farms with probability proportional to their surfaces areas in order to determine the amount of fertilizer used (each district would have a list of its farms organized by sizes). Those farms are then the population elements. In this example, note that the traveling costs will be smaller for a given number of sampled farms than it would be if one had to sample directly from the frame of all farms within Switzerland. This idea can be generalized to more stages: sample of districts, then sample of schools, then sample of classes, and, finally sample of students. In a forest inventory context we shall consider a two-stage sampling procedure in which the volume of the first-stage trees is by definition a simple approximation based on the diameter (DBH) and tree species alone. In contrast, for the second-stage trees the volume is estimated more accurately by taking costly additional measurements of height and diameter at 7m above the ground.

Chapter 3 provides some results for three-stage sampling and two-stage cluster sampling. General multi-stage theory has been previously discussed by Särndal et al. (2003) and Kendall et al. (1983). Let us now introduce the formalism.

The population of elements $\mathcal{P} = \{1, 2, \ldots, k, \ldots N\}$ is partitioned in N_I **PSU**

The whole set of PSUs is represented by $U_I = \{1, 2 \ldots, i, \ldots, N_I\}$. Clearly, $N = \sum_{i \in U_I} N_i$. The sampling procedure is as follows:

1. *First stage*: a sample s_I of PSUs is drawn from U_I according to the design $p_I(\cdot)$.

2. *Second stage*: for every $i \in s_I$, a sample s_i of elements is drawn from $(s_i \subset U_i)$ according to the design $p_i(\cdot \mid s_I)$

The resulting sample of elements, denoted s, is composed as $s = \cup_{i \in s_I}$
We narrow down the class of second-stage designs, by requiring that every time the ith PSU is included in a first-stage sample, the same second-stage sampling design must be used, i.e. $p_i(\cdot \mid s_I) = p_i(\cdot)$ (**invariance of sub-sampling**). Furthermore, we require that sub-sampling in a given PSU is carried out independent of sub-sampling in any other PSU (**independence of sub-sampling**). The number of PSUs in s_I is denoted by n_{s_I}, or simply n_I, if $p_I(\cdot)$ is of a fixed sample size. The number of elements in s_i is denoted n_{s_i}, or simply n_i, if $p_i(\cdot)$ is of a fixed size. The total number of elements in is $n_s = \sum_{i \in s_i} n_{s_i}$.

For the first-stage design $p_i(\cdot)$, the inclusion probabilities are π_{Ii}, π_{Iij}. We set as usual $\Delta_{Iij} = \pi_{Iij} - \pi_{Ii}\pi_{Ij}$ with $\Delta_{Iii} = \pi_{Ii}(1 - \pi_{Ii})$ and $\check{\Delta}_{Iij} = \frac{\Delta}{\pi_{Iij}}$
Likewise, for the second-stage design, we use the notation $\pi_{k|i}$ and $\pi_{kl|i}$
the conditional inclusion probabilities within the ith PSU.

The Δ-quantities are $\Delta_{kl|i} = \pi_{kl|i} - \pi_{k|i}\pi_{l|i}$, $\Delta_{kk|i} = \pi_{k|i}(1 - \pi_{k|i})$ and finally
$\check{\Delta}_{kl|i} = \frac{\Delta_{kl|i}}{\pi_{kl|i}}$.

The unconditional inclusion probabilities are then

$$\pi_k = \pi_{Ii}\pi_{k|i} \text{ if } k \in U_i \tag{2.47}$$

$$\pi_{kl} = \begin{cases} \pi_{Ii}\pi_{k|i} & \text{if } k = l \in U_i \\ \pi_{Ii}\pi_{kl|i} & \text{if } k, l \in U_i \text{ and } k \neq l \\ \pi_{Iij}\pi_{k|i}\pi_{l|j} & \text{if } k \in U_i \text{ and } l \in U_j \text{ and } i \neq j \end{cases} \tag{2.48}$$

One could derive an HT point estimate as well as theoretical and estimated variances from the first principles by applying Theorems 2.2.1, 2.2.2, and Eq. 2.5, 2.47, 2.48. However, it is much easier to use the hierarchical structure of the design and the decomposition rule for conditional expectation and variance given in Appendix B. First, let us note that

$$\hat{Y}_{i\pi} = \sum_{k \in s_i} \frac{Y_k}{\pi_{k|i}} = \sum_{k \in s_i} \check{Y}_{k|i} \tag{2.49}$$

is the unbiased HT estimate of the total $T_{Y,i} = \sum_{k \in U_i} Y_k$ of the ith PSU. The

$$V_i = \sum_{k,l \in U_i} \Delta_{kl|i} \check{Y}_{k|i} \check{Y}_{l|i} \tag{2.50}$$

$$\hat{V}_i = \sum_{k,l \in s_i} \check{\Delta}_{kl|i} \check{Y}_{k|i} \check{Y}_{l|i} \tag{2.51}$$

In two-stage sampling, the HT estimator of the population total $Y = \sum_{k \in \mathcal{P}} Y_k$ **is given by**
Theorem 2.8.1.

$$\hat{Y}_{2st\pi} = \sum_{i \in s_I} \frac{\hat{Y}_{i\pi}}{\pi_{Ii}}$$

Using B.3 one sees that $\hat{Y}_\pi$ is an unbiased estimate of the population total and that its variance is given by
Theorem 2.8.2.

$$\mathbb{V}(\hat{Y}_{2st\pi}) = V_{PSU} + V_{SSU}$$

where the variance between PSU and within SSU are

$$V_{PSU} = \sum_{i,j \in U_I} \Delta_{Iij} \check{Y}_i \check{Y}_j \tag{2.52}$$

$$V_{SSU} = \sum_{i \in U_I} \frac{V_i}{\pi_{Ii}} \tag{2.53}$$

Note the difference between $\check{Y}_i = \frac{T_{Y,i}}{\pi_{Ii}}$ containing the unknown total of the PSU and $\hat{Y}_{i\pi}$ defined by Eq. 2.49. A naive approach to estimating V_{PSU} would be to replace in Eq. 2.52 $\check{Y}_i \check{Y}_j$ with $\hat{Y}_{i\pi} \hat{Y}_{j\pi}$. Because of the independence this is correct for $i \neq j$ but not for $i = j$, as $\mathbb{E}(Y^2) = \mathbb{E}^2(Y) + \mathbb{V}(Y)$ for any random variable Y. However, using the decomposition rule in B.3 for the variance it is tedious but easy to show, see Särndal et al. (2003), that V_{PSU} can be unbiasedly estimated by

$$\hat{V}_{PSU} = \sum_{i,j \in s_I} \check{\Delta}_{Iij} \frac{\hat{Y}_{i\pi} \hat{Y}_{j\pi}}{\pi_{Ii} \pi_{Ij}} - \sum_{i \in s_I} \frac{(1 - \pi_{Ii})}{\pi_{Ii}^2} \hat{V}_i \tag{2.54}$$

Finally, from Eq. 2.5, 2.50 and B.3 one has at once the following unbiased estimate of V_{SSU}:

$$\hat{V}_{SSU} = \sum_{i \in s_I} \frac{\hat{V}_i}{\pi_{Ii}^2} \tag{2.55}$$

Then, combining all these pieces one can conclude that **the variance of the two-stage HT estimator is unbiasedly estimated by**
Theorem 2.8.3.

$$\hat{\mathbb{V}}(\hat{Y}_{2st\pi}) = \sum_{i,j \in s_I} \check{\Delta}_{Iij} \frac{\hat{Y}_{i\pi} \hat{Y}_{j\pi}}{\pi_{Ii} \pi_{Ij}} + \sum_{i \in s_I} \frac{\hat{V}_i}{\pi_{Ii}}$$

because a variance estimate $\hat{V}_i$ must be calculated for every $i \in s$. However, it can be shown that the first term alone gives a much simpler alternative estimate:

$$\hat{\mathbb{V}}^*(\hat{Y}_{II\pi}) = \sum_{i,j \in s_I} \check{\Delta}_{Iij} \frac{\hat{Y}_{i\pi} \hat{Y}_{j\pi}}{\pi_{Ii} \pi_{Ij}}$$

which usually underestimates, but only slightly, the variance.

Remarks:

1. If the first stage is a census of the PSUs then two-stage sampling is equivalent to stratified sampling and $V_{PSU} = 0$.

2. If the second stage is a census of the sampled PSUs then two-stage sampling is called **single-stage cluster-sampling**, which is described in more detail in section 2.9. In this case $V_{SSU} = 0$.

3. Estimation of the mean is straightforward if N is known. Even so, it is usually better, particularly with random sample sizes, to estimate it as a ratio, i.e. via

$$\hat{\bar{Y}} = \frac{\sum_{i \in S_I} \frac{\hat{Y}_{i\pi}}{\pi_{Ii}}}{\sum_{i \in S_I} \frac{\hat{N}_{i\pi}}{\pi_{Ii}}}$$

with

$$\hat{N}_i = \sum_{k \in s_i} \frac{1}{\pi_{k|i}}$$

In addition the residual technique explained in Section 2.6 can be used to get an estimate of the variance.

We can now specifically apply the above general result to simple random sampling without replacement at both stages. In stage 1, we sample n_I out of the N_I PSUs, in stage 2, we sample for each $i \in s_I$ n_i out of the N_i elements in U_i. Simple algebra leads to the following result:

$$\hat{Y}_{II\pi} = \frac{N_I}{n_I} \sum_{i \in s_I} N_i \bar{Y}_{s_i}$$

$$\mathbb{V}(\hat{Y}_{II\pi}) = N_I^2 \frac{1 - f_I}{n_I} S_{U_I}^2 + \frac{N_I}{n_I} \sum_{i \in U_I} N_i^2 \frac{1 - f_i}{n_i} S_{Y,U_i}^2 \qquad (2.56)$$

where we have set $\bar{Y}_{s_i} = \frac{\sum_{k \in s_i} Y_k}{n_i}$, $f_I = \frac{n_I}{N_I}$, $f_i = \frac{n_i}{N_i}$.

$$S_{U_I}^2 = \frac{1}{N_I - 1} \sum_{i \in U_i} (T_{Y,i} - \bar{T}_{Y,i})^2$$

is the variance in U_I of the total for the PSU with $T_{Y,i}$ $\quad N_I \quad$ and

$$S^2_{Y,U_i} = \frac{1}{N_i - 1} \sum_{k \in U_i} (Y_k - \bar{Y}_{U_i})^2$$

with $\bar{Y}_{U_i} = \frac{\sum_{k \in U_i} Y_k}{N_i}$. The unbiased variance estimator is then

$$\hat{V}(\hat{Y}_{II\pi}) = N_I^2 \frac{1 - f_I}{n_I} \hat{S}^2_{U_I} + \frac{N_I}{n_I} \sum_{i \in U_I} N_i^2 \frac{1 - f_i}{n_i} \hat{S}^2_{Y,U_i} \qquad (2.57)$$

where

$$\hat{S}^2_{U_I} = \frac{1}{n_I - 1} \sum_{i \in s_I} \left[\hat{Y}_{i\pi} - \left(\frac{\sum_{i \in s_I} \hat{Y}_{i\pi}}{n_I} \right) \right]^2$$

is the variance in s_I of the estimated PSU totals $\hat{Y}_{i\pi} = N_i \bar{Y}_{s_i}$, and

$$\hat{S}^2_{Y,U_i} = \frac{1}{n_i - 1} \sum_{k \in s_i} (Y_k - \bar{Y}_{s_i})^2$$

Note that the second term in Eq. 2.57 is not unbiased for the second term in Eq. 2.56.

2.9 Single-stage cluster-sampling

This scheme is a special case of two-stage sampling in which the second stage is a full census. For easier reference let us recall the main features in this unique context. A finite population $\mathcal{P} = 1, 2, \ldots, N$ is partitioned into subpopulation (PSUs) called clusters, which are denoted as $U_1, U_2, \ldots, U$ The set of clusters is shown by $U_I = \{1, 2, \ldots, i, \ldots, N_I\}$. This represents a population of clusters from which a sample of clusters is selected. The number of population elements in the ith cluster U_i is N_i. Thus, $\mathcal{P} = \bigcup_{i \in U_I} U_i$ $N = \sum_{i \in U_I} N_i$. Single-stage cluster-sampling (or simply cluster-sampling) is defined as follows:

1. A probability sample s_I of clusters is drawn from U_I according to the design $p_I(\cdot)$, with size n_I if fixed, or n_{s_I} if random.

2. Every population element in the cluster is observed.

Note that the sample s is the set $\bigcup_{i \in s_I} U_i$. Likewise, the sample size is n $\sum_{i \in s_I} N_i$, which is generally random, even if $p_I(\cdot)$ is a fixed-size design (equal to n_I), because the cluster size N_i may vary. The first- and second-order cluster inclusion probabilities are π_{Ii} and π_{Iij}, with $\pi_{Iii} = \pi_{Ii}$. At the element level we have $\pi_k = \pi_{Ii}$ for every element k in the cluster U_i, $\pi_{kl} = \pi_{Ii}$ if both and l belong to the same cluster U_i or $\pi_{kl} = \pi_{Iij}$, if k and l belong to different cluster U_i and U_j. To simplify the notation we set $T_i = T_{Y,i} = \sum_{k \in U}$

th cluster total (which is equal to
sampling because the second stage is a census).

Our HT estimate of the population total $Y = \sum_{k \in \mathcal{P}} Y_k$ is

$$\hat{Y}_\pi = \sum_{i \in s_I} \frac{T_i}{\pi_{Ii}} = \sum_{i \in s_I} \check{T}_i \qquad (2.58)$$

with variance

$$\mathbb{V}(\hat{Y}_\pi) = \sum_{i,j \in U_I} \Delta_{Iij} \check{T}_i \check{T}_j \qquad (2.59)$$

and estimated variance

$$\hat{\mathbb{V}}(\hat{Y}_\pi) = \sum_{i,j \in s_I} \check{\Delta}_{Iij} \check{T}_i \check{T}_j \qquad (2.60)$$

For a fixed-size design the Yates-Grundy formula from Eq. 2.12 applies and one can draw the following conclusions, valid also for random-size design:

1. If π_{Ii} are approximately proportional to the cluster totals T_i, then cluster sampling will be highly efficient.
2. If the cluster sizes are known at the planning stage, one can choose a design with $\pi_{Ii} \propto N_i$. Because $T_i = N_i \bar{T}_i$, this is efficient if there is little variation among the cluster means $\bar{T}_i = \frac{1}{N_i} \sum_{k \in U_i} Y_k$.
3. An equal probability cluster-sampling design is often a poor choice when the clusters are of different sizes.

In simple random sampling without replacement one has $\pi_{Ii} = \frac{n_I}{N_I}$. The HT estimate of the total $Y = \sum_{k \in \mathcal{P}} Y_k$ is then

$$\hat{Y} = \frac{N_I}{n_I} \sum_{i \in s_I} T_i \qquad (2.61)$$

By 2.16 the estimated variance is given by

$$\hat{\mathbb{V}}(\hat{Y}) = N_I^2 (1 - \frac{n_I}{N_I}) \frac{1}{n_I} \frac{\sum_{i \in s_I} (T_i - \bar{T}_{s_I})^2}{n_I - 1} \qquad (2.62)$$

where $\bar{T}_{s_I} = \frac{\sum_{i \in s_I} T_i}{n_I}$ is the estimated mean per cluster (not per element!).

An intuitive estimate of the mean per element is the ratio

$$\bar{Y}_{s_I} = \frac{\sum_{i \in s_I} T_i}{\sum_{i \in s_I} N_i} \qquad (2.63)$$

which is simply the mean over all elements selected, independent of the cluster structure. According to the results presented in Section 2.6 it is asymptotically $(n \to \infty)$ unbiased for $\bar{Y} = \frac{\sum_{k \in \mathcal{P}} Y_k}{N}$, the mean per element in the population. An alternative estimate for the total is therefore

$$\hat{Y}_R = N \frac{\sum_{i \in s_I} T_i}{\sum_{i \in s_I} N_i} \qquad (2.64)$$

$$\hat{\mathbb{V}}(\hat{Y}_R) = N_I^2 (1 - \frac{n_i}{N_I}) \frac{1}{n_I} \frac{\sum_{i \in s_I} N_i^2 (\bar{T}_i - \bar{Y}_{s_I})^2}{n_I - 1} \tag{2.65}$$

where $\bar{T}_i$ is the mean per element in the ith cluster. It turns out that the estimated variance from Eq. 2.65 is usually much smaller than that from 2.62. To estimate the mean per element in the population one can choose

$$\hat{\bar{Y}} = \frac{\hat{Y}}{N} \tag{2.66}$$

if the total number of elements N is known, with an estimated variance

$$\hat{\mathbb{V}}(\hat{\bar{Y}}) = \frac{N_I^2}{N^2} (1 - \frac{n_I}{N_I}) \frac{1}{n_I} \frac{\sum_{i \in s_I} (\bar{T}_i - \bar{T}_{s_I})^2}{n_I - 1} \tag{2.67}$$

If N is unknown one must take the ratio estimate $\bar{Y}_{s_I} = \frac{\sum_{i \in s_I} T_i}{\sum_{i \in s_I} N_i}$ with the estimated variance

$$\hat{\mathbb{V}}(\bar{Y}_{s_I}) = (1 - \frac{n_I}{N_I}) \frac{1}{n_I} \frac{\sum_{i \in s_I} (\frac{N_i}{\bar{N}_{s_I}})^2 (\bar{T}_i - \bar{Y}_{s_I})^2}{n_I - 1} \tag{2.68}$$

where $\bar{N}_{s_I} = \frac{1}{n_i} \sum_{i \in s_I} N_i$ is the mean cluster size in the sample. Note that the ratio estimate $\bar{Y}_{s_I}$ requires us to know only the cluster sizes N_i of the selected clusters and that its variance is usually smaller than the variance of $\hat{\bar{Y}}$ given by Eq. 2.67.

Another estimate is occasionally used to determine the mean per element namely

$$\hat{\bar{Y}}^* = \frac{1}{n_i} \sum_{i \in s_I} \bar{T}_i$$

i.e. the mean of the cluster means. **When the N_i vary, this estimate is not only biased but also inconsistent**. That is, it does not converge towards the true mean as $n \to \infty$. The bias may be unimportant if the $\bar{T}_i$ are not correlated with the N_i.

To better understand the impact of clustering on the variance we briefly consider a very special case in which the population $\mathcal{P}$ of N elements is partitioned into N_I clusters all of the same size $N_i = M$, i.e. $N = N_I M$. Then, one obtains

$$\bar{Y}_{s_I} = \frac{\sum_{i \in s_I} T_i}{n_I M} = \frac{\bar{T}_{s_I}}{M} \tag{2.69}$$

with $\bar{T}_{s_I} = \frac{\sum_{i \in s_I} T_i}{n_I}$. The variance of $\bar{Y}_{s_I}$ is therefore given by

$$\mathbb{V}(\bar{Y}_{s_I}) = \frac{1}{M^2} (1 - \frac{n_I}{N_I}) \frac{1}{n_I} \frac{\sum_{i \in U_I} (\bar{T}_i - \bar{T})^2}{N_I - 1} \tag{2.70}$$

where T_{N_I} $N_I Y$ MY. Now, one has

$$T_i - \bar{T} = \sum_{k \in U_i} Y_k - M\bar{Y} = \sum_{k \in U_i} (Y_k - \bar{Y})$$

because the number of terms in the sum is a constant equal $N_i \equiv M$. Hence, one arrives at

$$\sum_{i \in U_I} (T_i - \bar{T})^2 = \sum_{i \in U_I} \sum_{k \in U_i} (Y_k - \bar{Y})^2 + \sum_{i \in U_I} \sum_{k \neq l \in U_i} (Y_k - \bar{Y})(Y_l - \bar{Y})$$

Let us now define the **intra-cluster correlation coefficient** ρ as

$$\rho = \frac{\frac{\sum_{i \in U_I} \sum_{k \neq l \in U_i} (Y_k - \bar{Y})(Y_l - \bar{Y})}{N_I M (M-1)}}{\frac{\sum_{i \in U_I} \sum_{k \in U_i} (Y_k - \bar{Y})^2}{N_I M}}$$

in agreement with the classical interpretation. This term represents the average of cross-products for deviations from the mean divided by the variance. Recall that the population variance at the element level is calculated as

$$S^2 = \frac{\sum_{i \in U_I} \sum_{k \in U_i} (Y_k - \bar{Y})^2}{N_I M - 1}$$

so that the intra-cluster correlation can be rewritten as

$$\rho = \frac{\sum_{i \in U_I} \sum_{k \neq l \in U_i} (Y_k - \bar{Y})(Y_l - \bar{Y})}{(M-1)(N_I M - 1)S^2} \tag{2.71}$$

Finally, one obtains the asymptotic variance under cluster-sampling as

$$\mathbb{V}(\bar{Y}_{s_I}) = \frac{1 - f_I}{n_I} \frac{N_I M - 1}{M^2 (N_I - 1)} S^2 \left[1 + (M-1)\rho\right] \tag{2.72}$$

with $f_I = \frac{n_I}{N_I}$. For N_I large this can be further simplified to

$$\mathbb{V}(\bar{Y}_{s_I}) \approx \frac{1 - f_I}{n_I M} S^2 \left[1 + (M-1)\rho\right] \tag{2.73}$$

This is the same expression as the variance under simple random sampling of $n_I M$ elements, up to the inflation factor $1 + (M-1)\rho$. **In most applications the correlation of elements within the same cluster is positive, so that cluster-sampling is less precise for a given overall size.** Of course, one should also take costs into account (see Cochran, 1977 for details) and also recall that PPS sampling is in this case usually better than simple random sampling. In a forest inventory scenario, we shall see that the term cluster-sampling is used in a slightly different context and the Swedish word "trakt" is occasionally used instead. It is of interest to note that under stratified sampling the strata should be as homogenous as possible and markedly different from each other, while in cluster-sampling, the clusters should be, if feasible, as heterogenous as possible.

We consider only the simplest version. A first element is drawn at random and with equal probability among the first a elements in the population list. The positive integer a is fixed in advance and is called the sampling interval. The rest of the sample is determined by systematically taking every ath element thereafter, through the entire list. Let us assume for simplicity that $N =$ We then select, with equal probability $\frac{1}{a}$, a random integer r between 1 and a inclusively. The selected sample of fixed size n consists of the elements $s_r = \{k := r + (j-1)a \leq N; \quad j = 1, 2, \ldots n\}$, while the set of all possible samples is $\mathcal{S} = \{s_1, s_2, \ldots, s_r\}$. Our inclusion probability is $\pi_i = \frac{1}{a}$, $\pi_{ij} =$ i and j belong to the same sample s_r and zero otherwise. Simple calculations lead to

$$\hat{Y}_{\pi,r} = a \sum_{i \in s_r} Y_i \tag{2.74}$$

$$\mathbb{V}(\hat{Y}_{\pi,r}) = a \sum_{r=1}^{a} (\hat{Y}_{\pi,r} - \bar{Y})^2$$

Because of its inherent simplicity systematic sampling is very popular, especially in field work, but it has one drawback. The condition $\pi_{ij} > 0$ $\forall$ in Theorem 2.2.2 is not fulfilled so that **there exists, in design-based inference, no unbiased estimate of the variance based on a single systematic sample.** Systematic sampling can also be viewed as the drawing of one cluster only. If the intra-cluster correlation coefficient ρ defined in Eq. 2.71 is zero, this could be true if the elements are in a "random order", then the variance estimate given by the formula for simple random sampling would be approximately correct. In contrast, it would be conservative (i.e. overestimating the variance) if ρ is negative. Cochran (1977) and Särndal et al. (2003) provide further details in this regard.

2.11 Exercises

Problem 2.1. *Consider the population $\mathcal{P} = \{1, 2, 3\}$ endowed with the sampling scheme $p(\{1,2\}) = \frac{1}{2}$, $p(\{1,3\}) = \frac{1}{4}$ and $p(\{2,3\}) = \frac{1}{4}$. Determine the inclusion probabilities π_i, $i = 1, 2, 3$ and the positive definite $(3,3)$ matrix defined by*

$$\Delta_{kl} = \mathbb{COV}(I_k, I_l) = \pi_{kl} - \pi_k \pi_l, \Delta_{kk} = \pi_k(1 - \pi_k)$$

Problem 2.2. *Show that for a design of fixed sample size*

$$n_s = \sum_{k \in \mathcal{P}} I_k \equiv n$$

$$\sum_{k \in \mathcal{P}} \Delta_{kl} = 0 \ \forall l \in \mathcal{P}$$

Conversely, show that if

$$\sum_{k,l \in \mathcal{P}} \Delta_{kl} = 0$$

then the design is of a fixed size.

Problem 2.3. *Consider the $\boldsymbol{\Delta}$ matrix in a population of five elements $\mathcal{P}$ $\{1, 2, 3, 4, 5\}$ given by*

$$\boldsymbol{\Delta} = \begin{pmatrix} 1 & 1 & 1 & -1 & -1 \\ 1 & 1 & 1 & -1 & -1 \\ 1 & 1 & 1 & -1 & -1 \\ -1 & -1 & -1 & 1 & 1 \\ -1 & -1 & -1 & 1 & 1 \end{pmatrix}$$

1. *Is the design of a fixed size?*
2. *Calculate the inclusion probabilities knowing that $\pi_1 = \pi_2 = \pi_3 > \pi_4 =$*
3. *Give the pair-wise inclusion probability matrix $\boldsymbol{\Pi}$.*
4. *Determine the probabilities for all possible samples.*

Problems 2.1, 2.2 and 2.3 originate from the writing of Ardilly and Tillé (2006), which contains many more exercises that are generally rather difficult.

Problem 2.4. *Consider the population $\mathcal{P} = \{Y_1, Y_2, \ldots Y_N\}$, with mean $\bar{Y}$ $\frac{1}{N}\sum_{i=1}^{N} Y_i$ and variance $S_Y^2 = \frac{1}{N-1}\sum_{i=1}^{N}(Y_i - \bar{Y})^2$.*

1. *Express the population variance in terms of the quantities N, $\sum_{i=1}^{N} Y_i$ and $\sum_{i=1}^{N} Y_i^2$.*

2. *Give the theoretical variance $\hat{\mathbb{V}}_{srs}(\hat{\bar{Y}}_s)$, under simple random sampling of n out of N units without replacement, for the Horwitz-Thompson-based estimate of the population mean*

$$\hat{\bar{Y}}_s = \frac{1}{N}\sum_{i \in s} \frac{Y_i}{\pi_i} = \frac{1}{n}\sum_{i \in s} Y_i$$

Note that because $n_s \equiv n$ the estimate $\hat{\bar{Y}}_s$ is the ordinary sample mean!

3. *Calculate the variance $\mathbb{V}_{BE}(\hat{\bar{Y}}_s)$ of $\hat{\bar{Y}}_s$ under Bernoulli sampling with constant inclusion probabilities $\pi_i \equiv \frac{n}{N}$. Note that $\hat{\bar{Y}}_s$ is no longer the ordinary sample mean because $n_s \neq n$ in general.*

4. *Likewise, calculate the variance $\mathbb{V}_{BE}(\tilde{Y}_s)$ of the weighted sample mean*

$$\tilde{Y}_s = \frac{\sum_{i=1}^{N} \frac{I_i Y_i}{\pi_i}}{\sum_{i=1}^{N} \frac{I_i}{\pi_i}} = \frac{1}{n_s}\sum_{i \in s} Y_i$$

equal to the estimate $\hat{\bar{Y}}_{alt}$ given in Eq. 2.31.

5. Determine also the corresponding variance estimates $\hat{\mathbb{V}}_{BE}(\hat{\bar{Y}}_s)$ and $\hat{\mathbb{V}}_{BE}$

6. Compare the theoretical variances of $\hat{\bar{Y}}_s$ and $\tilde{Y}_s$ under simple random sampling and Bernoulli sampling. Explain your conclusions.

7. Perform the above calculations for a population of the first 100 integers, i.e. $\mathcal{P} = \{Y_1, Y_2, \ldots Y_{100}\}$ with $Y_i = i$, for $n = \mathbb{E}(n_s) = 10$. What is the exact probability that n_s is between 5 and 15? What is the approximate value of this probability based on the normal approximation?

Problem 2.5. We consider the same population as in Problem 2.4, namely $\mathcal{P} = \{1, 2, 3 \ldots 100\}$. The population is partitioned into systematic samples defined according to

$$s_k = \{i \mid i = k + 10(L - 1), L = 1, 2, 3 \ldots 10\}, \ k = 1, 2, \ldots 10$$

This sampling scheme selects one of the subset s_k with probability $p(s_k) =$

1. Calculate the means $\bar{Y}_{s_k}$ for all systematic sample. What are the inclusion probabilities π_i and π_{ij}? What are the consequences?

2. Define $\hat{\bar{Y}}_{sys} = \bar{Y}_{s_k}$ if sample s_k is selected and calculate

$$\mathbb{E}(\hat{\bar{Y}}_{sys}) = \sum_{k=1}^{10} p(s_k)\bar{Y}_{s_k}$$

3. Determine the theoretical variance

$$\mathbb{V}(\hat{\bar{Y}}_{sys}) = \sum_{k=1}^{10} p(s_k)(\hat{\bar{Y}}_{sys} - \bar{Y})^2$$

and compare it with the values obtained in Problem 2.4.

4. Calculate the estimated variance by treating the systematic sample as a random sample.

5. By considering systematic sampling as cluster sampling calculate the intra-cluster correlation coefficient ρ (use Eq. 2.72).

6. Can you think of other partitions leading to systematic sampling schemes with higher or lower variances? What are the extreme values for the intra-cluster correlation coefficient? What do you expect to happen if prior to these calculations one performs a random permutation of the integers while keeping the allocation of the labels i to the samples?

7. How can you estimate the variance under replication of the samples?

Problem 2.6. Consider a population $\mathcal{P}$ of size N and a subset or domain D $\mathcal{P}$ of size N_D. Define the indicator variable $Y_i = 1$ if $i \in D$ and $Y_i = 0$ if $i \notin$ We want to estimate $p_d = \frac{N_D}{N} = \frac{\sum_{i=1}^{N} Y_i}{N}$ under simple random sampling without replacement on a fixed sample size n. Find the Horvitz-Thompson-based estimate $\hat{p}_D$, together with its theoretical and estimated variances.

Problem 2.8. *Consider a population* $\mathcal{P}$ *of* N *units with response variable* Y_i *and a Poisson sampling scheme with inclusion probabilities* π_i. *Determine the inclusion probability* π_i *in order to minimize the variance of the Horvitz-Thompson estimate* $\hat{Y} = \sum_{i=1}^{N} \frac{Y_i}{\pi}$ *under the cost constraint* $\sum_{i=1}^{N} \pi = n$. *Use the Lagrange Multiplier technique.*

Problem 2.9. *We again consider the population* $\mathcal{P} = \{1, 2, 3 \ldots 100\}$ *partitioned into* $H = 5$ *strata* $\mathcal{P}_k = \{(k-1)20 + 1, \ldots k20\}$, $k = 1, 2, 3, 4, 5$, N_k 20. *We use simple random sampling with* $n_k = 2$ *in each stratum. Find the variance for the stratified estimate of the mean given by*

$$\hat{\bar{Y}}_{strat} = \sum_{k=1}^{5} \frac{N_h}{N} \bar{Y}_k$$

Compare this result with those from Problems 2.4 and 2.5.

Problem 2.10. *For* $\mathcal{P} = \{1, 2, 3 \ldots 100\}$ *we define* 10 *primary units* U_i $\{(i-1)10, \ldots i10\}$ *of sizes* $N_i \equiv 10$. *We select a sample of* $n_I = 5$ *primary units by simple random sampling without replacement with* $f_i = \frac{n_I}{N_I} =$ *In each selected primary unit we choose, via simple random sampling without replacement* $n_i = 2$ *secondary units. Calculate the variance for the two-stage estimate of the population mean according to Eq. 2.56. Summarize the findings for this, albeit artificial, population.*

CHAPTER 3

Sampling finite populations: advanced topics

3.1 Three-stage element sampling

Going beyond two stages of sampling is straightforward as long as the invariance and independence properties hold. However, the notation for general r-stage sampling is cumbersome. Three types of sampling units are described below. The term element sampling means that the third-stage sampling units are the populations element. A typical example would be to designate schools as PSUs, classes as SSUs, and the students in each class as elements. Let us now introduce the notation:

1. The N elements of $\mathcal{P}$ are partitioned into PSUs $U_1, U_2 \ldots, U_i, \ldots U_{N_I}$. The set of PSUs is denoted by $U_I = \{1, \ldots, i, \ldots, N_I\}$. With N_i being the size of U_i we have $N = \sum_{i \in U_I} N_i$.
2. The N_i elements in U_i $(i = 1, \ldots N_i)$ are partitioned into N_{IIi} secondary sampling units
$$U_{i1}, \ldots, U_{iq}, \ldots, U_{iN_{IIi}}$$
The set of SSUs formed by the partitioning of U_i is denoted by
$$U_{IIi} = \{1, \ldots, q, \ldots, N_{IIi}\}$$
With N_{iq} being the size of U_{iq} we have $N_i = \sum_{q \in U_{IIi}} N_{iq}$.
3. **The tertiary sampling units are the population elements**

Our sampling scheme (with parameters summarized in Table 3.1) proceeds as follows:

1. *First stage* (I): a sample s_I of PSUs is drawn from U_I $(s_I \subset U_I)$ according to the design $p_I(\cdot)$.
2. *Second stage* (II): for $i \in s_I$, a sample s_{IIi} of SSUs is drawn from U_{IIi} $(s_{IIi} \subset U_{IIi})$ according to the design $p_{IIi}(\cdot)$.
3. *Third stage* (III): for $q \in s_{IIi}$, a sample s_{iq} of elements is drawn from U_{iq} $(s_{iq} \subset U_{iq})$ according to the design $p_{IIIiq}(\cdot)$ (abbreviated $p_{iq}(\cdot)$). Those ultimately selected are the elements $k \in s$, where
$$s = \cup_{i \in s_I} \cup_{q \in s_{IIi}} s_{iq}$$

Table 3.1 *Inclusion probabilities*

Stage	Design	First-order	Second-order	Δ Parameters
I	$p_I(\cdot)$	π_{Ii}	π_{Iij}	$\Delta_{Iij} = \pi_{Iij} - \pi_{Ii}\pi_{Ij}$
II	$p_{IIi}(\cdot)$	$\pi_{IIq\|i}$	$\pi_{IIqr\|i}$	$\Delta_{IIqr\|i} = \pi_{IIqr\|i} - \pi_{IIq\|i}\pi$
III	$p_{iq}(\cdot)$	$\pi_{k\|iq}$	$\pi_{kl\|iq}$	$\Delta_{kl\|iq} = \pi_{kl\|iq} - \pi_{k\|iq}\pi_{l\|iq}$

Here, i and j denote distinct PSUs, q and r distinct SSUs, and k and l distinct tertiary units, or TSUs (recall that they are the elements of the population).

$$\pi_{Iii} = \pi_{Ii}, \quad \pi_{IIqq\|i} = \pi_{IIq\|i}, \quad \pi_{kk\|iq} = \pi_{k\|iq}$$

$$\check{\Delta}_{Iij} = \frac{\Delta_{Iij}}{\pi_{Iij}}, \quad \check{\Delta}_{IIqr\|i} = \frac{\Delta_{IIqr\|i}}{\pi_{IIqr\|i}}, \quad \check{\Delta}_{kl\|iq} = \frac{\Delta_{kl\|iq}}{\pi_{kl\|iq}}$$

We introduce the totals

$$T_{Y,iq} = \sum_{k \in U_{iq}} Y_k, \quad T_{Y,i} = \sum_{q \in U_{IIi}} T_{Y,iq}, \quad T_Y = \sum_{i \in U_I} T_{Y,i}$$

Proceeding backwards from the last to the first stage, we obtain the following HT estimators:

1. *Stage III*

$$\hat{Y}_{iq\pi} = \sum_{k \in s_{iq}} \frac{Y_k}{\pi_{k\|iq}} \quad \text{with} \quad \mathbb{E}_{(III\|I,II)}(\hat{Y}_{iq\pi}) = T_{Y,iq} \tag{3.1}$$

2. *Stage II*

$$\hat{Y}_{i\pi} = \sum_{q \in s_{IIi}} \frac{\hat{Y}_{iq\pi}}{\pi_{IIq\|i}} \quad \text{with} \quad \mathbb{E}_{(II,III\|I)}(\hat{Y}_{i\pi}) = T_{Y,i} \tag{3.2}$$

3. *Stage I*

$$\hat{Y}_{3st\pi} = \sum_{i \in s_I} \frac{\hat{Y}_{i\pi}}{\pi_{Ii}} \quad \text{with} \quad \mathbb{E}_{(I,II,III)}(\hat{Y}_\pi) = T_Y \tag{3.3}$$

To calculate the variance we use Eq. B.5. First, the variance of $\hat{Y}_{iq\pi}$ in repeated sampling from U_{iq} is by 2.8

$$\mathbb{V}_{(III\|I,II)}(\hat{Y}_{iq\pi}) = V_{iq} = \sum_{k,l \in U_{iq}} \Delta_{kl\|iq} \frac{Y_k Y_l}{\pi_{k\|iq}\pi_{l\|iq}} \tag{3.4}$$

The variance of

$$\sum_{q \in s_{IIi}} \frac{T_{Y,iq}}{\pi_{IIq\|i}}$$

$$V_{IIi} = \sum_{q,r \in U_{IIi}} \Delta_{IIqr|i} \frac{T_{Y,iq}T_{Y,ir}}{\pi_{IIq|i}\pi_{IIr|i}} \tag{3.5}$$

Consequently, the variance of $\hat{Y}_{i\pi}$ due to third and second stage sampling is, according to the decomposition Theorem B.2, calculated by

$$V_i = V_{IIi} + \sum_{q \in U_{IIi}} \frac{V_{iq}}{\pi_{IIq|i}} \tag{3.6}$$

Collecting the pieces together and using Eq. B.5 we obtain the following result for the variance of the tree-stage HT estimator:

Theorem 3.1.1.

$$\mathbb{V}(\hat{Y}_{3st\pi}) = V_{PSU} + V_{SSU} + V_{TSU}$$

where

$$V_{PSU} = \sum_{i,j \in U_I} \Delta_{Iij} \check{T}_{Y,i} \check{T}_{Y,j} \tag{3.7}$$

with $\check{T}_{Y,i} = \frac{T_{Y,i}}{\pi_{Ii}}$ gives the variance component due to first-stage sampling,

$$V_{SSU} = \sum_{i \in U_I} \frac{V_{IIi}}{\pi_{Ii}} \tag{3.8}$$

gives the variance contribution due to second stage-sampling, and

$$V_{TSU} = \sum_{i \in U_I} \frac{\left(\sum_{q \in U_{IIi}} \frac{V_{iq}}{\pi_{IIq|i}} \right)}{\pi_{Ii}} \tag{3.9}$$

gives the variance component due to the third-stage sampling.

Using the same arguments as for two-stage sampling we can conclude that the variance of the three-stage Horvitz-Thompson estimator can be unbiasedly estimated by

Theorem 3.1.2.

$$\hat{\mathbb{V}}_{3st}(\hat{Y}_\pi) = \sum_{i,j \in s_I} \check{\Delta}_{Iij} \frac{\hat{Y}_{i\pi}\hat{Y}_{j\pi}}{\pi_{Ii}\pi_{Ij}} + \sum_{i \in s_I} \frac{\hat{V}_i}{\pi_{Ii}}$$

where

$$\hat{V}_i = \sum_{q,r \in s_{IIi}} \check{\Delta}_{IIqr|i} \frac{\hat{Y}_{iq\pi}\hat{Y}_{ir\pi}}{\pi_{IIq|i}\pi_{IIr|i}} + \sum_{q \in s_{IIi}} \frac{\hat{V}_{iq}}{\pi_{IIq|i}}$$

and

$$\hat{V}_{iq} = \sum_{k,l \in s_{iq}} \check{\Delta}_{kl|iq} \frac{Y_k Y_l}{\pi_{k|iq}\pi_{l|iq}}$$

Table 3.2 *Inclusion probabilities for simple random sampling*

Stage	First-order	Second-order
I	$\pi_{Ii} = \frac{n_i}{N_i}$	$\pi_{Iij} = \frac{n_I(n_I-1)}{N_I(N_I-1)}$
II	$\pi_{IIq\mid i} = \frac{n_{IIi}}{N_{IIi}}$	$\pi_{IIqr\mid i} = \frac{n_{IIi}(n_{IIi}-1)}{N_{IIi}(N_{IIi}-1)}$
III	$\pi_{k\mid iq} = \frac{n_{iq}}{N_{iq}}$	$\pi_{kl\mid iq} = \frac{n_{iq}(n_{iq}-1)}{N_{iq}(N_{iq}-1)}$

For planning purposes it might be useful to obtain separate estimates for the three components of variance:

$$\hat{V}_{TSU} = \sum_{i \in s_I} \frac{\left(\sum_{q \in s_{IIi}} \frac{\hat{V}_{iq}}{\pi_{IIq\mid i}^2} \right)}{\pi_{Ii}^2} \tag{3.10}$$

$$\hat{V}_{SSU} = \sum_{i \in s_I} \frac{\hat{V}_i}{\pi_{Ii}^2} - \hat{V}_{TSU} \tag{3.11}$$

$$\hat{V}_{PSU} = \hat{\mathbb{V}}_{3st}(\hat{Y}_\pi) - \hat{V}_{SSU} - \hat{V}_{TSU} \tag{3.12}$$

Remarks:

1. If the first stage is a census three-stage element sampling is also called **stratified two-stage sampling**, then $V_{PSU} = 0$.

2. If the third stage is a census three-stage element sampling is also called **two-stage cluster sampling**, then $V_{TSU} = 0$.

As for two-stage sampling we adapt the above general results to simple random sampling at each stage according to Table 3.2. The algebra is simple but extremely tedious so only the end results are given.

The estimate of the total in the iq-th SSU is

$$\hat{Y}_{iq\pi} = N_{iq} \bar{Y}_{s_{iq}} \quad \text{with} \quad \bar{Y}_{s_{iq}} = \frac{1}{n_{iq}} \sum_{k \in s_{iq}} Y_k$$

For the i-th PSU it is

$$\hat{Y}_{i\pi} = N_{IIi} \bar{\hat{Y}}_{i\cdot\pi} \quad \text{with} \quad \bar{\hat{Y}}_{i\cdot\pi} = \frac{1}{n_{IIi}} \sum_{q \in s_{IIi}} \hat{Y}_{iq\pi}$$

Finally, for the overall total, it is

$$\hat{Y}_\pi = N_I \bar{\hat{Y}}_{\cdot\cdot\pi} \quad \text{with} \quad \bar{\hat{Y}}_{\cdot\cdot\pi} = \frac{1}{n_I} \sum_{i \in s_I} \hat{Y}_{i\pi}$$

These calculations will involve therefore only standard sample means (denoted according to the convention used in a classical ANOVA). To present the formulae required for the variances in a relatively compact form we introduce the following notation:

$$s_I^2 = \frac{1}{n_I - 1} \sum_{i \in s_I} (\hat{Y}_{i\pi} - \bar{\hat{Y}}_{\cdot\cdot\pi})^2 \tag{3.13}$$

$$s_{IIi}^2 = \frac{1}{n_{IIi} - 1} \sum_{q \in s_{IIi}} (\hat{Y}_{iq\pi} - \bar{\hat{Y}}_{i\cdot\pi})^2$$

$$s_{iq}^2 = \frac{1}{n_{iq} - 1} \sum_{k \in s_{iq}} (Y_k - \bar{Y}_{s_{iq}})^2 \tag{3.14}$$

An estimated variance of the three-stage Horvitz-Thompson estimate can then be written as

$$\hat{\mathbb{V}}_{3st}(\hat{Y}_\pi) = N_I^2 \left(1 - \frac{n_I}{N_I}\right) \frac{1}{n_I} s_I^2 + \frac{N_I}{n_i} \sum_{i \in s_i} N_{IIi}^2 \left(1 - \frac{n_{IIi}}{N_{IIi}}\right) \frac{1}{n_{IIi}} s_{IIi}^2$$

$$+ \frac{N_I}{n_I} \sum_{i \in s_I} \frac{N_{IIi}}{n_{IIi}} \sum_{q \in s_{IIi}} N_{iq}^2 \left(1 - \frac{n_{iq}}{N_{iq}}\right) \frac{1}{n_{iq}} s_{iq}^2 \tag{3.15}$$

The components of variance can be individually estimated according to

$$\hat{V}_{TSU} = \frac{N_I^2}{n_i^2} \sum_{i \in s_I} \frac{N_{IIi}^2}{n_{IIi}^2} \sum_{q \in s_{IIi}} N_{iq}^2 \left(1 - \frac{n_{iq}}{N_{iq}}\right) \frac{1}{n_{iq}} s_{iq}^2 \tag{3.16}$$

$$\hat{V}_{SSU} = \frac{N_I^2}{n_i^2} \sum_{i \in s_I} \left[N_{IIi}^2 \left(1 - \frac{n_{IIi}}{N_{IIi}}\right) \frac{1}{n_{IIi}} s_{IIi}^2 \right] - \hat{V}_{TSU}$$

$$\hat{V}_{PSU} = \hat{\mathbb{V}}_{3st}(\hat{Y}_\pi) - \hat{V}_{SSU} - \hat{V}_{TSU} \tag{3.17}$$

In general, the optimal allocation of resources is analytically extremely difficult. To illustrate we assume that $N_I = N_1$, $N_{IIi} \equiv N_2$ and $N_{iq} \equiv N_3$. That is, we have N_1 PSUs, all containing the same number N_2 of SSUs, and each comprising exactly N_3 elements. The overall population size is then $N = N_1 N_2$ Similarly, we perform simple random sampling without replacement with PSUs, n_2 SSUs in each of the selected PSUs and, finally, n_3 elements in each selected SSU. The following notation is necessary:

$$\bar{Y}_{iq} = \frac{1}{N_3} \sum_{k \in U_{iq}} Y_k$$

is the mean per element in the iqth SSU

$$\bar{Y}_{i\cdot} = \frac{1}{N_2 N_3} \sum_{q \in U_{IIi}} \sum_{k \in U_{iq}} Y_k$$

$$\bar{Y}_{..} = \frac{1}{N_1 N_2 N_3} \sum_{i \in U_I} \sum_{q \in U_{IIi}} \sum_{k \in U_{iq}} Y_k$$

is the overall mean per element in the population. Here, the overall estimate of the population mean will obviously be the ordinary mean of the sample of the $n_1 n_2 n_3$ selected elements. That is

$$\bar{Y}_s = \frac{1}{n_1 n_2 n_3} \sum_{i \in s_I} \sum_{q \in s_{IIi}} \sum_{k \in s_{iq}} Y_k$$

To express the variance more intuitively, we define the components of variance in terms of mean per element.

$$\sigma_1^2 = \frac{1}{N_1 - 1} \sum_{i \in U_I} (\bar{Y}_{i.} - \bar{Y}_{..})^2$$

$$\sigma_2^2 = \frac{1}{N_1(N_2 - 1)} \sum_{i \in U_I} \sum_{q \in U_{IIi}} (\bar{Y}_{iq} - \bar{Y}_{i.})^2$$

$$\sigma_3^2 = \frac{1}{N_1 N_2 (N_3 - 1)} \sum_{i \in U_I} \sum_{q \in U_{IIi}} \sum_{k \in U_{iq}} (Y_k - \bar{Y}_{iq})^2$$

The theoretical variance is then given by

$$\mathbb{V}(\bar{Y}_s) = \frac{1 - f_1}{n_1} \sigma_1^2 + \frac{1 - f_2}{n_1 n_2} \sigma_2^2 + \frac{1 - f_3}{n_1 n_2 n_3} \sigma_3^2 \qquad (3.18)$$

where $f_i = \frac{n_i}{N_i}$ are the sampling fractions. The estimated variance is

$$\hat{\mathbb{V}}(\bar{Y}_s) = \frac{1 - f_1}{n_1} \hat{\sigma}_1^2 + \frac{f_1(1 - f_2)}{n_1 n_2} \hat{\sigma}_2^2 + \frac{f_1 f_2(1 - f_3)}{n_1 n_2 n_3} \hat{\sigma}_3^2$$

We obtain the $\hat{\sigma}_i^2$ from these theoretical counterparts by taking sample copies. However, they are not the corresponding unbiased estimates but, instead, they satisfy the following relationships:

$$\mathbb{E}(\hat{\sigma}_1^2) = \sigma_1^2 + \frac{1 - f_2}{n_2} \sigma_2^2 + \frac{1 - f_3}{n_2 n_3} \sigma_3^2$$

$$\mathbb{E}(\hat{\sigma}_2^2) = \sigma_2^2 + \frac{1 - f_3}{n_3} \sigma_3^2$$

$$\mathbb{E}(\hat{\sigma}_3^2) = \sigma_3^2$$

(see Cochran (1977) pp. 286 and ff. for more details).

The theoretical variance can be rewritten as

$$\mathbb{V}(\bar{Y}_s) = -\frac{\sigma_1^2}{N_1} + \frac{1}{n_1}\left(\sigma_1^2 - \frac{\sigma_2^2}{N_2}\right) + \frac{1}{n_1 n_2}\left(\sigma_2^2 - \frac{\sigma_3^2}{N_3}\right) + \frac{1}{n_1 n_2 n_3} \sigma_3^2$$

$$\mathbb{V}(\bar{Y}_s) = v_0 + \frac{v_1}{n_1} + \frac{v_2}{n_1 n_2} + \frac{v_3}{n_1 n_2 n_3}$$

with weights v_i that depend only on the population parameters and not on the sample sizes. A reasonable model for costs would require taking a function of the form

$$C = c_0 + c_1 n_1 + c_2 n_1 n_2 + c_3 n_1 n_2 n_3$$

where c_0 is overhead cost and c_l is the cost of a sampling unit at the lth stage. Setting $w_1 = n_1$, $w_2 = n_1 n_2$ and $w_3 = n_1 n_2 n_3$ the optimal solution minimizes the variance

$$\mathbb{V}(\bar{Y}_s) = V = v_0 + \sum_{l=1}^{3} \frac{v_l}{w_l} \tag{3.19}$$

under given costs

$$C = c_0 + \sum_{l=1}^{3} w_l c_l \tag{3.20}$$

or, conversely, minimizes the costs for a given variance. Before solving this problem, however, we recall the famous **Cauchy-Schwarz** inequality which states that

$$\left(\sum_{i=1}^{n} a_i b_i \right)^2 \le \sum_{i=1}^{n} a_i^2 \sum_{i=1}^{n} b_i^2 \tag{3.21}$$

for all real numbers a_i, b_i. The equality holds if and only if we have proportionality $a_i = c b_i$, for all i and some real number c. This inequality plays a key role in many optimization problems. Minimizing the costs for a given variance or conversely leads to minimizing

$$(V - v_0)(C - c_0)$$

Using the Cauchy-Schwarz inequality with $a_i = \sqrt{\frac{v_i}{c_i}}$, $b_i = \sqrt{v_i c_i}$, we get

$$(V - v_0)(C - c_0) \ge \left(\sum_{l=1}^{3} \sqrt{v_l c_l} \right)^2$$

the lower bound being achieved when $\frac{v_l}{w_l} \propto w_l c_l$, that is, whenever

$$w_l^2 \propto \frac{v_l}{c_l} \tag{3.22}$$

or, equivalently, $(n_1 n_2 n_3)^2 = \frac{v_3}{c_3}$, $n_2^2 = \frac{v_2 c_1}{c_2 v_1}$ and $n_3^2 = \frac{v_3 c_2}{c_3 v_2}$. Our n_1 is therefore determined by the constraint on either the cost or on the variance. This is a notable result, for it implies that later-stage sample sizes are determined by variances and costs irrespective of the total sample size $n_1 n_2 n_3$. If, for example, the available budget changes, only n_1 should be altered. See Cochran (1977) for more details on cost optimization and Kendall et al. (1983) for more general results from multi-stage sampling. It is rare for multi-stage schemes

is almost invariably to lower costs rather then to reduce variance directly; economized resources can, of course, be applied to increasing the sample sizes.

It is worthwhile to note that the structure of the variance and cost functions given by Eq. 3.19 and Eq. 3.20 occur in many different problems, obviously also for $l > 3$, so that the optimality condition from Eq. 3.22 holds. This is especially the case with optimal stratification.

3.2 Abstract nonsense and elephants

Students may have heard in lectures on classical statistics that standard estimation procedures, e.g. least squares or maximum likelihood, enjoy some optimality properties, at least asymptotically. Therefore, it may come as a surprise that in sampling theory the situation is indeed very different, in the sense that there is simply no best estimator. This subject is very subtle and still controversial. Here, we present only the gist of the main arguments; Cassel et al. (1977) provide further details on this topic. We consider a **given population** $\mathcal{P}$ endowed **with all possible response variables** $\vec{Y} = (Y_1, Y_2, \ldots, Y_N)$, the Y_i being real numbers (i.e. $\vec{Y} \in \Re^N$), a **fixed sampling design** $p(\cdot)$ and the class $\mathcal{C}$ of all **unbiased estimators** $\hat{T}(\vec{Y})$ of the total $Y = \sum_{I=1}^{N} Y_i$, that is $\mathbb{E}_{p(\cdot)}\hat{T}(\vec{Y}) = Y$ for all $\vec{Y}$. The HT is one possible choice and, in some sense, the simplest.

One estimator $T_1 \in \mathcal{C}$ is said to be better than another $T_2 \in \mathcal{C}$ if

$$\mathbb{V}(\hat{T}_1(\vec{Y})) \leq \mathbb{V}(\hat{T}_2(\vec{Y}))$$

for all $\vec{Y}$ and with strict inequality for at least one $\vec{Y}$.

Consider now an arbitrarily chosen response variable $\vec{Y}_0$, with total Y_0. Let $\hat{T}(\vec{Y}_0)$ be any unbiased estimator of Y_0. A new unbiased estimator of the total for the actual response $\vec{Y}$ is

$$\hat{T}^*(\vec{Y}) = \hat{T}(\vec{Y}) + Y_0 - \hat{T}(\vec{Y}_0)$$

because $\mathbb{E}\hat{T}(Y) = Y$ and $\mathbb{E}\hat{T}(Y_0) = Y_0$. Now, when $\vec{Y} = \vec{Y}_0$, that is when the response variable being sampled in the population and the arbitrary response variable are identical, then

$$\hat{T}^*(\vec{Y}_0) = \hat{T}(\vec{Y}_0) + Y_0 - \hat{T}(\vec{Y}_0) = Y_0$$

and $\hat{T}^*(\vec{Y}_0)$ has zero variance. Hence for an estimator to be uniformly best for all $\vec{Y}$ it must have zero variance everywhere because $\vec{Y}_0$ has been chosen arbitrarily. This is impossible (unless of course the design is a census), so no uniformly best estimator can exist. This astounding negative result was first determined by V.P Godambe in 1955 and the above enlightening proof was given in 1991 by D. Basu (see Godambe, 1955; Basu, 1991). The problem

can the population of N elements consist of mice, the 10 biggest sequoia trees in California and, of course, elephants, but the response variable can be something such as the weight, the volume or the indicator variables that define the $2^N - 1$ possible sub-populations. Meaningful optimality criteria are possible only if some structure is imposed on the $\vec{Y}$. One consequence of this non-existence result is that no empirical conclusion can ever be conclusive.

If the very best is not available in this world one could at least hope that the estimators used in everyday life are not that bad after all. Fortunately, it has been proved, Godambe and Joshi (1965), that the Horvitz-Thompson estimator, among others, is admissible in the class of all unbiased estimators. This means that no other estimator has a smaller variance for all $\vec{Y}$s (and strictly smaller for at least one $\vec{Y}$). As we shall see later in forestry sampling, the HT estimator is practically the only one with a direct physical interpretation. In short, one may do better but also worse!

Let us consider a natural extension of the foregoing problem, namely: can we, by changing both design $p(\cdot)$ and estimator $\hat{T}(\vec{Y})$, find one strategy that is better than a particular strategy (p, T). The strategy (p_1, T_1) with unbiased T_1 is said to be at least as good as a competing one (p_2, T_2) with unbiased if $\mathbb{V}_{p_1}(T_1(\vec{Y})) \leq \mathbb{V}_{p_2}(T_2(\vec{Y}))$ for all $\vec{Y} \in \Re^N$. If the strict inequality holds for at least one $\vec{Y}$, then the strategy (p_1, T_1) is said to be better than $(p_2, T_2$ strategy (p, T) belonging to some class of strategies $\mathcal{C}$ is considered admissible in $\mathcal{C}$, if and only if no strategy in $\mathcal{C}$ is better than (p, T). It has been proved, Ramakrishnan (1975a), that **any** Horvitz-Thompson strategy $(p(\cdot), T_{HT})$ with expected sample size n is admissible in the class of all unbiased strategies with an expected sample size fixed at n. In particular the sample mean under simple random sampling without replacement is admissible.

Although unbiasedness, admissibility and other concepts are certainly important, common sense should not be dismissed. To illustrate that point let us examine this famous example from Basu (1991), that deals with elephants in a circus (note that Indian statisticians and elephants have made substantial contributions to sampling theory!).

The manager of a circus plans the dispatch of his 50 elephants to the next city the circus will visit. He needs an estimate of the total weight of the herd $Y = \sum_{i=1}^{50} Y_i$, where Y_i is the current weight of the ith elephant. He knows their individual weights $X_1, X_2, \ldots X_{50}$ from the previous year. Instead of re-weighting all the elephants, the manager proposes the following intuitive procedure: after having ranked in ascending order the X_i the manager notices that the elephant named Sambo corresponds roughly to the median of the X_i values. He suggests that Sambo be weighted again, and that value then multiplied by 50 as the estimation of the total Y. The statistician of the circus is opposed to that idea, arguing it is not rigorous, particularly because of the absolute lack of randomization. The manager admits that his knowledge of

the manager's idea has a lot of common sense. They agree on a compromise, which involves randomly selecting a single elephant. The sampling scheme is as follows: the inclusion probabilities are $\pi_i = 0.99$ if $i = i_0$ is Sambo and $\pi_i = \frac{1}{4900}$ for all other elephants, ensuring $\sum_{I=1}^{50} \pi_i = \frac{49}{4900} + 0.99 = 1 =$ The sampling takes place with a random number generator and, not inevitably, Sambo is selected. Instead of using the method suggested by the manager the statistician insists that the Horvitz-Thompson estimator be applied because it is unbiased and many scientific papers praise its properties. It is equal to $\frac{}{0}$ if Sambo is selected and $4900Y_i$ if any other elephant is chosen. Because Sambo has been chosen the estimated total weight calculated by the statistician is $1.01 \times Y_{\text{Sambo}}$, which is obviously and totally absurd. The dismayed manager then asked what the estimate would be if Jumbo, the largest elephant, were chosen. Even worse, the statistician then must admit that his new estimate would be equal to $4900 \times Y_{\text{Jumbo}}$. The career of our statistician, at least in this circus, stopped there. Although the HT is unbiased and admissible its variance is incredibly large in the present case, whereas the manager's estimate

$$\hat{Y}_M = 50Y_{\text{Sambo}}$$

would most likely be very close to the true value. Note that

$$\mathbb{E}(\hat{Y}_M) = \mathbb{E}50\sum_{i=1}^{50} Y_i I_i \approx 50(0.98Y_{\text{Sambo}} + 0.01\bar{Y})$$

and that the variance is much smaller than that of the HT estimate. Although this scheme with sample size $n = 1$ is so absurd that no reasonable person would apply it, it raises the question as to whether by using auxiliary information known prior to the sampling (i.e. weights in the previous year $X_1, X_2, \ldots, X_{50}$), one can construct a much better estimate. For instance, the intuitively appealing ratio estimator

$$\hat{Y}_R = \left(\sum_{i=1}^{50} X_i\right) \frac{Y_{i_0}}{X_{i_0}}$$

which has a small bias but a much smaller variance than the HT estimator. Generally speaking the HT estimator performs well if the inclusion probabilities π_i are roughly proportional to the response variable Y_i. Otherwise the estimator can be very poor, as in our elephant example, where obviously the π_i are totally uncorrelated with the Y_i.

We shall see in the next section how estimators like $\hat{Y}_M$ and $\hat{Y}_R$, can be logically obtained through linear modeling that uses auxiliary information.

3.3 Model-assisted estimation procedures

Sampling theory is characterized by an emphasis laid on using auxiliary information to improve the precision of estimates. We have already seen how such

and we shall now learn how to apply it in deriving more efficient estimators.

Generally speaking, an auxiliary variable is any variable about which information is available prior to sampling. Usually, we assume that the values of the variable are known for each of the N population elements before sampling begins. However it actually suffices to know the values for the elements in the sample and the sum over the population. We can describe the auxiliary information with p dimensional column vectors $\boldsymbol{x}_i = (x_{i1}, x_{i2}, \ldots, x_{ip})^t \in \Re^p$, i $1, 2 \ldots, N$. **Vectors are written in bold small characters and matrices in bold capital characters.** As an upper index the symbol t denotes the transposition operator for vectors and matrices. To simplify the notation small characters are also adopted for the individual components, i.e. the auxiliary variables, of vectors. Recall that qualitative auxiliary information can be represented by sets of $0 - 1$ indicator variables.

In the previous section we learned that no reasonable optimality criteria can be developed without imposing a structure on the response variable and the population. To do so we introduce the concept of a **super-population model** The response variable Y_i, which so far has been considered fixed in contrast to the random indicator variable I_i induced by the sampling scheme, is assumed to be the realization y_i of a hypothetical model M. More precisely we shall develop a multiple linear regression model that relates the response variable to the auxiliary information in the following sense:

1. The observed values $y_1, y_2, \ldots, y_N$ of the response variables are realized values of independent random variables $Y_1, Y_2 \ldots, Y_N$,

2.
$$\mathbb{E}_M(Y_i) = \sum_{k=1}^{p} \beta_k x_{ik} = \boldsymbol{x}_i^t \boldsymbol{\beta}, \quad i = 1, 2 \ldots N$$
where $\boldsymbol{\beta}^t = (\beta_1, \beta_2, \ldots, \beta_p) \in \Re^p$ is a p-dimensional vector of parameter.

3.
$$\mathbb{V}_M(Y_i) = \sigma_i^2, \quad i = 1, 2 \ldots N$$

Here, M introduces a new type of randomness. So far the only haphazard component has been that different sample outcomes s can occur under a given sampling design. A new randomness assumption is postulated, but has nothing to do with random selection. Therefore, the role of Model M is to describe the finite population point scatter $(y_i, \boldsymbol{x}_i)$. We may believe that our finite population looks as if it might have been generated in accordance Model However, the assumption is never made that the population is really generated this way. As we shall see, our conclusions about the finite population parameters, are by construction, independent of the model assumptions. That model serves as a tool to find an appropriate estimate of $\boldsymbol{\beta}$ by which to calculate the so-called regression estimator defined below. The efficiency of this estimator

of the regression estimator (e.g. approximate unbiasedness or validity of the variance formulas) do not depend on whether Model M holds or not. Thus the procedure is **model-assisted** and not **model-dependent**, a concept we shall encounter later.

Let us consider the estimation problem of the unknown parameter $\beta \in \Re$ Suppose that we can observe all the elements of the population, so that the following quantities are available

- $\boldsymbol{y} = (y_1, y_2, \ldots, y_N)^t$ the N-dimensional vector of all observed value of the response variable

- The (N, p) matrix $\boldsymbol{X}$ whose ith row is precisely the auxiliary vector $\boldsymbol{x}_i^t$.

- The diagonal (N, N) variance covariance matrix $\boldsymbol{\Sigma}$ with $\sigma_{ii} = \sigma_i^2$ and σ_{ij} 0 for all $i \neq j$

- The diagonal (N, N) matrix of inclusion probabilities $\boldsymbol{\Pi}$ with $\pi_{ii} = \pi_i$ and $\pi_{ij} = 0$ for all $i \neq j$

- The least squares estimate of the regression coefficient vector $\boldsymbol{\beta}_{\mathcal{P}}$ based on the entire population, which is given by

$$\boldsymbol{\beta}_{\mathcal{P}} = \left(\boldsymbol{X}^t \boldsymbol{\Sigma}^{-1} \boldsymbol{X}\right)^{-1} \boldsymbol{X}^t \boldsymbol{\Sigma}^{-1} \boldsymbol{y}$$

Note that one has

$$\boldsymbol{T} = \boldsymbol{X}^t \boldsymbol{\Sigma}^{-1} \boldsymbol{X} = \sum_{i=1}^{N} \frac{\boldsymbol{x}_i \boldsymbol{x}_i^t}{\sigma_i^2}$$

and

$$\boldsymbol{t} = \boldsymbol{X}^t \boldsymbol{\Sigma}^{-1} \boldsymbol{y} = \sum_{i=1}^{N} \frac{y_i \boldsymbol{x}_i}{\sigma_i^2}$$

and therefore

$$\boldsymbol{\beta}_{\mathcal{P}} = \boldsymbol{T}^{-1} \boldsymbol{t}$$

Also note that, if we had a perfect linear relationship $\boldsymbol{y} = \boldsymbol{X}\boldsymbol{\beta}$ for some $\boldsymbol{\beta}$ then, by the above formula, $\boldsymbol{\beta}_{\mathcal{P}} = \boldsymbol{\beta}$. Of course, $\boldsymbol{T}$ and $\boldsymbol{t}$ are not available but they can be estimated with the sample drawn s according to the Horvitz-Thompson procedure. That is we set

$$\hat{\boldsymbol{T}}_s = \sum_{i \in s} \frac{\boldsymbol{x}_i \boldsymbol{x}_i^t}{\pi_i \sigma_i^2} \qquad (3.23)$$

$$\hat{\boldsymbol{t}}_s = \sum_{i \in s} \frac{y_i \boldsymbol{x}_i}{\pi_i \sigma_i^2}$$

$$\hat{\boldsymbol{\beta}}_s = \hat{\boldsymbol{T}}_s^{-1} \hat{\boldsymbol{t}}_s$$

Accounting for the restriction of the matrices $\boldsymbol{X}$, $\boldsymbol{\Sigma}$ and $\boldsymbol{\Pi}$, as well as the ordering of the label $i \in s$) we can rewrite

$$\hat{\boldsymbol{\beta}}_s = \left(\boldsymbol{X}_s^t(\boldsymbol{\Sigma}_s\boldsymbol{\Pi}_s)^{-1}\boldsymbol{X}_s\right)^{-1}\boldsymbol{X}_s^t(\boldsymbol{\Sigma}_s\boldsymbol{\Pi}_s)^{-1}\boldsymbol{Y}_s \qquad (3.24)$$

We have assumed implicitly that the inverse matrices all exist. Depending on the model this may not be true for all realized samples s, e.g. with analysis of variance models if one of the group is not present in the sample. In such a case one must simplify the model by merging some of the groups.

For fixed realizations y_i the sample regression coefficient vector $\hat{\boldsymbol{\beta}}_s$ is an asymptotically design-unbiased estimate of the population regression coefficient vector $\boldsymbol{\beta_P}$. ($\hat{\boldsymbol{t}}_s$ is exactly unbiased for $\boldsymbol{t}$, but, because of the matrix inversion, $\boldsymbol{T}$ is only approximately unbiased for $\boldsymbol{T}^{-1}$).

Again, let us note that a perfect linear relationship $\boldsymbol{y} = \boldsymbol{X}\boldsymbol{\beta}$ for some is inherited by the sample in the sense that $\boldsymbol{y}_s = \boldsymbol{X}_s\boldsymbol{\beta}$, thereby implying $\boldsymbol{\beta_P} = \boldsymbol{\beta} = \hat{\boldsymbol{\beta}}_s$.

The theoretical, or non-observable, predictions are defined as $\tilde{y}_i = \boldsymbol{x_i}^t\boldsymbol{\beta_P}$ the theoretical residuals as $\tilde{r}_i = y_i - \tilde{y}_i$. Because $y_i = \tilde{y}_i + \tilde{r}_i$ we can write

$$y = \sum_{i=1}^N y_i = \sum_{i=1}^N \tilde{y}_i + \sum_{i=1}^N \tilde{r}_i$$

In words, the sum of the observed values is the sum of the predictions plus the sum of the residuals. Because the auxiliary vector is known for all elements in the population $\sum_{i=1}^N \tilde{y}_i$ is known exactly if $\boldsymbol{\beta_P}$ were known. In contrast, the residuals are known only for the elements in the sample s. To estimate $\sum_{i=1}^N$ we use the Horvitz-Thompson technique and obtain an unbiased theoretical regression estimate as

$$\hat{y}_{treg} = \sum_{i=1}^N \tilde{y}_i + \sum_{i \in s} \frac{\tilde{r}_i}{\pi_i} \qquad (3.25)$$

We do not assume that the sum of the residuals over the entire population is zero, or, equivalently, that the model does not have a systematic error. Correcting the predictions by the residuals ensures that the regression estimate is unbiased even when the model is wrong. Actually, the above arguments remain valid for any arbitrary choice $\boldsymbol{\beta_0}$ in place of To calculate the design-based variance, i.e. for fixed y_i, note that $\sum_{i=1}^N$ is non-random and that the only contribution to the variance comes from $\sum_{i \in s} \frac{\tilde{r}_i}{\pi_i}$, for which we can use Eq. 2.8 to obtain

$$\mathbb{V}(\hat{y}_{treg}) = \sum_{i,j=1}^N \check{\tilde{r}}_i\check{\tilde{r}}_j\Delta_{ij} \qquad (3.26)$$

given in Eq. 3.23. Empirical predictions are then defined as $\hat{y}_i = \boldsymbol{x}_i^t \hat{\boldsymbol{\beta}}_s$ and the empirical residuals as $r_{is} = y_i - \hat{y}_i$. We use the notation r_{is} instead of just because the empirical residual of the ith element also depends, through $\hat{\boldsymbol{\beta}}_s$, on the other elements in the sample. The regression estimator is then defined as

$$\hat{y}_{reg} = \sum_{i=1}^{N} \hat{y}_i + \sum_{i \in s} \frac{r_{is}}{\pi_i} \tag{3.27}$$

It is intuitively plausible that, with a large sample size, the effect of replacing $\boldsymbol{\beta}_p$ with $\hat{\boldsymbol{\beta}}_s$ is negligible so that $\hat{y}_{reg}$ is approximately unbiased. Furthermore, the variance from Eq. 3.26 will remain approximately valid if we replace the theoretical residuals with their empirical counterparts. This is likewise true for the estimated variance given by Eq. 2.9 (Särndal et al., 2003). Setting $r_{is} = y_i - \hat{y}_i = y_i - \boldsymbol{x}_i^t \hat{\boldsymbol{\beta}}_s$ in 3.27, one obtains the equivalent version $\hat{y}_{reg}$ $\hat{y}_\pi + (\boldsymbol{x} - \boldsymbol{x}_\pi)^t \hat{\boldsymbol{\beta}}_s$, where $\hat{y}_\pi$ is the ordinary HT estimate of the total of the response variable, $\boldsymbol{x}$ is the known total for the vector of auxiliary variables and $\boldsymbol{x}_\pi$ is the HT estimate of that total. Our regression estimate adjusts therefore the ordinary HT estimate for differences in the auxiliary variables between the population and the sample. These results can be summarized by

Theorem 3.3.1.

$$
\begin{aligned}
\hat{y}_{reg} &= \sum_{i=1}^{N} \hat{y}_i + \sum_{i=1}^{N} \frac{r_{is}}{\pi_i} \\
\hat{y}_{reg} &= \hat{y}_\pi + (\boldsymbol{x} - \boldsymbol{x}_\pi)^t \hat{\boldsymbol{\beta}}_s \\
\mathbb{V}(\hat{y}_{reg}) &= \sum_{i,j=1}^{N} \breve{r}_{is} \breve{r}_{js} \Delta_{ij} \\
\hat{\mathbb{V}}(\hat{y}_{reg}) &= \sum_{i,j \in s} \breve{r}_{is} \breve{r}_{js} \breve{\Delta}_{ij}
\end{aligned}
$$

Remarks:

1. Although we are working in pure design-based inference we use small letters rather than capitals for the various variables to emphasize the fact that they were originally considered to be realizations. For example, one can replace in 3.3.1 y_i with Y_i, $\boldsymbol{x}_\pi$ with $\boldsymbol{X}_\pi$ etc., so that the notation will be the same as in previous sections.

2. Many models entail an intercept term, i.e. $x_{i1} \equiv 1$. Therefore, one must be careful with qualitative variables that are described by indicators in order to ensure that the matrix $\hat{\boldsymbol{T}}_s$ is regular.

3. We have assumed that the σ_i^2 are known. Nevertheless, in many problems, one models the variance through a relationship of the form $\sigma_i^2 = \sigma^2 f$

$(\;)$ (e.g. $_i$ $_{ij}$ $)$. In doing so, the unknown scale factor σ^2 cancels out in $\hat{\boldsymbol{\beta}}_s$.

4. Under a perfect linear relationship $\boldsymbol{y} = \boldsymbol{X}\boldsymbol{\beta}$ one arrives at $\sum_{i\in s} \frac{r_{is}}{\pi_i} = 0$. This follows from the fact that $r_{is} = y_i - \boldsymbol{x}_i^t\hat{\boldsymbol{\beta}}_s = \boldsymbol{x}_i^t\boldsymbol{\beta} - \boldsymbol{x}_i^t\boldsymbol{\beta} = 0$, because, as previously shown, $\hat{\boldsymbol{\beta}}_s = \boldsymbol{\beta}$ in this case. Hence, all the residuals vanish.

5. One sufficient condition under which the residual adjustment might disappear, i.e. $\sum_{i\in s} \frac{r_{is}}{\pi_i} = 0$, would be that

$$\sigma_i^2 = \boldsymbol{\lambda}^t\boldsymbol{x}_i \tag{3.28}$$

for some constant column vector $\boldsymbol{\lambda}$. This condition can always be fulfilled: if necessary one simply extends $\boldsymbol{x}_i$ with one further component, namely The proof follows from the definitions (Särndal et al., 2003).

Finally, let us consider another alternative form of the regression estimate. Using $\hat{y}_{reg} = \hat{y}_{\pi} + (\boldsymbol{x} - \boldsymbol{x}_{\pi})^t\hat{\boldsymbol{\beta}}_s$ and the definition of $\hat{\boldsymbol{\beta}}_s$ we obtain, with the so-called g-weights,

$$g_{is} = 1 + (\boldsymbol{x} - \boldsymbol{x}_{\pi})^t\hat{\boldsymbol{T}}_s^{-1} \frac{\boldsymbol{x}_i}{\sigma_i^2} \tag{3.29}$$

where the notation g_{is} indicates that the weight depends not only on the element but also, via the matrix $\hat{\boldsymbol{T}}_s$, on all other elements in sample s. The regression estimator can be written as

$$\hat{y}_{reg} = \hat{y}_{greg} = \sum_{i\in s} g_{is} \frac{y_i}{\pi_i} \tag{3.30}$$

The total weight attached to the observation y_i is the product of the sampling weight $\frac{1}{\pi_i}$ and the g-weight g_{is}. This is in contrast to the classic HT estimator, in which the weight depends only on the label i and not on the whole sample. An important property of the g-weights is

$$\sum_{i\in s} g_{is}\check{\boldsymbol{x}}_i^{\,t} = \sum_{i=1}^{N} \boldsymbol{x}_i^t = \boldsymbol{x}^t \tag{3.31}$$

The proof is as follows:

$$
\begin{aligned}
\sum_{i\in s} g_{is}\check{\boldsymbol{x}}_i^{\,t} &= \sum_{i\in s} \left(1 + (\boldsymbol{x} - \boldsymbol{x}_{\pi})^t\hat{\boldsymbol{T}}_s^{-1} \frac{\boldsymbol{x}_i}{\sigma_i^2}\right) \frac{\boldsymbol{x}_i^t}{\pi_i} \\
&= \sum_{i\in s} \check{\boldsymbol{x}}_i^{\,t} + (\boldsymbol{x} - \boldsymbol{x}_{\pi})^t\hat{\boldsymbol{T}}_s^{-1}\hat{\boldsymbol{T}}_s \\
&= \boldsymbol{x}_{\pi}^t + \boldsymbol{x}^t - \boldsymbol{x}_{\pi}^t \\
&= \sum_{i=1}^{N} \boldsymbol{x}_i^t
\end{aligned}
$$

Intuitively, because the g-weights provide perfect estimates for the total of

model is adequate. We note that under the condition $\sigma_i^2 = \boldsymbol{\lambda}^t \boldsymbol{x}_i$ the g-weight expression can be simplified to

$$g_{is} = \boldsymbol{x}^t \hat{\boldsymbol{T}}^{-1} \frac{\boldsymbol{x}_i}{\sigma_i^2} \qquad (3.32)$$

Under this condition one has

$$\hat{y}_{reg} = \sum_{i \in s} g_{is} \frac{y_i}{\pi_i} = \sum_{i=1}^{N} \hat{y}_i = \boldsymbol{x}^t \hat{\boldsymbol{T}}_s^{-1} \sum_{i \in s} \frac{y_i \boldsymbol{x}_i}{\pi_i \sigma_i^2}$$

which is equal to $\sum_{i \in s} (\boldsymbol{x}^t \hat{\boldsymbol{T}}_s^{-1} \frac{\boldsymbol{x}_i}{\sigma_i^2}) \frac{y_i}{\pi_i}$. Because this equality holds for all values y_i and the g_{is} are independent of the y_i, Eq. 3.32 follows. To calculate the variance of $\hat{y}_{greg}$ we set in 3.30 $y_i = \tilde{y}_i + \tilde{r}_i = \boldsymbol{x}_i{}^t \boldsymbol{\beta_P} + \tilde{r}_i$ and use 3.31 to get $\hat{y}_{greg} = \boldsymbol{x}^t \boldsymbol{\beta_P} + \sum_{i \in s} g_{is} \frac{\tilde{r}_i}{\pi_i}$. The first term is a constant so that $\mathbb{V}(\hat{y}_{greg}) = \mathbb{V}(\sum_{i \in s} g_{is} \frac{\tilde{r}_i}{\pi_i})$. This looks like the variance of an HT estimator, with the difference being that the weights g_{is} are now attached to the $\tilde{r}_i$. However, disregarding that the weights are sample-dependent, and inserting the empirical residual r_i in place of the theoretical $\tilde{r}_i$ we obtain by Theorem 2.2.2 the asymptotically unbiased estimate of variance that was first proposed by Särndal in 1982 (the arguments developed in Särndal et al. (2003) are easier to understand).

Theorem 3.3.2.

$$\hat{\mathbb{V}}(\hat{y}_{greg}) = \sum_{i,j \in s} \check{\Delta}_{ij} g_{is} \check{r}_{is} g_{js} \check{r}_{js}$$

There is convincing theoretical and empirical evidence that 3.3.2 is better than the formula given in theorem 3.3.1 (see Särndal et al. (2003) for details and further references).

Let us examine some examples of regression estimators.

1. **The common mean model:**

   This model specifies that

   $$\begin{aligned} \mathbb{E}_M(Y_i) &= \beta \qquad (3.33) \\ \mathbb{V}_M(Y_i) &= \sigma^2 \end{aligned}$$

   There is a single auxiliary variable $x_i \equiv 1$, which although apparently not providing any information, does in fact do so. Condition 3.28 obviously holds. Calculations are, in this case, trivial and one obtains : $\hat{\boldsymbol{T}}_s$ $\sum_{i \in s} \frac{1}{\pi_i \sigma^2}$, $\hat{\boldsymbol{t}}_s = \sum_{i \in s} \frac{y_i}{\pi_i \sigma^2}$,

   $$\hat{\beta}_s = \frac{\sum_{i \in s} \frac{y_i}{\pi_i}}{\sum_{i \in s} \frac{1}{\pi_i}} = \tilde{y}_s$$

$$N \qquad \sum_{i \in s} \pi_i$$

$$\hat{y}_{reg} = N\tilde{y}_s \tag{3.34}$$

$$\hat{V}(\hat{y}_{reg}) = \left(\frac{N}{\tilde{N}}\right)^2 \sum_{i,j \in s} \check{\Delta}_{ij} \check{r}_{is} \check{r}_{js}$$

with the residuals $r_{is} = y_i - \tilde{y}$. This is the estimator we presented in Eq. 2.29. It is also the one suggested by that circus manager for the elephant herd discussed in Section 3.2. Hence, when we consider the observations as the realization of random variables with the same mean and the same variance, we find they indeed contain a lot of information!

2. **The ratio estimator:**

Our model specifies that

$$\mathbb{E}_M(Y_i) = \beta x_i \tag{3.35}$$
$$\mathbb{V}_M(Y_i) = \sigma^2 x_i$$

The response variable as well as its variability are proportional to a single auxiliary variable, a situation that frequently occurs in practice, at least approximately. Note that Eq. 3.28 holds. A simple calculation leads to $\hat{T}_s = \frac{1}{\sigma^2} \sum_{i \in s} \frac{x_i}{\pi_i}$, $\hat{t}_s = \frac{1}{\sigma^2} \sum_{i \in s} \frac{y_i}{\pi_i}$, $\hat{\beta}_s = \frac{\sum_{i \in s} \frac{y_i}{\pi_i}}{\sum_{i \in s} \frac{x_i}{\pi_i}} = \frac{\hat{y}_{HT}}{\hat{x}_{HT}}$, $g_{is} = \frac{x}{\hat{x}_{HT}}$ and, hence, to

$$\hat{y}_{ratio} = \left(\frac{x}{\hat{x}_{HT}}\right) \hat{y}_{HT} \tag{3.36}$$

$$\hat{V}(\hat{y}_{ratio}) = \left(\frac{x}{\hat{x}_{HT}}\right)^2 \sum_{i,j \in s} \check{\Delta}_{ij} \check{r}_{is} \check{r}_{js}$$

Using simple random sampling with $\pi_i \equiv \frac{n}{N} = f$ one obtains $\hat{\beta}_s = \frac{\bar{y}_s}{\bar{x}_s}$ and

$$\hat{y}_{ratio} = N\bar{x}\frac{\bar{y}_s}{\bar{x}_s} \tag{3.37}$$

$$\hat{V}(\hat{y}_{ratio}) = \left(\frac{\bar{x}}{\bar{x}_s}\right)^2 N^2 \frac{1-f}{n} \frac{\sum_{i \in s}(y_i - \hat{\beta}_s x_i)^2}{n-1}$$

where $\bar{y}_s$ and $\bar{x}_s$ are the ordinary sample means.

It turns out that the ratio estimator has a smaller variance than the ordinary HT estimator if the coefficients of variations for x_i and y_i are nearly equal and if the correlation between them is greater than 0.5. If the regression of y_is on x_is show a distinct positive intercept the HT estimator is better. In such a case the model should include an intercept term, which leads to the simple linear regression estimator presented in the next example.

This model reads

$$
\begin{aligned}
\mathbb{E}_M(Y_i) &= \alpha + \beta x_i \\
\mathbb{V}_M(Y_i) &= \sigma^2
\end{aligned}
\tag{3.38}
$$

The calculations are simple but tedious (inversion of a $(2,2)$ matrix) so we present only the results here. One obtains with $\hat{N} = \sum_{i \in s} \frac{1}{\pi_i}$, $\tilde{y}_s = \frac{\sum_{i \in s} \frac{\tilde{y}_i}{\pi_i}}{\hat{N}}$, $\tilde{x}_s = \frac{\sum_{i \in s} \frac{\tilde{x}_i}{\pi_i}}{\hat{N}}$, and

$$
\hat{\beta}_s = \frac{\sum_{i \in s} \frac{(x_i - \tilde{x}_s)(y_i - \tilde{y}_s)}{\pi_i}}{\sum_{i \in s} \frac{(x_i - \tilde{x}_s)^2}{\pi_i}}
$$

$$
\hat{\alpha}_s = \tilde{y}_s - \hat{\beta}_s \tilde{x}_s
$$

The regression estimator is then given by

$$
\hat{y}_{reg} = N \left(\tilde{y}_s - \hat{\beta}_s (\bar{x} - \tilde{x}_s) \right)
\tag{3.39}
$$

Setting $\tilde{S}_{xs}^2 = \frac{\sum_{i \in s} \frac{(x_i - \tilde{x}_s)^2}{\pi_i}}{\hat{N}}$ and $a_s = \frac{(\bar{x} - \tilde{x}_s)}{\tilde{S}_{xs}^2}$. The g-weights are

$$
g_{is} = \frac{N}{\tilde{N}} (1 + a_s(x_i - \tilde{x}_s))
$$

and the residuals

$$
r_{is} = y_i - \tilde{y}_s - \hat{\beta}_s(x_i - \tilde{x}_s)
$$

The variance can be then calculated with Theorem 3.3.2. Under simple random sampling one obtains $\tilde{y}_s = \bar{y}_s$, $\tilde{x}_s = \bar{x}_s$, $\hat{N} = N$ and

$$
\hat{\mathbb{V}}(\hat{y}_{reg}) = \frac{N^2(1-f)}{n(n-1)} \sum_{i \in s} (1 + a_s(x_i - \bar{x}_s))^2 r_{is}^2
$$

where $a_s = \frac{n(\bar{x} - \bar{x}_s)}{\sum_{i \in s}(x_i - \bar{x}_s)^2}$.

For simple random sampling the simple regression estimator will, for large samples, usually perform better than the HT and the ratio estimator. In small samples, say, of size ≤ 12, the situation is not as clear, particularly because the bias can be a source of concern.

More complex models such as the group mean model (which corresponds to a one-way ANOVA) and the group ratio model (i.e. separate ratio estimators are calculated within G different groups) have been described by Särndal et al. (2003). Likewise, it can be generalized to cluster-sampling and two-phase sampling where the auxiliary information is unavailable for the entire population but exists only for a large sample s_1 that contains a smaller sample $s_2 \subset s_1$, in which the response variable y_i is known for all elements. We shall consider such advanced procedures later, in the context of a forest inventory.

has no satisfactory solution from a pure design-based approach. For unbiased or asymptotically unbiased estimators under so-called non-informative designs, the π_i can depend on the auxiliary variables $\boldsymbol{x}_i$ but not explicitly on the response variable y_i. In this case, the efficiency of an estimator $\hat{Y}$ be assessed by its **anticipated variance** defined as $\mathbb{E}_M \mathbb{V}_{p(\cdot)|M}(\hat{Y})$, i.e. the average (over all realizations of the Y_i) under Model M of the design-based variance (for a given realization). Minimizing the anticipated variance under a constraint for the expected sample size lead to inclusion probabilities π_i that are proportional to the standard deviations σ_i and, therefore, to the prediction error. In many applications this prediction is itself roughly proportional to some x_i, so that we develop a **PPS**-like procedure. Let us briefly mention the model-dependent approach in finite populations. The basic idea is to postulate a model M, such that the y_i can be viewed as the realization of random variables Y_i and $\mathbb{E}_M Y_i = \boldsymbol{x}_i^t \boldsymbol{\beta}$ with a given covariance structure and an auxiliary vector $\boldsymbol{x}_i$ known for each unit in $\mathcal{P}$. The goal is to find an $\hat{\boldsymbol{\beta}}$ for an optimal model-dependent estimator of the form $\hat{y}_{pred} = \sum_{i \in s} y_i + (\sum_{i \notin s} \boldsymbol{x}_i^t \hat{\boldsymbol{\beta}})$, see e.g. Chaudhuri and Stenger (1992) for details. We shall use this approach again in the forest inventory context.

So far our emphasis has been placed on estimating totals, means, ratios, and eventually correlations between pairs of variables. The objective is to attain an estimate of such quantities, together with estimated variances and approximate confidence intervals, primarily by following a design-based approach (model-assisted estimation included). In this sense we have **descriptive studies**. Survey data are increasingly being used to model complex relationships among variables and to analyze those data and draw inferences in a super-population setup. We shall summarily state here that we have **analytical studies**. These survey data can be employed either to estimate parameters and test hypotheses in the design-based approach or to infer about parameters of super-population models. This subject is rather complex, but the reader can consult Särndal et al. (2003) and Chaudhuri and Stenger (1992) for details. Nevertheless, a word of warning is necessary here. Very often categorical survey data are presented in tables. One might be tempted to use standard statistical techniques, such as chi-squares, to test for homogeneity or independence. However, the validity of those tests rests upon the assumption that units have been drawn by simple random sampling, which is frequently not true in survey sampling (and practically never the case in forest inventory). **These classical tests are therefore not generally valid and can be very misleading**. We shall briefly address this topic in the context of a forest inventory in Subsections 6.4.1 and 6.4.2.

Problem 3.1. *We consider a population $\mathcal{P} = \{1, 2, \ldots, 100\}$, partitioned into 10 primary units of 10 consecutive integers $U_1, U_2, \ldots U_{10}$. Each primary unit is further partitioned into 5 secondary units of pairs of 2 consecutive integers. For instance $U_2 = \{11, 12, 13, 14, 15, 16, 17, 18, 19, 20\}$ and U_{21} $\{11, 12\}, U_{22} = \{13, 14\}, U_{23} = \{15, 16\}, U_{24} = \{17, 18\}, U_{25} = \{19, 20\}$. We consider the following three-stage sampling scheme: $n_1 = 5$ out of $N_1 = 10$ primary units at the first stage, $n_2 = 2$ out of $N_2 = 5$ secondary units at the second stage and $n_3 = 1$ out of $N_3 = 2$ elements (tertiary units) at the third stage. At each stage we use simple random sampling without replacement. Hence $N_1 N_2 N_3 = N = 100$ and $n = n_1 n_2 n_3 = 10$ as in Problems 2.4, 2.5, 2.9, 2.10. Find the tree-stage Horvitz-Thompson estimate of the population mean and its variance by applying Eq. 3.18.*

Problem 3.2. *We consider a population of $N = 2010$ farms for which the mean of the total land cultivated (auxiliary variable X_i) is 118.32ha. We want to estimate the mean surface area of cereal crops being grown (response variable Y_i). A sample of $n = 100$ farms is selected by simple random sampling. The data are*

$$\bar{X}_s = 131.25ha, \ \bar{Y}_s = 29.07ha$$

The estimated variance and covariances are

$$\hat{S}_X^2 = 9173ha^2, \ \hat{S}_Y^2 = 708ha^2, \ \hat{S}_{X,Y} = \frac{1}{n-1}\sum_{i\in s}(X_i - \bar{X}_s)(Y_i - \bar{Y}_s) = 1452.60$$

Determine the ratio estimate of $\bar{Y}$ and its 95% confidence interval by using Theorem 3.3.1 and Eq. 3.37.

Problem 3.3. *For which condition is the ratio estimator better than the ordinary sample mean under simple random sampling?*

Problem 3.4. *This problem concerns a generalization of the ratio estimator given in Eq. 3.36, the so-called post-stratified ratio estimator. Population $\mathcal{P}$ size N is decomposed into G sub-populations $\mathcal{P}_g$, $g = 1, 2 \ldots G$ ($\mathcal{P} = \cup_{g=1}^{G}\mathcal{P}$ of respective sizes N_g, and likewise samples s_g of sizes n_g ($s = \cup_{g=1}^{G}s_g$). By definition $N = \sum_{g=1}^{G} N_g$ and $n = \sum_{g=1}^{G} n_g$. The model is*

$$\mathbb{E}_M(Y_i) = \beta_g x_i \ \text{if} \ i \in \mathcal{P}_g$$

$$\mathbb{V}(Y_i) = \sigma_g^2 x_i \ \text{if} \ i \in \mathcal{P}_g$$

This can be rewritten as

$$\mathbb{E}_m(Y_i) = \boldsymbol{x}_i^t \boldsymbol{\beta}$$

with the notation $\boldsymbol{x}_i = (x_{1i}, x_{2i}, \ldots x_{gi}, \ldots x_{Gi})^t$, where $x_{gi} = 1$ if $i \in \mathcal{P}_g$ $x_{gi} = 0$ if $i \notin \mathcal{P}_g$ for $g = 1, 2, \ldots G$. We consider simple random sampling without replacement of n out of the N individuals from $\mathcal{P}$. Because the allocation of the units to the sub-population is known after sampling we are working

$\hat{\boldsymbol{\beta}}_s = (\hat{\beta}_1, \hat{\beta}_2, \ldots \hat{\beta}_G)^t$ *where*

$$\hat{\beta}_g = \frac{\sum_{i \in s_g} y_i}{\sum_{i \in s_g} x_{gi}} := \frac{\bar{y}_{s_g}}{\bar{x}_{s_g}}$$

Setting $t_g = \sum_{i \in P_g} x_i =: N_g \bar{x}_g$ *show that the regression estimator can be written as*

$$\hat{Y}_{reg} = \sum_{g=1}^{G} t_g \hat{\beta}_g$$

and that its variance is given by

$$\hat{\mathbb{V}}(\hat{Y}_{reg}) = (1 - f) \sum_{g=1}^{G} \left(\frac{\bar{x}_g}{\bar{x}_{s_g}}\right)^2 N_g^2 \frac{\hat{\sigma}_g^2}{n_g}$$

$$\hat{\sigma}_g^2 \approx \frac{1}{n_g - 1} \sum_{i \in s_g} (y_i - \hat{\beta}_g x_i)^2$$

by using the approximation

$$\frac{n_g - 1}{n_1} \frac{n}{n_g} = 1$$

and where $f = \frac{n}{N}$.

CHAPTER 4

Forest Inventory: one-phase sampling schemes

4.1 Generalities

In this chapter we shall present the most important concepts and tools available for forest sampling in a framework general enough to cover the majority of the situations encountered in practice. Unfortunately, the terminology often used is not uniform, so the reader should carefully look at the various definitions provided in the literature.

Most modern forest inventories are combined forest inventories, i.e. the information obtained directly in the forest, via terrestrial plots, is enhanced by auxiliary information gained through remote sensing data, such as aerial photographs or satellites images, or any other sources, like thematic maps or previous inventories. In short, and for the purpose of this book, we shall say that auxiliary information is compiled in the **first phase**, usually associated with a very large sample size (may be even infinite if thematic maps are used). Then, the **second phase** collects terrestrial information from a **sub-sample of the first phase sample**. Those terrestrial data are gathered either through **one-stage** procedures, in which trees are selected to obtain directly the response variable of interest, or by **two-stage** procedures, in which the **first-stage** trees are chosen to determine an approximation of the response variable. A sub-sample of the first-stage trees, the **second-stage** trees, is drawn to obtain the exact response. For each of the 4 scenarios described here, one can use 1) **simple random sampling**, in which the information is collected from single points uniformly and independently distributed within the forest area, or 2) **cluster random sampling**, in which the information is gathered in clusters (a set of points with a fixed geometrical structure), whose origins are drawn by simple random sampling. The number of points per cluster that fall within the forest area is also a random variable. In that situation, cluster-sampling is sometimes also called **trakt sampling**.

In practice, random sampling is rarely if ever used. Instead, most inventories rely on systematic grids that have random start or orientation. In two-phase systematic sampling the second-phase grid is a sub-grid of the first-phase grid. **It is impossible to construct with one single systematic sample a**

forest inventorists will treat points lying on systematic grids as if they were random. To a large extent this is acceptable for point estimates, but less so for variance estimates, which, as suggested by empirical evidence, are usually overestimated (theoretically the converse is also possible). If one suspects a periodicity of some kind in the forest structure it is essential that it does not coincide with the periodic structure of the grid. A mathematically coherent variance estimation from a single systematic sample is only feasible within the model-dependent approach.

Here, sampling theory for forest inventories will be developed systematically within the framework of the infinite population model, as proposed by Mandallaz (1991). Furthermore, some ideas from the pioneering work of C.E. Saerndal have been adapted to this model (Särndal et al., 2003). The approach presented in this book differs from the classical references such as de Vries (1986) and Schreuder et al. (1993), which primarily rely on the paradigm of classical sampling theory for finite populations as formulated by Cochran (1977).

4.1.1 Terminology and formulation of the inventory problem

We now proceed to define the terminology and notation that will be used throughout this book.

Consider a forest area F that is assumed to be a subset of the Euclidean plane $\Re^2$. In practice this implies that F is already the suitable projection of a real forest, e.g. a parallel projection whenever the earth curvature can be neglected. We shall not delve into great detail about the approximate practical solutions, like slope correction, to achieve this (Kangas and Maltamo, Eds, 2006). The surface area of F is denoted by $\lambda(F)$, usually in ha.

We also consider a well-defined population $\mathcal{P}$ of N trees lying in F; these trees are identified by their labels $i = 1, 2 \ldots N$. Their positional vectors are indicated by $u_i \in \Re^2$, $i = 1 \ldots N$. To simplify this notation we shall write $i \in G$ instead of $u_i \in G \subseteq F$; the surface area of an arbitrary set G is always denoted by $\lambda(G)$.

The response variables of interest measured or observed at a given time point for each tree in $\mathcal{P}$ are presented as $Y_i^{(m)}, m = 1, \ldots p$. **These response are assumed to be error-free.** Whenever confusion can be excluded from the context we shall drop the upper index (m). The response $Y_i^{(m)}$ can be a real or integer number in the usual sense, e.g. the timber volume, the basal area, $Y_i \equiv 1$ for the number of stem, or a set of binary $0, 1$ variables that code categorical variables such as species, state of health or any variable defined with the previous ones. Here, we take the obvious abstract view that trees are dimensionless points in the plane at which the response variables $Y_i^{(m)}$ defined. It is worth noting that a variable, e.g. $Y_i^{(2)}$, can illustrate a change in another variable, say $_i$, over a given time period. This is particularly useful in the context of continuous forest inventories with permanent plots.

In most instances the population $\mathcal{P}$ comprises trees with a minimum DBH This is called the **inventory threshold**. It is tacitly assumed that all response variables are by definition equal to zero if the tree does not belong to this population, i.e. if its DBH is below that threshold.

Given any set $G \subseteq F$ the objectives of a forest inventory are, in the restricted sense, to gather information on spatial means (densities), totals or ratios, that are defined according to

$$\bar{Y}_G^{(m)} = \frac{1}{\lambda(G)} \sum_{i \in G} Y_i^{(m)}$$

$$Y_G^{(m)} = \sum_{i \in G} Y_i^{(m)}$$

$$R_{l,m} = \frac{\bar{Y}_G^{(l)}}{\bar{Y}_G^{(m)}} = \frac{Y_G^{(l)}}{Y_G^{(m)}} \tag{4.1}$$

possibly also for many different sets G and over several time points.

If the set G is small a full census is feasible and often preferable to a sample survey. However, a full census usually is not practical for several and almost never for an entire forest area F. Hence, one must use sampling techniques that generally perform well for F but not necessarily so for small $G \subseteq F$. This is known as the problem of small-area estimation.

4.2 One-phase one-stage simple random sampling scheme

With respect to a particular design, we shall use the symbols $\mathbb{P}$, $\mathbb{E}$, $\mathbb{V}$ and $\mathbb{COV}$ for probability, expectation, variance and covariance. For a more intuitive understanding of the general procedure let us consider the simplest case first.

We draw a random point x uniformly in F, which means that for any set $B \subseteq \Re^2$ the probability that point x falls within B is given by

$$\mathbb{P}(x \in B) = \frac{\lambda(B \cap F)}{\lambda(F)} \tag{4.2}$$

Though point x is random it is denoted hereafter with a small letter and not with a capital X, according to the convention for random variables. The reason is purely typographical. For illustration Figure 4.1 displays a simple random sampling drawn in the Zürichberg forest that will be described in chapter 8. In contrast, Figure 4.2 presents a systematic random sample (i.e. the starting point and orientation of the systematic grid are random) with the same density of points per surface area.

Trees are selected if they are within the circle $K_r(x) = \{y \in \Re^2 \mid d(y,x) \le$

Figure 4.1 *Simple random sampling using 17 random plots in a* 17*ha* ***forest***

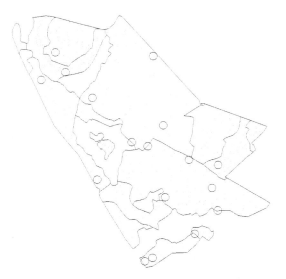

with its radius r and with its center in point x. $d(x, y)$ is the euclidian distance between the points x and y. We now define the random indicator variables

$$I_i(x) = \begin{cases} 1 \text{ if } i \in K_r(x) \\ 0 \text{ if } i \notin K_r(x) \end{cases}$$

and the N circles $K_i(r) = K_r(u_i)$ with constant radius r centered on the trees. By symmetry, the ith tree is in the circle $K_r(x)$ if and only if the random point x is in the ith circle $K_i(r)$. Hence, we have the obvious but important **duality principle**:

$$I_i(x) = 1 \Leftrightarrow x \in K_i(r)$$

The inclusion probability of the ith tree is consequently given by

$$\pi_i = \mathbb{P}(I_i(x) = 1) = \mathbb{E}_x\left(I_i(x)\right) = \frac{\lambda\left(K_i(r) \cap F\right)}{\lambda(F)}$$

Up to boundary effects at the forest edges this inclusion probability is constant. For a given variable Y, and neglecting those boundary effects, it is natural to define the **local density** $Y(x)$ at the point x as the sum of the Y_i over the trees selected, divided by the constant surface area of the circle $K_r(x)$ that is to set

$$Y(x) = \frac{1}{\lambda(F)} \sum_{i=1}^{N} \frac{I_i(x)Y_i}{\pi_i}$$

This dual consideration leads immediately to a far-reaching generalization. We can assign to each tree its circle $K_i = K_{r_i}(u_i)$, whose radius depends on the

Figure 4.2 *Systematic random sampling*

label i and therefore, generally, also on the $Y_i^{(m)}$ and u_i; for instance it might depend on diameter, species and location of the tree. In facts, circles are convenient but not compulsory: squares, rectangles or any other shape with a fixed orientation are suitable. Figure 4.3 illustrates this duality principle.

Figure 4.3 *Duality principle with two concentric circles*

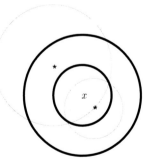

Legend: x sample point, $*$ trees, circles centered on sample point (bold) and on the trees.

The famous **angle count method** with limit angle α assigns to each tree of diameter (in cm) at breast height D_i a circle of radius (in m) $R_i = \frac{D_i}{2\sqrt{k}}$. Here,

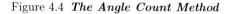 ($_2$) is the angle count factor (or basal area factor) in $_{ha}$ values $k = 2$ and $k = 4$ are frequently used.

Figure 4.4 *The Angle Count Method*

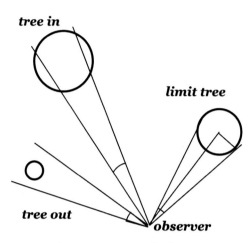

Legend: tree in $\alpha_i > \alpha$, tree out $\alpha_i < \alpha$, limit tree $\alpha_i = \alpha$ = critical angle

When a field survey is being conducted, a particular tree is included in the sample if its apparent diameter, viewed from point x (angle α_i) is larger than the limit angle α (see Figure 4.4). This task is usually performed with an optical instrument, the **Relaskop**. This method was invented in the 1930s, by the Austrian forester Bitterlich , and was published only in 1947. Inclusion probabilities are given as $\pi_i = \frac{G_i}{k\lambda(F)}$ and are proportional to the basal area $G_i = \frac{\pi D_i^2 10^{-4}}{4}$ (in m^2). If the response variable is the **basal area** of the tree in m^2, i.e. $Y_i = G_i$, then one obtains $Y(x) = kn(x)$, where $n(x) = \sum_{i=1}^{N} I_i$ is simply the number of trees included in the sample at point x. By doing so, the density of the basal area (in m^2 per ha) can be obtained purely optically, without measurements. This explains the popularity of the method. However, the above property is no longer true for other variables, especially densities of stems or timber volume density. In dense stands, the problem of hidden trees can be a serious drawback. This angle count method is also known as the **variable radius plot sampling**, or **VRP**.

The widely used technique involving concentric circles can be viewed as a discrete approximation of the angle count: trees are included in the sample if they lie within the corresponding circles that are centered on point x and if their diameters are larger than the corresponding thresholds. Usually, the thresholds for the radii of the circles take $1, 2$, eventually 3 but rarely more values.

$$I_i(x) = \begin{cases} 1 \text{ if } x \in K_i \\ 0 \text{ if } x \notin K_i \end{cases} \qquad (4.3)$$

and the **inclusion probabilities** for a single tree as well as for any pair of trees according to

$$\pi_i \quad = \quad \frac{\lambda\,(K_i \cap F)}{\lambda(F)}$$

$$\pi_{ij} \quad = \quad \frac{\lambda\,(K_i \cap K_j \cap F)}{\lambda(F)} \qquad (4.4)$$

The **local density** is a random variable defined by

$$Y(x) = \frac{1}{\lambda(F)} \sum_{i=1}^{N} \frac{I_i(x) Y_i}{\pi_i} \qquad (4.5)$$

Except for the constant $\lambda(F)$, the local density $Y(x)$ is actually the Horvitz-Thompson estimator. It is interesting to note that Bitterlich would not have been familiar with the work of Horvitz and Thompson (1952), because his method was devised prior to their research. The first probabilistic interpretation of his method was described by Grosenbaugh (1952).

Remarks:

1. The surface area of the forest $\lambda(F)$ appears only formally in Eq. 4.5. For effective calculations we only need to know $\lambda(F)\pi_i$, which, according to Eq. 4.4, is equal to the surface area of the circle and is eventually corrected for boundary effects, i.e. $\lambda(K_i \cap F)$.

2. A tree can rightly be included in the sample even if its DBH is below the inventory threshold (e.g. using the angle count method one can place a very small tree in the sample if it is very near the sample point). Recall that, according to our definition of the population $\mathcal{P}$, all the response variables of that tree are set to zero so that it does not contribute to the local density in Eq. 4.5.

3. In field work it is important to remember that the inclusion of a tree in the sample is based on the distance of **the sample point** x **to the center point of the tree** u_i, and not to its outer edge at $1.3m$ above ground. For limit trees, i.e. those with $d(x, u_i) \approx r_i$, it is sometimes recommended that one flips a coin to determine their eligibility for inclusion in order to avoid systematic errors.

4. The only universal, yet still approximate, method for calculating π boundary trees, with respect to the forest edge, is to record the polar co-ordinates of all trees included in the sample as well as to determine the geometry of the forest boundary (assuming there is a clear definition thereof). In most instances, the boundary will be a straight line or a simple

(   ) with a
line boundary one could use the famous reflection principle. Most other
procedures are biased. **Boundary adjustments are crucial**, and failing
to perform them will usually generate a bias greater than the sampling
error, see Schreuder et al. (1993) and Kangas and Maltamo (Eds, 2006) for
more details.

5. If point x, i.e. the sample plot, is located on a slope, one must perform a
slope correction. Because the inclusion circles are in the horizontal plane
containing F, one must determine their elliptic inverse projections onto the
slope. To simplify field work the slope correction is usually performed at
the plot level and not at the tree level. That is, one can consider concentric
ellipses in the slope as having their semi-axis equal to the nominal circle
radius r_i in the direction perpendicular to the slope, and their semi-axis
$\frac{r_i}{cos(\beta)}$ in the direction parallel to the slope, where β is the slope angle. Then,
the vertical projections of these ellipses onto the horizontal plane will be
circles with the correct nominal surface area. In angle count sampling, the
Relaskop measures the horizontal distances of the sample point to the trees,
so that any slope correction is automatic. See Kangas and Maltamo (Eds,
2006) for details and further references.

Given the π_i the local density $Y(x)$ is a function $Y(.) : x \in F \mapsto Y(x)$ which
by construction satisfies

$$\mathbb{E}_x\left(Y(x)\right) = \frac{1}{\lambda(F)} \int_F Y(x) dx = \frac{1}{\lambda(F)} \sum_{i=1}^{N} Y_i = \bar{Y}_F \qquad (4.6)$$

In other words, the spatial average of the local density is equal to the true
density of the response variable. **The above formulation transforms the
problem of estimating a sum over a finite population $\mathcal{P}$ of trees
into one of estimating the integral of a function over a domain.
This allows one to view $Y(x), x \in F$ as an infinite population $\mathcal{Y}$.** By
construction the function $Y(x)$ is constant around each point x. The set of
discontinuities of $Y(x)$ then consists of circle arcs of zero surface area, so that
the above integral is well defined.

When expressing the local density $Y(x)$ the inclusion probabilities π_i do already take into account boundary effects. An alternative approach is to draw
the random point x uniformly in a domain $\tilde{F} \supset F$ such that $\forall i \; K_i \subset \tilde{F}$. No
boundary effects occur at the forest edge, therefore the inclusion probabilities
are $\tilde{\pi}_i = \frac{\lambda(K_i)}{\lambda(\tilde{F})}$ and the resulting local density

$$\tilde{Y}(x) = \frac{1}{\lambda(\tilde{F})} \sum_{i=1}^{N} \frac{I_i(x) Y_i}{\tilde{\pi}_i}$$

$$\mathbb{E}_{x \in \tilde{F}} \left(Y(x) \right) = \frac{1}{\lambda(\tilde{F})} \int_{\tilde{F}} Y(x) dx = \frac{1}{\lambda(\tilde{F})} \sum_{i=1}^{N} Y_i$$

However, this method is not always feasible when the point $x \in \tilde{F} \setminus F$ inaccessible, e.g. at the boundary of a lake or a cliff.

Not all local densities derive from an Horvitz-Thompson estimator. Consider for example the angle count technique. Define the **critical height** $H_i(x$ the *ith* tree, viewed from sample point x, as the height at which that tree has angular diameter α (the tree is a limit tree at its critical height). If $I_i(x) = 0$, i.e. the angular diameter is smaller than α at breast height, then $H_i(x) = 0$. Set

$$Y(x) = k \sum_{i=1}^{N} I_i(x) H_i(x)$$

Then

$$\mathbb{E}Y(x) = \frac{1}{\lambda(F)} \int_{F} Y(x) dx = \frac{1}{\lambda(F)} \sum_{I=1}^{N} Y_i$$

where Y_i is the volume of the tree. See Schreuder et al. (1993) for the proof and further details.

The infinite population framework is simpler and better suited for a forest inventory than the other approaches based on various finite populations. For instance, when one samples with a constant circle, it is impossible to partition the plane into a finite population of circles. They either overlap or do not fill out the plane. Things get even worse with concentric circles or the angle count technique. In contrast, the infinite population model is very elegant for cluster-sampling and is unavoidable when working with model-dependent techniques, in particular with geostatistics. **A further mathematical advantage concerns asymptotic considerations**, which are cumbersome with finite populations. These proofs require the artificial introduction of nested increasing sequences of populations. In the infinite population $\mathcal{Y}$ this problem is solved from the onset.

Because one samples the function $Y(x)$ to estimate its integral over a plane, the infinite population model is now sometimes called the **Monte Carlo approach**, in reference to the simulation algorithms used in physics to evaluate complex multidimensional integrals.

It is worth noting that the techniques presented in chapters 4 and 5 can also be used as alternatives to the mathematically and numerically more sophisticated geostatistical methods, now applied in various fields of research. The latter were originally developed for estimation problems in the mining and oil industries, where one must predict the integral of a function over a spatial domain (for example, evaluating the total oil resource in a field amounts to estimating the 3-D integral of the oil density, measured at relatively few and expensive drill holes). Here, geostatistical techniques will be presented in chapter 7.

Given the inclusion circles K_i, the function $Y(.)$ is well-defined and suffices to construct all the statistical quantities required for an estimation of $\bar{Y}_F$. In that case, the trees can vanish behind the scene. However, if the question is which function $Y(.)$ to use and when, then we have to go back to the K_i and therefore to the trees. This we shall do for designing optimal sampling schemes. It is unfortunate that standard sampling theory has been, up to a certain degree, misused as a corset for the forest inventory. Finite populations are simply not well-tailored to the intrinsically geometrical nature of the problem.

The crucial difference between the **design-based** and the **model-dependent** approaches is that, in the former, $Y(x)$ is a random variable because point is random, while the forest is fixed, i.e. $N, Y_i^{(m)}, u_i$ are fixed. With the latter framework x is given and the actual forest is considered to be the realization of a complex stochastic process, making $Y(x)$ random because the $N, Y_i^{(m)}$ are random.

One important advantage of the schemes described above and of the Horvitz-Thompson estimator in Eq. 4.5 is that the inclusion probabilities are known for the units (trees) drawn, and that we do not have to know them for those units not drawn. Other schemes do not have this property, such as all approaches that rely on nearest-neighbors methods (in which one requires a full census to determine the inclusion probabilities, thereby completely defeating the aim of forest sampling, at least in the design-based framework). These techniques, though popular, should be used only in very special situations, such as those focused on rare events.

The individual inclusion probabilities π_i are completely arbitrary. Up to boundary effects, they do not depend on the location of the trees. In contrast, the π_{ij} depend by 4.4 on the geometrical pattern of the trees in the forest. The number of trees drawn from point x is the random variable:

$$n(x) = \sum_{i=1}^{N} I_i(x) \qquad \mathbb{E}_x(n(x)) = \sum_{i=1}^{N} \pi_i \qquad (4.7)$$

One obtains from the definition and Theorem 2.1.2 the following relationships between the inclusion probabilities and the expected number of trees drawn

$$\mathbb{E}_x(n^2(x)) = \sum_{i,j=1}^{N} \pi_{ij} \ , \qquad \sum_{j=1, j\neq i}^{N} \pi_{ij} = \pi_i \left(\mathbb{E}_x(n(x)|I_i(x) = 1) - 1\right) \qquad (4.8)$$

Now we consider a set s_2 of n_2 points drawn uniformly and independent of each other in the forest area F. This **one-phase one-stage estimate for simple random sampling** is defined as

$$\hat{Y} = \frac{1}{n_2} \sum_{x \in s_2} Y(x) \qquad (4.9)$$

variance is easily found to be

$$\frac{1}{n_2\lambda^2(F)}\left(\sum_{i=1}^{N}\frac{Y_i^2(1-\pi_i)}{\pi_i}+\sum_{i\neq j}^{N}\frac{Y_iY_j(\pi_{ij}-\pi_i\pi_j)}{\pi_i\pi_j}\right)=\frac{1}{n_2}V_s \qquad (4.10)$$

where

$$V_s=\frac{1}{\lambda(F)}\int_{F}\left(Y(x)-\bar{Y}\right)^2 dx$$

is the variance of the local density under simple random sampling.

Because the $Y(x)$ are identically and independently distributed we immediately obtain the following unbiased estimate of variance:

$$\hat{\mathbb{V}}(\hat{Y})=\frac{1}{n_2}\frac{1}{(n_2-1)}\sum_{x\in s_2}(Y(x)-\bar{Y})^2 \qquad (4.11)$$

Let us recapitulate the **main points of the infinite population model for local density** $Y(x)$.

Theorem 4.2.1.

$$Y(x) \;=\; \frac{1}{\lambda(F)}\sum_{i=1}^{N}\frac{I_i(x)Y_i}{\pi_i}$$

$$\mathbb{E}(Y(x)) \;=\; \frac{1}{\lambda(F)}\int_{F}Y(x)dx=\frac{1}{\lambda(F)}\sum_{i=1}^{N}Y_i=\bar{Y}_F$$

$$\mathbb{V}(Y(x)) \;=\; \frac{1}{\lambda(F)}\int_{F}\left(Y(x)-\bar{Y}\right)^2 dx$$

$$\mathbb{V}(Y(x)) \;=\; \frac{1}{\lambda^2(F)}\left\{\sum_{i=1}^{N}\frac{Y_i^2(1-\pi_i)}{\pi_i}+\sum_{i\neq j}^{N}\frac{Y_iY_j(\pi_{ij}-\pi_i\pi_j)}{\pi_i\pi_j}\right\}$$

$$\hat{Y} \;=\; \frac{1}{n_2}\sum_{x\in s_2}Y(x)$$

$$\mathbb{E}(\hat{Y}) \;=\; \bar{Y}_F$$

$$\mathbb{V}(\hat{Y}) \;=\; \frac{1}{n_2}\mathbb{V}(Y(x))$$

$$\hat{\mathbb{V}}(\hat{Y}) \;=\; \frac{1}{n_2}\frac{1}{n_2-1}\sum_{x\in s_2}\left(Y(x)-\hat{Y}\right)^2$$

In many application one must estimate a ratio $R_{1,2}=\frac{\bar{Y}^{(1)}}{\bar{Y}^{(2)}}$, e.g. the proportion of spruce or the mean timber volume per tree. Using the rules defined in Section 2.6 we can then make the following asymptotically unbiased estimates:

$$\hat{R}_{1,2} = \frac{\hat{Y}^{(1)}}{\hat{Y}^{(2)}}$$

$$\hat{\mathbb{V}}(\hat{R}_{1,2}) = \frac{1}{n_2(\hat{Y}^{(2)})^2} \frac{1}{n_2 - 1} \sum_{x \in s_2} (Y^{(1)}(x) - \hat{R}_{1,2} Y^{(2)}(x))^2$$

A common mistake by beginners is to take the mean over s_2 of the ratios obtained at each point x. For instance, the percentage of spruce is calculated as the mean of the proportion of spruce obtained at each $x \in s_2$. **This is totally wrong** and the resulting estimates can be severely biased, even when using the simplest method of a single circle of constant radius.

The theoretical variance from Eq. 4.10 depends via the π_{ij} on the spatial structure of the forest. That is, all other things being equal, displacing the trees will change the variance, which, as we shall see, is the source of great difficulties, and one of the main reasons for introducing the concept of anticipated variance.

For completeness we emphasize some subtle but important differences between the formalisms for finite populations and a forest inventory. Given a population $\mathcal{P}$ of N individuals a sampling design p is a probability function defined on all 2^N subsets of $\mathcal{P}$. Obviously, in practice, $p(s) = 0$ for most subsets s and the empty set has a probability of zero in this setup, i.e. $p(\emptyset) = 0$. In forest sampling, however, it may well happen that no tree is drawn from point x. We have then selected the empty set $\emptyset$, which occurs with a probability of $p(\emptyset) = \mathbb{P}\{I_i(x) = 0, \forall i\}$. In such a case $Y(x) = 0$. Note also that, in standard sampling theory, N is known (most sampling schemes used rest upon a complete list of the individuals), whereas it is generally unknown, but fixed, in forest sampling.

Almost all designs used in finite-population sampling are **non-informative** i.e. $p(s)$ does not depend explicitly on the response variable Y_i for $i \in s$ $p(s) > 0$. This concept is important when expectations with respect to model and design probabilities are required. For non-informative designs the order of the expectations is irrelevant. The angle count method described above is an example of an informative design, at the tree level, for determining the basal area. In the infinite population setup, however, we are sampling the function $Y(x)$ with uniform sampling and the design is non-informative because the design-probability density function $\frac{dx}{\lambda(F)}$ does not depend on the value $Y($ In contrast to classical theory, where one draws one single reasonably large sample of units for which the π_i, π_{ij} are usually known prior to sampling (i.e. list sampling), forest sampling draws a sufficiently large number of independent points $x \in s_2$, each of them selecting a relatively small random number of trees, for which the π_i can be exactly determined a posteriori. In principle, π_{ij} could be calculated, although this is unnecessary because one can obtain an unbiased estimate of the theoretical variance by first principles.

We can now proceed further to an important generalization of simple random inventories over large areas. Before doing so the reader can have a brief look at the simulation examples given in Appendix A.2.

4.3 One-phase one-stage cluster random sampling scheme

A **cluster**, sometimes called **trakt**, of (nominal) size M is determined by a fixed set of M vectors $e_l \in \mathbb{R}^2$, $l = 1, \ldots M$. Without loss of generality we shall assume that one of the e_l, say e_1 is the null vector.

A correct definition of cluster-sampling requires some technical details (which are crucial when designing simulations!). For any set $A \subset \mathbb{R}^2$ let us denote by A_l the set $A + e_l = \{x | \exists a \in A, x = a + e_l\}$. We make the key assumption that set A is large enough to ensure $F \subset A_l$ $l = 1, \ldots M$. For instance one could take any set A containing the set $\cup_{l=1}^{M}\{F - e_l\}$.

We now draw a random point x uniformly in A. Obviously, the points $x + e_l$ are uniformly distributed in A_l. Following the above convention, the origin x of the cluster is always the point x_1. We first note that given $x_l \in x_l$ is uniformly distributed in F. Indeed, for any set $B \in \mathbb{R}^2$, because $F \subset$ and $\lambda(A_l) = \lambda(A)$ $\forall l$ we have

$$\mathbb{P}\left(x_l \in B | x_l \in F\right) = \frac{\mathbb{P}\left(x_l \in B \cap F\right)}{\mathbb{P}\left(x_l \in F\right)} = \frac{\frac{\lambda(B \cap F \cap A_l)}{\lambda(A_l)}}{\frac{\lambda(F \cap A_l)}{\lambda(A_l)}} = \frac{\lambda\left(B \cap F\right)}{\lambda\left(F\right)} \qquad (4.12)$$

By Eq. 4.6 and the above we also get $\mathbb{E}_x\left(Y(x_l) | x_l \in F\right) = \bar{Y}$. We introduce the indicator variable of set F (the definition is valid for any set) as

$$I_F(x) = \begin{cases} 1 \text{ if } x \in F \\ 0 \text{ if } x \notin F \end{cases}$$

The number of points per cluster falling into the forest area is the random variable defined by

$$M(x) = \sum_{l=1}^{M} I_F(x_l) \qquad (4.13)$$

whereas the local density found at the cluster level is given by

$$Y_c(x) = \frac{\sum_{l=1}^{M} I_F(x_l) Y(x_l)}{M(x)} \qquad (4.14)$$

By using Eq. 4.12 and 4.6 we get the following important relationships:

$$\mathbb{E}_x M(x) = \sum_{l=1}^{M} \mathbb{P}\left(x_l \in F\right) = \sum_{l=1}^{M} \frac{\lambda\left(F \cap A_l\right)}{\lambda\left(A_l\right)} = M \frac{\lambda(F)}{\lambda(A)} \qquad (4.15)$$

$$\mathbb{E}_x \left(\sum_{l=1} I_F(x_l) Y(x_l) \right) = \sum_{l=1} \mathbb{P}\left(I_F(x_l) = 1\right) \mathbb{E}_{x_l}\left(Y(x_l)|x_l \in F\right)$$

$$= \sum_{l=1}^{M} \frac{\lambda\left(F \cap A_l\right)}{\lambda\left(A_l\right)} \bar{Y} = M \frac{\lambda(F)}{\lambda(A)} \bar{Y} \qquad (4.16)$$

If we now draw n_2 points $x \in s_2$ independently and uniformly in set A generate n_2 clusters of effective sizes $M(x)$.

For illustration Figure 4.5 displays the positioning of points for 17 non-void clusters. Each cluster, nominally, comprises 4 points on a $50m \times 50m$ square. The origin (lower left corner) of the cluster is uniformly distributed in Likewise, Figure 4.6 presents the positions of points in non-void clusters, when their origins are distributed according to a systematic $100m \times 100m$ grid.

Figure 4.5 *Random cluster sampling*

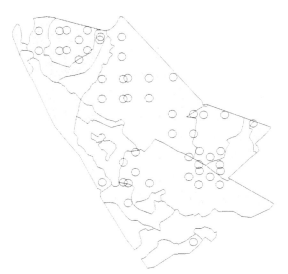

Cluster: $50m \times 50m$ **square**

The **one-phase one-stage estimate for cluster random sampling** defined by

$$\widehat{Y}_c = \frac{\sum_{x \in s_2} M(x) Y_c(x)}{\sum_{x \in s_2} M(x)} \qquad (4.17)$$

Note that $\widehat{Y}_c$ is the average of all $Y(x_l)$ ignoring the cluster structure and is the ratio of two random variables. These facts explain why cluster-sampling is slightly more complicated technically than random sampling. Dividing the

Figure 4.6 *Systematic cluster sampling*

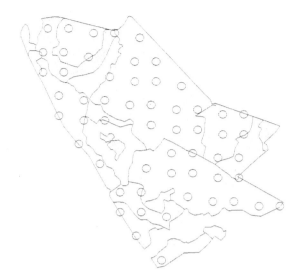

Cluster: $50m \times 50m$ **square, origin** $100m \times 100m$ **grid**

numerator and the denominator in Eq. 4.17 by n_2 we arrive by Eq. 4.16 and 4.15 at

$$\lim_{n_2 \to \infty} \mathbb{E}_{x \in A}(\widehat{Y}_c) = \frac{\mathbb{E}_x \left(\sum_{l=1}^{M} I_F(x_l) Y(x_l) \right)}{\mathbb{E}_x M(x)} = \bar{Y}$$

At this point, **it is important to realize that** $Y_c(x)$ **does not yield generally an unbiased estimate**, not even asymptotically, i.e.,

$$\lim_{n_2 \to \infty} \mathbb{E}_{x \in A} \frac{1}{n_2} \sum_{x \in s_2} Y_c(x) \neq \bar{Y}$$

More precisely, one can show that the cluster sampling point estimate via Eq. 4.17 is asymptotically unbiased (Mandallaz, 1991), in the sense that

$$\mathbb{E}_{x \in A} \widehat{Y}_c = \bar{Y} + O\left(n_2^{-1}\right) \tag{4.18}$$

and that its theoretical asymptotic variance is given by

$$\mathbb{V}(\widehat{Y}_c) = \frac{1}{n_2} \frac{\mathbb{E}_x M^2(x)(Y_c(x) - \bar{Y})^2}{\mathbb{E}_x^2 M(x)} + O\left(n_2^{-2}\right) \tag{4.19}$$

Using the techniques illustrated in Section 2.6 for the variance of a ratio, we can summarize the results for one-phase one-stage cluster sampling in the following theorem:

$$\hat{Y}_c = \frac{\sum_{x \in s_2} M(x) Y_c(x)}{\sum_{x \in s_2} M(x)}$$

$$\widehat{\mathbb{V}}(\hat{Y}_c) = \frac{1}{n_2(n_2 - 1)} \sum_{x \in s_2} \left(\frac{M(x)}{\bar{M}_2}\right)^2 (Y_c(x) - \hat{Y}_c)^2$$

where $\bar{M}_2 = \frac{1}{n_2} \sum_{x \in s_2} M(x)$ is the average number of points per cluster falling within a forest area F. Note that void clusters, i.e. $M(x) = 0$, do not contribute to either the point estimate or the variance. The point estimate has a bias of order n_2^{-1} and its variance estimate is correct up to n_2^{-2}.

Remarks:
The above framework for cluster sampling differs markedly from the traditional approach primarily because the number of points per cluster is treated as a random variable and also because the random origin of the cluster may fall outside of the forest area. Therefore, adjustments for boundary effects are performed only at the point (plot) level and not at the cluster level (see also Gregoire and Valentine, 2007). This approach is simpler for both statistical estimation and for field work. Loss of efficiency can occur if the variance for the number of points per cluster is large. This will be the case when the geometry of the clusters and of the forest are incompatible, e.g. with a very patchy forested area and clusters with points far apart.

The estimation of a ratio is a straightforward generalization of Theorem 4.2.2 and is given by

Theorem 4.3.2.

$$\hat{R}_{1,2} = \frac{\hat{Y}_c^{(1)}}{\hat{Y}_c^{(2)}}$$

$$\widehat{\mathbb{V}}(\hat{R}_{1,2}) = \frac{1}{n_2} \frac{1}{(\hat{Y}_c^{(2)})^2} \frac{1}{(n_2 - 1)} \sum_{x \in s_2} \left(\frac{M(x)}{\bar{M}_2}\right)^2 (Y_c^{(1)}(x) - \hat{R}_{1,2} Y_c^{(2)}(x))$$

For a better intuitive understanding we quote a description by Mandallaz (1991), which expresses the variance under cluster sampling as a function of the variance under simple random sampling V_s, the intra-cluster correlation coefficient ρ and a term describing the topological characteristics of the forest. Namely:

$$\mathbb{V}(\hat{Y}_c) = \frac{1}{n_2 \mathbb{E}_x M(x)} V_s \left(1 + \rho \left(\mathbb{E}_x M(x) - 1\right) + \rho \frac{V_x M(x)}{\mathbb{E}_x M(x)}\right) + O(n_2^{-2}) \quad (4.20)$$

where

$$\rho = \frac{\mathbb{E}_x \sum_{l \neq m}^{M} I_F(x_l) I_F(x_m) \left(Y(x_l) - \bar{Y}\right) \left(Y(x_m) - \bar{Y}\right)}{V_s \mathbb{E}_x M(x) \left(M(x) - 1\right)} \quad (4.21)$$

One can estimate ρ by equating the sample versions of Eq. 4.20 with $\widehat{\mathbb{V}}($

sums-of-squares formula, while ignoring the cluster structure of the data.

The intra-cluster correlation coefficient ρ is usually positive and the variance under cluster-sampling is larger than under simple random sampling, obviously while keeping constant the function $Y(.)$ as well as the total number of points, i.e. $n_{2,simple} = n_{2,cluster}\mathbb{E}_x M(x)$. The inflation factor is determined by Eq. 4.20. That particular formula generalizes the classical result reported by Cochran (1977), in which the cluster size is held constant, and also according to Eq. 2.72, which, of course, is not suitable for forest inventories because the cluster size is random. Nevertheless, the reason for using cluster sampling is to reduce transport costs, a matter to be discussed later. Finally, let us note that one can view the points of a systematic grid over a forest area as a single very large cluster. This depiction then explains why one cannot estimate, strictly speaking, the design-based variance with systematic grids as used in forest inventories. It also implies that density estimates are only asymptotically unbiased, in contrast to estimates of totals (Mandallaz, 1991). In practice the finite sample bias seems to be negligible. The reader can now have a brief look at the simulation example given in Appendix A.3.

4.4 One-phase two-stage simple random sampling

In many applications costs to measure the response variable Y_i are high. For instance, a good determination of the volume may require that one records DBH, as well as the diameter at $7m$ above ground and total height in order to utilize a three-way yield table. However, one could rely on a coarser, but cheaper, approximation of the volume based only on DBH. Nonetheless, it may be most sensible to assess those three parameters only on a sub-sample of trees. We now formalize this simple idea. For each point $x \in s_2$ trees are drawn with probabilities π_i. The set of selected trees is denoted by $s_2(x)$. From each of the selected trees $i \in s_2(x)$ one gets an approximation Y_i^* of the exact value Y_i. From the finite set $s_2(x)$ one draws a sub-sample $s_3(x) \subset s_2(x$ trees. For each tree $i \in s_3(x)$ one then measures the exact variable Y_i. Let us now define the second stage indicator variable

$$J_i(x) = \begin{cases} 1 \text{ if } i \in s_3(x) \\ 0 \text{ if } i \notin s_3(x) \end{cases} \tag{4.22}$$

Note that by construction $I_i(x)J_i(x) = J_i(x)$. Before going further, we must introduce some notation for expectations and variances. In our general context, the sub-index 1 refers to the first phase (in which one collects the auxiliary information in the large sample s_1 (this will be thoroughly defined later on). The sub-index 2 refers to the second phase, when one gathers the terrestrial information from first-stage trees. Finally, the sub-index 3 refers to the second-stage trees. All three sub-indices are necessary because generally three

tation $\mathbb{E}_{2,3}(.), \mathbb{V}_{2,3}(.), \mathbb{E}_{3|2}(.), \mathbb{V}_{3|2}(.)$ for the overall expectation and variance under the random selections $(2,3)$, as well as for the conditional expectation and variance of the second stage procedure, given the second-phase and first-stage selections. According to Eq. B.1 and B.2 we can calculate the expected value and variance of an arbitrary random variable Z depending on random selection $(2,3)$ with

$$\begin{aligned}
\mathbb{E}_{2,3}(Z) &= \mathbb{E}_2\left(\mathbb{E}_{3|2}(Z)\right) \\
\mathbb{V}_{2,3}(Z) &= \mathbb{E}_2\left(\mathbb{V}_{3|2}(Z)\right) + \mathbb{V}_2\left(\mathbb{E}_{3|2}(Z)\right)
\end{aligned} \tag{4.23}$$

Hence, we arrive at

$$\begin{aligned}
\mathbb{E}_{2,3}(J_i(x)) &= \mathbb{E}_2\mathbb{E}_{3|2}(J_i(x)I_i(x)) \\
&= \mathbb{E}_2 I_i(x)\mathbb{E}_{3|2}(J_i(x)|I_i(x)) \\
&= \mathbb{P}(J_i(x) = 1|I_i(x) = 1)\mathbb{P}(I_i(x) = 1) := p_i\pi_i
\end{aligned} \tag{4.24}$$

where we have introduced the **second-stage** conditional inclusion probability $p_i = \mathbb{P}(J_i(x) = 1|I_i(x) = 1)$ that, so far, depends on the label i and is, therefore, completely arbitrary.

We now assume that trees in $s_2(x)$ are sampled **independently of each other**, so that $p_{ij} = \mathbb{P}(J_i(x)J_j(x) = 1|I_i(x)I_j(x) = 1) = p_i p_j$. Thus, we have **Poisson sampling** at the second stage, as defined in Section 2.4. The advantage of this proposed scheme is that a field crew can collect the required information on first-stage trees one by one, then enter these data in a portable computer. Using a software generating appropriate random numbers, the crew can then determine immediately whether further measurements will be taken. Of course, other schemes are possible, but these are not necessarily better nor as easily implemented (because one needs a list of all first-stage trees at point x and, possibly, at others, too).

To construct a good point estimate, we must have **the residual $R_i = Y_i -$** which is known only for trees $i \in s_3(x)$. The **generalized local density** $Y^*(x)$ is defined according to

$$\begin{aligned}
Y^*(x) &= \frac{1}{\lambda(F)}\left(\sum_{i=1}^N \frac{I_i(x)Y_i^*}{\pi_i} + \sum_{i=1}^N \frac{I_i(x)J_i(x)R_i}{\pi_i p_i}\right) \\
&= \frac{1}{\lambda(F)}\left(\sum_{i\in s_2(x)} \frac{Y_i^*}{\pi_i} + \sum_{i\in s_3(x)} \frac{R_i}{\pi_i p_i}\right)
\end{aligned} \tag{4.25}$$

Those formulae are simply the regression estimator from Theorem 3.3.1, as adapted to the forest inventory context.

We assume here that the prediction Y_i^* of Y_i is based on an **external model** That is, the corresponding model is not adjusted with the data collected by

$Y_i = Y_i^* + R_i$ and $\mathbb{E}_{3|2} J_i(x) = p_i$ we have

$$\mathbb{E}_{2,3}(Y^*(x)) = \mathbb{E}_2(Y(x)) = \bar{Y} \qquad (4.26)$$

so that $Y^*(x)$ from Eq. 4.25 is also unbiased. Because the second-stage trees are drawn independently of each other we end up with

$$
\begin{aligned}
V(x) &= \mathbb{V}_{3|2}(Y^*(x)) = \mathbb{V}_{3|2}\left(\frac{1}{\lambda(F)} \sum_{i \in s_3(x)} \frac{R_i}{\pi_i p_i}\right) \\
&= \frac{1}{\lambda^2(F)}\left(\sum_{i \in s_2(x)} \frac{R_i^2(1 - p_i)}{\pi_i^2 p_i}\right) \\
\mathbb{E}_2 \mathbb{V}_{3|2}(Y^*(x)) &= \mathbb{E}_x V(x) = \frac{1}{\lambda^2(F)}\left(\sum_{i=1}^N \frac{R_i^2}{\pi_i p_i} - \sum_{i=1}^N \frac{R_i^2}{\pi_i}\right) \qquad (4.27)
\end{aligned}
$$

The one-phase two-stage estimate for simple random sampling is defined by

$$\widehat{Y}^* = \frac{1}{n_2} \sum_{x \in s_2} Y^*(x) \qquad (4.28)$$

Its theoretical variance is easily found, according to Eq. 4.23, to be

$$\mathbb{V}(\widehat{Y}^*) = \frac{1}{n_2} \mathbb{V}_x(Y(x)) + \frac{1}{n_2} \mathbb{E}_x V(x) \qquad (4.29)$$

Then, using Eq. 4.10 and 4.27, this is equivalent to

$$
\begin{aligned}
\mathbb{V}(\widehat{Y}^*) &= \frac{1}{n_2 \lambda^2(F)}\left(\sum_{i=1}^N \frac{Y_i^2(1 - \pi_i)}{\pi} + \sum_{i \neq j} \frac{Y_i Y_j(\pi_{ij} - \pi_i \pi_j)}{\pi_i \pi_j}\right) \qquad (4.30) \\
&+ \frac{1}{n_2 \lambda^2(F)}\left(\sum_{i=1}^N \frac{R_i^2}{\pi_i p_i} - \sum_{i=1}^N \frac{R_i^2}{\pi_i}\right) \qquad (4.31)
\end{aligned}
$$

It is somewhat remarkable that the usual variance estimate can also be utilized in this case. More precisely one has

$$
\begin{aligned}
\widehat{\mathbb{V}}(\widehat{Y}^*) &= \frac{1}{n_2(n_2 - 1)} \sum_{x \in s_2}(Y^*(x) - \widehat{Y}^*)^2 \\
\mathbb{E}_x(\widehat{\mathbb{V}}(\widehat{Y}^*)) &= \mathbb{V}(\widehat{Y}^*) \qquad (4.32)
\end{aligned}
$$

$$n_2(n_2 - 1)\widehat{\mathbb{V}}(\widehat{Y}^*) \;=\; \sum_{x \in s_2} Y^*(x)^2 - \frac{\left(\sum_{x \in s_2} Y^*(x) \right)^2}{n_2}$$

$$=\; \sum_{x \in s_2} Y^*(x)^2 - \frac{1}{n_2} \left(\sum_{x \in s_2} Y^*(x)^2 + \sum_{x \neq y \in s_2} Y^*(x)Y^*(y) \right)$$

Taking the expectation according to Eq. 4.23, and using Eq. 4.26 and the independence of $Y^*(x), Y^*(y)$ for $x \neq y$ we obtain

$$n_2(n_2 - 1)\mathbb{E}_{2,3}(\widehat{\mathbb{V}}(\widehat{Y}^*)) = \frac{n_2 - 1}{n_2} \mathbb{E}_2 \left(\sum_{x \in s_2} (\mathbb{V}_{3|2} Y^*(x) + Y^2(x)) \right)$$

$$- \frac{1}{n_2} \mathbb{E}_2 \sum_{x \neq y \in s_2} Y(x)Y(y)$$

and the result. We can summarize this one-phase two-stage sampling scheme into the following theorem:

Theorem 4.4.1.

$$Y^*(x) \;=\; \frac{1}{\lambda(F)} \left(\sum_{i \in s_2(x)} \frac{Y_i^*}{\pi_i} + \sum_{i \in s_3(x)} \frac{R_i}{\pi_i p_i} \right)$$

$$\widehat{Y}^* \;=\; \frac{1}{n_2} \sum_{x \in s_2} Y^*(x)$$

$$\widehat{\mathbb{V}}(\widehat{Y}^*) \;=\; \frac{1}{n_2(n_2 - 1)} \sum_{x \in s_2} (Y^*(x) - \widehat{Y}^*)^2$$

It should be clear that, because of Eq. 4.32, Theorem 4.2.2 can also be applied to estimate the ratio under two-stage sampling. Here, it suffices to replace the $Y^{(k)}(x), k = 1, 2$ with the $Y^{(k)*}(x)$.

In some instances one may want to know the variance of the exact density $Y(x)$ when only $Y^*(x)$ is available. To this end, let us note first that

$$\widehat{V}(x) = \frac{1}{\lambda^2(F)} \sum_{i \in s_3(x)} \frac{R_i^2(1 - p_i)}{\pi_i^2 p_i^2}$$

is an unbiased estimate of $V(x)$, in the sense that

$$\mathbb{E}_{3|2} \widehat{V}(x) = V(x)$$

Hence, by Eq. 4.29, we get an unbiased estimate of $\mathbb{V}(\widehat{Y})$ through

$$\widehat{\mathbb{V}}(Y(x)) = n_2 \widehat{\mathbb{V}}(\widehat{Y}^*) - \frac{1}{n_2} \sum_{x \in s_2} \widehat{V}(x) \qquad (4.33)$$

here to conduct the second Swiss National Forest Inventory (SNFI2), based on data from the first Swiss National Forest Inventory (SNFI1). The earlier version selected roughly 120,000 first-stage trees from 10,000 points, plus 40,000 second-stage trees with a constant probability $p_i \equiv \frac{1}{3}$. Afterwards, it was found that one could reduce the number of second-stage trees to 10,000, while ensuring the same accuracy and even decreasing the bias in small areas. This was accomplished simply by optimizing the p_i (a topic we shall discuss later), and by using the point estimate from Eq. 4.25, which explicitly took the residuals into account.

4.5 One-phase two-stage cluster random sampling

Assuming that the **second-stage trees are drawn independently of each other in each point of each cluster**, we can generalize $Y_c(x)$ in a straightforward manner, i.e.

$$Y_c^*(x) = \frac{\sum_{l=1}^{M} I_F(x_l)Y^*(x_l)}{M(x)} \tag{4.34}$$

where $Y^*(x_l)$ is the generalized local density at point $x_l = x + e_l$. Note that

$$\mathbb{E}_{3|2}Y_c^*(x) = Y_c(x)$$

The **one-phase two-stage point estimate for cluster-sampling** is then defined in perfect analogy to $\hat{Y}_c$ as

$$\hat{Y}_c^* = \frac{\sum_{x \in s_2} M(x)Y_c^*(x)}{\sum_{x \in s_2} M(x)} \tag{4.35}$$

Because of Eq. 4.18 and the above remark, it is clear that Y_c^* is asymptotically unbiased, i.e.

$$\mathbb{E}_{x \in A}\hat{Y}_c^* = \bar{Y} + O\left(n_2^{-1}\right) \tag{4.36}$$

To calculate the theoretical variance we use the decomposition from Eq. 4.23. The second term is simply the variance of the one-phase one-stage estimate,

$$\mathbb{E}_2 \mathbb{V}_{3|2}(\widehat{Y}_c^*) = \mathbb{E}_2 \left(\mathbb{V}_{3|2} \frac{\sum_{x \in s_2} \sum_{l=1}^{M} I_F(x_l) Y^*(x_l)}{\sum_{x \in s_2} M(x)} \right)$$

$$= \mathbb{E}_2 \left(\frac{\sum_{x \in s_2} \sum_{l=1}^{M} I_F(x_l) V(x_l)}{(\sum_{x \in s_2} M(x))^2} \right)$$

$$= \frac{\sum_{x \in s_2} \sum_{l=1}^{M} \mathbb{E}_2(V(x_l)|x_l \in F) \mathbb{P}(x_l \in F)}{n_2^2 \mathbb{E}_2^2 M(x)} + O(n_2^{-2})$$

$$= \frac{n_2 \mathbb{E}_{x \in F} V(x) \mathbb{E}_{x \in A} M(x)}{n_2^2 \mathbb{E}_{x \in A}^2 M(x)} + O(n_2^{-2})$$

$$= \frac{\mathbb{E}_{x \in F} V(x)}{n_2 \mathbb{E}_{x \in A} M(x)} + O(n_2^{-2})$$

where we have applied Eq. 4.12. Hence, by using the above result plus Eq. 4.23 and 4.19, we can calculate the asymptotic theoretical variance of the one-phase two-stage point estimate under cluster-sampling as

$$\mathbb{V}(\widehat{Y}_c^*) = \frac{1}{n_2} \frac{\mathbb{E}_{x \in A} M^2(x)(Y_c(x) - \bar{Y})^2}{\mathbb{E}_{x \in A}^2 M(x)} + \frac{\mathbb{E}_{x \in F} V(x)}{n_2 \mathbb{E}_{x \in A} M(x)} + O(n_2^{-2}) \quad (4.37)$$

Analogous to the result from Eq. 4.32 for one-phase two-stage simple random sampling one can construct an asymptotically unbiased estimate of variance according to

$$\widehat{\mathbb{V}}(\widehat{Y}_c^*) = \frac{1}{n_2(n_2 - 1)} \sum_{x \in s_2} \left(\frac{M(x)}{\bar{M}_2} \right)^2 (Y_c^*(x) - \widehat{Y}_c^*)^2$$

$$\mathbb{E}_{2,3}(\widehat{\mathbb{V}}(\widehat{Y}_c^*)) = \mathbb{V}(\widehat{Y}_c^*) + O(n_2^{-2}) \quad (4.38)$$

We give only the flavor of the proof, which can be made rigorous, by relying on the weak law of large numbers (convergence in probability), which implies here convergence in the mean because all random variables are bounded in practice.

Proof:

Asymptotically, one can replace $\bar{M}_2$ and $\widehat{Y}_c^*$ with their true values; likewise, $n_2 - 1$ with n_2. We then multiply both sides of Eq. 4.38 by $n_2^2 \bar{M}_2^2$. It remains to show that

$$\mathbb{E}_{2,3} \sum_{x \in s_2} M^2(x)(Y_c(x)^* - \bar{Y})^2 = n_2 \mathbb{E}_{x \in A} M^2(x)(Y_c(x) - \bar{Y})^2$$

$$+ n_2 \mathbb{E}_{x \in A} M(x) \mathbb{E}_{x \in F} V(x)$$

Setting $V_c(x) = \mathbb{V}_{3|2} Y_c^*(x)$ and using the decomposition from Eq. 4.23 we see first that the left-hand side is equal to

$$n_2 \mathbb{E}_{2,3} \left(M^2(x) Y_c^*(x)^2 - 2M^2(x) Y_c^*(x) \bar{Y} + M^2(x) \bar{Y}^2 \right)$$

$$n_2(\mathbb{E}_2 M^2(x)(Y_c^2(x) + V_c(x))) + n_2(-2M^2(x)Y_c(x)\bar{Y} + M^2(x)\bar{Y}^2)$$

to finally obtain

$$n_2\mathbb{E}_{x\in A}M^2(x)(Y_c(x) - \bar{Y})^2 + n_2\mathbb{E}_{x\in A}\sum_{l=1}^{M}I_F(x_l)V(x_l)$$

Using Eq. 4.12 the second term can be rewritten as

$$n_2\sum_{l=1}^{M}\mathbb{E}_{x\in A}V(x_l|x_l \in F)\mathbb{P}(x_l \in F) = n_2\mathbb{E}_{x\in F}V(x)\mathbb{E}_{x\in A}M(x)$$

which ends the proof. We summarize the asymptotic results for one-phase two-stage cluster-sampling in the following theorem:

Theorem 4.5.1.

$$\widehat{Y}_c^* = \frac{\sum_{x\in s_2} M(x)Y_c^*(x)}{\sum_{x\in s_2} M(x)}$$

$$\mathbb{V}(\widehat{Y}_c^*) = \frac{1}{n_2}\frac{\mathbb{E}_{x\in A}M^2(x)(Y_c(x) - \bar{Y})^2}{\mathbb{E}_{x\in A}^2 M(x)} + \frac{\mathbb{E}_{x\in F}V(x)}{n_2\mathbb{E}_{x\in A}M(x)}$$

$$\widehat{\mathbb{V}}(\widehat{Y}_c^*) = \frac{1}{n_2(n_2 - 1)}\sum_{x\in s_2}\left(\frac{M(x)}{\bar{M}_2}\right)^2(Y^* - \widehat{Y}_c^*)^2$$

The estimation of ratio under one-phase two-stage cluster sampling can be obtained by applying the result for one-phase one-stage cluster sampling given in Theorem 4.3.2. This is done by replacing the exact local densities $Y(x_l), Y_c$ with the generalized local densities $Y^*(x_l), Y_c^*(x)$.

From a theoretical design-based point of view this is essentially all there is to know about one-phase terrestrial inventories performed at a single time point, and where one assumes no measurement errors. If those errors have a zero mean (i.e. in the absence of systematic errors) the previous estimates are unbiased or asymptotically unbiased and therefore consistent. However, the variance is underestimated. This subject is rather complex and Fuller (1995) provides further explanation. Most inventory reports present error estimates based only on the sampling errors.

The next chapter deals with two-phase sampling schemes used in combined forest inventories.

4.6 Exercises

Problem 4.1. *Can you think of a forest F in which, at every point $x \in$ the number of trees $n(x)$ sampled with an angle count procedure is constant, i.e. $n(x) = 1$?*

( )

angle count scheme with $k = 10^4 \sin^2(\frac{\alpha}{2}) = 4$ and a DBH inventory threshold of 20cm if, in the neighborhood of x, we have the following candidate trees:

Tree no	Distance to x (m)	DBH (cm)
1	6	20
2	8	40
3	2	16
4	8	30
5	10	60
6	5	22
7	1	10

Problem 4.3. *We consider a version of the generalized Yates-Grundy formula (Eq. 2.11) in the forest-inventory context:*

1. *Prove that the variance $\mathbb{V}(Y(x))$ of the local density $Y(x)$ can be written as*

$$\frac{1}{2\lambda^2(F)} \sum_{i\neq j=1}^{N} (\pi_i \pi_j - \pi_{ij}) \left(\frac{Y_i}{\pi_i} - \frac{Y_j}{\pi_j}\right)^2 + \frac{1}{\lambda^2(F)} \sum_{I=1}^{N} \frac{Y_i^2}{\pi_i} \left(\mathbb{E}(n(x) \mid I_i = 1) - \mathbb{E}n\right)$$

where $n(x) = \sum_{i=1}^{N} I_i(x)$ is the number of trees sampled at x.

2. *Give the physical/geometrical interpretation for the condition*

$$\mathbb{E}(n(x) \mid I_i(x) = 1) - \mathbb{E}(n(x)) = 1 \;\; \forall i$$

3. *If Y_i is the basal area G_i and the sampling scheme is the angle count technique with factor k, show that, under those conditions, the variance of the basal area local density satisfies*

$$\mathbb{V}(G(x)) = k\bar{G}$$

and that

$$\mathbb{V}(n(x)) = \mathbb{E}(n(x))$$

4. *From the above scenario deduce the distribution of the random variable $n(x)$.*

Problem 4.4. *Draw a simple-shape forest F, e.g. a rectangle, and place as you wish five to height points in F, which will be the locations of the trees. A point uniformly distributed in F will sample the ith tree if it is the nearest neighbor of the random point x. Determine geometrically the inclusion probability for each tree. Then, what is the Horvitz-Thompson estimator for a response variable Y_i? What is the draw-back of this method? Can you generalize the technique to k's nearest neighbors?*

Problem 4.5. *Rewrite the local density $Y(x)$ as a Hansen-Hurwitz estimator (Eq. 2.23) and show the equivalence of those two approaches. Which one is simpler?*

cross-sections, and the following α-sampling scheme: $I_i^\alpha(x) = 1$ if and only if the distance from point x to the **outer edge** of the tree at $1.30m$ above ground is less than or equal to a given value r_α (recall that the usual rule stipulates the distance from x to the **center** of the tree!), otherwise $I_i^\alpha(x) = 0$. Note that, by definition, the tree will be sample if point x lies inside of it. Set $n_\alpha(x) = \sum_{i=1}^N I_i^\alpha(x)$ and calculate, neglecting boundary and slope corrections, the expected values

$$\mathbb{E}(n_\alpha(x))$$

Suppose that at each of n_2 points $x \in s_2$ you sample with three different distances r_α, $\alpha = 1, 2, 3$, so that you have the quantities

$$\hat{Y}_\alpha = \frac{1}{n_2} \sum_{x \in s_2} n_\alpha(x)$$

Calculate the $\mathbb{E}(\hat{Y}_\alpha)$. Which dendrometric quantities could you estimate with the $\hat{Y}_\alpha$? What happens if, instead of assuming circular cross-sections, you assume only convexity? What practical conclusion can you draw for field work using standard methods?

Problem 4.7. We consider simple random sampling with constant inclusion probabilities (i.e. involving a single circle). In order to obtain simple analytical results we neglect any boundary effects. More precisely, we have the response variables $Y_i^{(1)} \equiv 1$ and $Y_i^{(2)} = 1$ if the i-th tree has a given characteristic or $Y_i^{(2)} = 0$ otherwise. Our objective is to estimate the proportion

$$P = \frac{\sum_{i=1}^N Y_i^{(2)}}{\sum_{i=1}^N Y_i^{(1)}}$$

Calculate the point estimate $\hat{P}$ and its estimated variance $\hat{\mathbb{V}}(\hat{P})$ as a function of the quantities

$$n(x) = \sum_{i=1}^N I_i(x) Y_i^{(1)}, \quad r(x) = \sum_{i=1}^N I_i(x) Y_i^{(2)}, \quad \hat{p}(x) = \frac{r(x)}{n(x)}$$

Give an intuitive interpretation of these results. See also Problem 6.1 to apply the above solution to a real data set.

Problem 4.8. As an assignment, a group of students conducted a small inventory based on simple random sampling with $n_2 = 20$ plots, each consisting of a single $500m^2$ circle. Their point and variance estimates for the stem density were $\hat{Y} = 279$ and $\hat{\mathbb{V}}(\hat{Y}) = 2412$. In the same forest, another group of students performed an exercise with cluster sampling. The cluster comprised $M = 4$ plots (also one circle of $500m^2$) arranged on a $50m \times 50m$ square. Their results for $n_2 = 8$ clusters were $\hat{Y}_c = 285$, $\hat{\mathbb{V}}(\hat{Y}_c) = 4028$, $\bar{M} = 2$ and $\hat{\mathbb{V}}(M(x)) = 0.15$. Based on this data give an estimate of the intra-cluster correlation.

CHAPTER 5

Forest Inventory: two-phase sampling schemes

5.1 Two-phase one-stage simple random sampling

The **first phase** draws a large sample s_1 of n_1 points $x_i \in s_1$ that are independently and uniformly distributed within the forest area F. At each of those points auxiliary information is collected, very often of a purely qualitative nature (e.g. following the interpretation of aerial photographs). The **second phase** draws a small sample $s_2 \subset s_1$ of n_2 points from s_1 according to **equal probability sampling without replacement**. At each point $x \in s_2$ the terrestrial inventory provides the local density $Y(x)$. For points $x \in s_1 \setminus s_2$, i.e. in the large but not in the small sample, only the auxiliary information is available. Nevertheless, this allows one to make a prediction $\widehat{Y}(x)$ of the true local density $Y(x)$ in the forest. Strictly speaking, we shall assume that this prediction is given by an **external model** and is not adjusted with the data from the actual inventory. Let us examine probably the most important example. Here, stand structure is determined through the interpretation of aerial photographs in s_1, which, via pre-existing yield tables allows the inventorist to make a reasonable prediction of timber volume per ha or the number of stems per ha. However, if an external model is not available, an **internal model** first has to be fitted. Usually this is done by coding the auxiliary information at point x into a vector $\mathbf{Z}(x) \in \Re^m$. A prediction is then obtained with a linear model, i.e. $\widehat{Y}(x) = \mathbf{Z}(x)^t \boldsymbol{\beta}$ (the upper index indicating the transposition of the vector). Estimating the unknown parameter vector $\boldsymbol{\beta}$ can be done in several ways, in particular by completely ignoring sampling theory and using standard statistical tools for a linear model, i.e. by performing a linear regression of $Y(x)$ on $\mathbf{Z}(x)$. Alternatively, one can estimate $\boldsymbol{\beta}$ within the framework of sampling theory. There is some evidence that the choice of the estimation procedures is of secondary importance and that internal models can be treated as external models if n_2 is sufficiently large (Mandallaz, 1991). However, predictions based on models, particularly external ones, should not be blindly trusted. Moreover, it is intuitively clear that deviations should be considered between model and reality. By analogy with the model-assisted procedures discussed in Section 3.3, we examine the **residual** $R(x) = Y(x) - \widehat{Y}(x)$ and define **the two-phase one-stage estimator**

$$\widehat{\bar{Y}}_{reg} = \frac{1}{n_1} \sum_{x \in s_1} \widehat{Y}(x) + \frac{1}{n_2} \sum_{x \in s_2} R(x) \tag{5.1}$$

The lower index reg indicates that the two-phase estimator is indeed a model-assisted regression estimator. It is simply the mean of the prediction plus the mean of the residuals, which is intuitively very appealing. Given s_1 the properties of 5.1 are governed by standard sampling theory for a finite population (see Eq. 2.16) and the conditioning rules given in Appendix B. This leads immediately to

$$\mathbb{E}_{2|1} \frac{1}{n_2} \sum_{x \in s_2} R(x) = \frac{1}{n_1} \sum_{x \in s_1} R(x)$$

Because $Y(x) = \widehat{Y}(x) + R(x)$ we have by Eq. B.3 and 4.6

$$\mathbb{E}_{1,2}(\widehat{\bar{Y}}_{reg}) = \mathbb{E}_1 \frac{1}{n_1} \sum_{x \in s_1} Y(x) = \bar{Y}$$

and therefore unbiasedness. Furthermore

$$\mathbb{V}_{2|1} \frac{1}{n_2} \sum_{x \in s_2} R(x) = \left(1 - \frac{n_2}{n_1}\right) \frac{1}{n_2} \frac{1}{n_1 - 1} \sum_{x \in s_1} (R(x) - \bar{R}_1)^2$$

where we set $\bar{R}_i = \frac{1}{n_i} \sum_{x \in s_i} R(x)$ for $i = 1, 2$. Consequently we have

$$\mathbb{E}_1 \mathbb{V}_{2|1}(\widehat{\bar{Y}}_{reg}) = \left(1 - \frac{n_2}{n_1}\right) \frac{1}{n_2} \frac{1}{\lambda(F)} \int_F (R(x) - \bar{R})^2 dx$$

where

$$\bar{R} = \frac{1}{\lambda(F)} \int_F R(x) dx$$

Let us emphasize once again that we do not assume that the mean residual is zero. This is never the case with external models. With an internal model and least squares estimation of $\boldsymbol{\beta}$, $\bar{R}_2$ is usually 0 by construction, or nearly 0. We can then state the main results for two-phase one-stage simple random sampling according to the following theorem

Theorem 5.1.1.

$$\widehat{\bar{Y}}_{reg} = \frac{1}{n_1} \sum_{x \in s_1} \widehat{Y}(x) + \frac{1}{n_2} \sum_{x \in s_2} R(x)$$

$$\mathbb{V}(\widehat{\bar{Y}}_{reg}) = \left(1 - \frac{n_2}{n_1}\right) \frac{1}{n_2} \frac{1}{\lambda(F)} \int_F (R(x) - \bar{R})^2 dx$$
$$+ \frac{1}{n_1 \lambda(F)} \int_F (Y(x) - \bar{Y})^2 dx$$

$$\widehat{\mathbb{V}}(\widehat{\bar{Y}}_{reg}) = \left(1 - \frac{n_2}{n_1}\right) \frac{1}{n_2} \frac{1}{n_2 - 1} \sum_{x \in s_2} (R(x) - \bar{R}_2)^2$$

$$+ \frac{1}{n_1} \frac{1}{n_2 - 1} \sum_{x \in s_2} (Y(x) - \bar{Y}_2)^2$$

Remarks:

- The point and variance estimates are exactly unbiased when external models are implemented and are asymptotically unbiased if internal models are used.

- We implicitly assume that points $x \in s_2$ are error-free, in the sense that the prediction $\widehat{Y}(x)$ and the observation $Y(x)$ corresponds exactly to the same point. This, of course, is only approximately true in practice.

- When auxiliary information is provided by **thematic maps** (often based on aerial photographs), we can formally let n_1 tends to infinity in Theorem 5.1.1. Then, the variance of $\widehat{Y}_{reg}$ depends then only on the residuals and its estimate is

$$\widehat{\mathbb{V}}(\widehat{\bar{Y}}_{reg}) = \frac{1}{n_2} \frac{1}{n_2 - 1} \sum_{x \in s_2} (R(x) - \bar{R}_2)^2$$

The estimation of ratios can be performed according to the rules given in Section 2.6, here splitting the prediction and residual portions. One then obtains:

Theorem 5.1.2.

$$\widehat{R}_{1,2} = \frac{\widehat{Y}_{reg}^{(1)}}{\widehat{Y}_{reg}^{(2)}}$$

$$\widehat{\mathbb{V}}(\widehat{R}_{1,2}) = \frac{1}{(\widehat{Y}_{reg}^{(2)})^2} \left(1 - \frac{n_2}{n_1}\right) \frac{1}{n_2} \frac{1}{n_2 - 1} \sum_{x \in s_2} (U(x) - \bar{U}_2)^2$$

$$+ \frac{1}{(\widehat{Y}_{reg}^{(2)})^2} \frac{1}{n_1} \frac{1}{n_2 - 1} \sum_{x \in s_2} (Y^{(1)}(x) - \widehat{R}_{1,2} Y^{(2)}(x))^2$$

where $U(x) = R^{(1)}(x) - \widehat{R}_{1,2} R^{(2)}(x)$ (in practice $\bar{Z}_2 \approx 0$). The above theorem is valid even if different auxiliary vectors are used for the numerator and denominator.

5.2 Two-phase two-stage simple random sampling

In this procedure, all three random selections $(1, 2, 3)$ are involved. The concept is exactly as in two-phase one-stage simple random sampling, except that

( ) at point
i.e. the generalized local density $Y^*(x)$ defined in Eq. 4.25. The parameter
can be estimated by multiple linear regression of $Y^*(x)$ on $Z(x)$.

The two-phase two-stage estimate for simple random sampling
defined by

$$\widehat{Y}_{reg}^* = \frac{1}{n_1} \sum_{x \in s_1} \widehat{Y}(x) + \frac{1}{n_2} \sum_{x \in s_2} R^*(x) \tag{5.2}$$

where we have set $R^*(x) = Y^*(x) - \widehat{Y}(x)$.

Using Eq. B.3 and B.5 we arrive at

$$\mathbb{E}_{1,2,3}\widehat{Y}_{reg}^* = \mathbb{E}_1\mathbb{E}_{2|1}\mathbb{E}_{3|1,2}\widehat{Y}_{reg}^* = \bar{Y}$$

so that the estimate is design-unbiased. Likewise, we obtain

$$\mathbb{V}_{2,3|1}\widehat{Y}_{reg}^* = \mathbb{E}_{2|1}\mathbb{V}_{3|1,2}\widehat{Y}_{reg}^* + \mathbb{V}_{2|1}\mathbb{E}_{3|1,2}\widehat{Y}_{reg}^*$$

After some elementary algebra this leads to

$$\mathbb{V}_{1,2,3}(\widehat{Y}_{reg}^*) = \frac{1}{n_1}\mathbb{V}_x Y(x) + \left(1 - \frac{n_2}{n_1}\right)\frac{1}{n_2}\mathbb{V}_x R(x) + \frac{1}{n_2}\mathbb{E}_x V(x)$$

Therefore, the overall variance is essentially the sum of 1) the variance of
the exact local density as if we had observed it in the large sample; 2) the
residual variance of the true residuals because we have replaced, at the plot
level, observations with predictions; and 3) the second-stage variance, due to
replacing at the tree level exact measurements with predictions. This is indeed,
a very intuitive result.

Using exactly the same arguments as in the proof for Eq. 4.32 it is easy to see
that

$$\mathbb{E}_{1,2,3}\frac{1}{n_2 - 1}\sum_{x \in s_2}(\widehat{Y}^*(x) - \bar{Y}_2^*)^2 = \mathbb{E}_x V(x) + \mathbb{V}_x Y(x)$$

and that

$$\mathbb{E}_{1,2,3}\frac{1}{n_2 - 1}\sum_{x \in s_2}(\widehat{R}^*(x) - \bar{R}_2^*)^2 = \mathbb{E}_x V(x) + \mathbb{V}_x R(x)$$

We then obtain, after some algebra, the following result:

$$\widehat{Y}^*_{reg} = \frac{1}{n_1} \sum_{x \in s_1} \widehat{Y}(x) + \frac{1}{n_2} \sum_{x \in s_2} R^*(x)$$

$$\mathbb{V}(\widehat{Y}^*_{reg}) = \frac{1}{n_1} \mathbb{V}_x Y(x) + \left(1 - \frac{n_2}{n_1}\right) \frac{1}{n_2} \mathbb{V}_x R(x) +$$
$$+ \frac{1}{n_2} \mathbb{E}_x V(x)$$

$$\widehat{\mathbb{V}}(\widehat{Y}^*_{reg}) = \left(1 - \frac{n_2}{n_1}\right) \frac{1}{n_2} \frac{1}{n_2 - 1} \sum_{x \in s_2} (R^*(x) - \bar{R}^*_2)^2 +$$
$$+ \frac{1}{n_1} \frac{1}{n_2 - 1} \sum_{x \in s_2} (Y^*(x) - \bar{Y}^*_2)^2$$

The generalization of Theorem 5.1.2 in estimating the ratios through two-phase two-stage simple random sampling is straightforward. One simply replaces the $Y(x), R(x)$ with their estimates $Y^*(x), R^*(x)$.

Let us briefly consider the relative efficiency of two-phase one-stage versus one-phase one-stage sampling. The coefficient of determination is defined as

$$R^2 = \frac{\mathbb{V}_x(\widehat{Y}(x))}{\mathbb{V}_x(Y(x))} \tag{5.3}$$

Provided that the residuals and predictions are uncorrelated under the design (and this is usually the case, at least approximately) one obtains $\mathbb{V}_x R(x)$ $(1 - R^2)\mathbb{V}_x Y(x)$. Rewriting in Theorem 5.2.1 with $V(X) \equiv 0$

$$\mathbb{V}(\widehat{Y}_{reg}) = \frac{\mathbb{V}_x Y(x)(1 - R^2)}{n_2} + \frac{\mathbb{V}_x Y(x) R^2}{n_1} \tag{5.4}$$

Suppose that the cost function is given by

$$C = n_2 c_2 + n_1 c_1 \tag{5.5}$$

where c_1 and c_2 are the unitary first-phase and second-phase costs, respectively. It can be shown, see e.g. Cochran (1977), that, under the constraint of a given total cost C, two-phase sampling yields a smaller variance than does one-phase sampling if and only if the coefficient of determination is sufficiently large, i.e.,

$$R^2 > \frac{4c_1 c_2}{(c_1 + c_2)^2} \tag{5.6}$$

The case of thematic maps with $n_1 \to \infty$, $c_1 \to 0$, $n_1 c_1 \to C_{map}$ leads to the condition

$$R^2 > \frac{C_{map}}{C}$$

We apply the results obtained so far to the simplest but most important case encountered in practice. Here, a linear model consists of a simple ANOVA model, with one group variable representing the strata of a forest.

To illustrate the general theory we consider the important special case of post-stratification. We omit most of the tedious but elementary algebraic manipulations and provide the results for two-stage sampling at the tree level. For one-stage sampling, results can be obtained by setting $Y^*(x) = Y(x)$ and $V(x) = 0$.

Our forest area F is partitioned into L disjoined strata $F = \cup_{k=1}^{L} F_k$. The ideal prediction $\widehat{Y}(x)$ for $x \in F_k$ would be the true mean $\bar{Y}_k$ of the kth stratum. Because it is unknown, we estimate it via the corresponding internal linear model, which in this case is a simple one-way analysis of variance. The result is intuitively obvious and $\widehat{Y}(x)$ is simply the empirical mean of the stratum. That is $\widehat{Y}(x) = \bar{Y}_{2,k}^*$ for $x \in F_k$, where

$$\bar{Y}_{2,k}^* = \frac{1}{n_{2,k}} \sum_{x \in F_k \cap s_2} Y^*(x)$$

$n_{2,k}$ is the number of points in s_2 that fall into the kth stratum. In this case, as with almost all internal linear models, the residuals add up globally to 0, $\sum_{x \in F_k \cap s_2} R^*(x) = 0$. We finally arrive at

$$\widehat{Y}_{strat}^* = \sum_{k=1}^{L} \hat{p}_{1,k} \bar{Y}_{2,k}^* \qquad (5.7)$$

where the $\hat{p}_{1,k} = \frac{n_{1,k}}{n_1}$ are the proportions of the surface areas in the strata as estimated from the large sample. The usual variance estimate within the k stratum is defined as

$$\widehat{V}_k^* = \frac{1}{n_{2,k} - 1} \sum_{x \in F_k \cap s_2} (Y^*(x) - \bar{Y}_{2,k}^*)^2$$

Tedious but simple algebra shows that, here, Theorem 5.2.1 yields

$$\widehat{V}(\widehat{Y}_{strat}^*) = \frac{1}{n_2} \sum_{k=1}^{L} \left(\frac{n_{2,k} - 1}{n_2 - 1} \right) \widehat{V}_k^* + \frac{1}{n_1} \sum_{k=1}^{L} \left(\frac{n_{2,k}}{n_2 - 1} \right) (\bar{Y}_{2,k}^* - \bar{Y}_2^*)^2 \qquad (5.8)$$

where we have set

$$\bar{Y}_2^* = \frac{1}{n_2} \sum_{x \in s_2} Y^*(x)$$

The above estimate has some disturbing features: 1) it does not use the strata weights estimated from the large sample and 2) the variance is not inversely proportional to the $n_{2,k}$. Using a g-weight technique (see Eq. 6.29) it is possible to derive a better variance estimate, namely

$$\widehat{V}(\widehat{Y}_{strat}^*) = (1 - \frac{n_2}{n_1}) \sum_{k=1}^{L} \hat{p}_{1,k}^2 \frac{\widehat{V}_k^*}{n_{2,k}} + \frac{1}{n_1} \frac{1}{n_2 - 1} \sum_{x \in s_2} (Y^*(x) - \bar{Y}_2^*)^2 \qquad (5.9)$$

,k n_1
known exactly and Eq. 5.9 yields

$$\widehat{\mathbb{V}}(\widehat{Y}^*_{strat}) = \sum_{k=1}^{L} p_{1,k}^2 \frac{\widehat{V}^*_k}{n_{2,k}}$$

with $p_{1,k} = \frac{\lambda(F_k)}{\lambda(F)}$. Double sampling for stratification is then equivalent to exact stratification as defined in Section 2.7. This g-weight technique will be presented in the next chapter.

Because we have a simple analytical expression for the point estimate we can directly calculate the variance. All we must do is repeatedly apply Eq. B.3 (conditioning on given $n_{1,k}$) and exploit the fact that the $n_{1,k}$ have a multinomial distribution (Mandallaz, 1991). One obtains

$$\mathbb{V}(\widehat{Y}^*_{reg}) = \frac{1}{n_2}\sum_{k=1}^{L} p_k V_k + \frac{1}{n_1}\sum_{j=1}^{L} p_j(\bar{Y}_j - \bar{Y})^2 + \frac{1}{n_2}\mathbb{E}_x V(x) + O(n_2^{-2}) \quad (5.10)$$

The variance of the unstratified estimate $\widehat{Y}^* = \frac{1}{n_2}\sum_{x \in s_2} Y^*(x)$ is then approximated by

$$\mathbb{V}(\widehat{Y}^*) \approx \frac{1}{n_2}\sum_{j=1}^{L} p_j V_j + \frac{1}{n_2}\sum_{j=1}^{L} p_j(\bar{Y}_j - \bar{Y})^2 + \frac{1}{n_2}\mathbb{E}_x V(x)$$

This follows at once by using the decomposition

$$\int_F (Y(x) - \bar{Y})^2 dx = \sum_{j=1}^{L}\int_{F_j} (Y(x) - \bar{Y}_j + \bar{Y}_j - \bar{Y})^2 dx$$

Because a point $x \in F_k$ may sample trees from an adjacent stratum and edge effects are usually adjusted for only at the forest boundary $Y(x)$ is not exactly unbiased within each F_k. Nevertheless, we see that post-stratification can substantially reduce the variance when the among-strata variance, i.e. the term $\sum_{j=1}^{L} p_j(\bar{Y}_j - \bar{Y})^2$, is large.

Let us now calculate the expected value of the variance estimate from Eq. 5.8. One first conditions on the $n_{2,k}$ and then uses $\mathbb{E}_{2,3}(V_k^*|n_{2,k}) = V_k + \mathbb{E}_{x \in F_k} V$ (which follows by the same arguments as in the proof for Eq. 4.32). Now let $n_{2,k}n_2^{-1} \to p_k$, and replace in the last squared term the random variables with their expected values. We then obtain

$$\mathbb{E}_{1,2,3}\widehat{\mathbb{V}}(\widehat{Y}^*_{reg}) \approx \frac{1}{n_2}\sum_{j=1}^{L} p_j V_j + \frac{1}{n_1}\sum_{j=1}^{L} p_j(\bar{Y}_j - \bar{Y})^2 + \frac{1}{n_2}\mathbb{E}_x V(x) \quad (5.11)$$

so that, as expected, Eq. 5.8, and consequently also Theorem 5.2.1, yield asymptotically unbiased estimates of variance. In this case, we have shown that we can treat the internal model $\widehat{Y}(x) = \bar{Y}^*_{2,k}$ for $x \in F_k$ as the external model $\widehat{Y}(x) = \bar{Y}_k$, $x \in F_k$. This result holds for a large class of linear models.

lated (which is the case for internal linear models adjusted by ordinary least squares), then $\mathbb{V}_x(Y(x)) = \mathbb{V}_x(R(x)) + \mathbb{V}_x(\hat{Y}(x))$ and we can rewrite Theorem 5.2.1 as

$$\mathbb{V}(\hat{Y}^*_{reg}) = \frac{1}{n_2}\mathbb{V}_x(Y(x)) + \frac{1}{n_2}\mathbb{E}_x V(x) - \frac{1}{n_2}\left(1 - \frac{n_2}{n_1}\right)\mathbb{V}_x(\hat{Y}(x)) \qquad (5.12)$$

For post-stratification one must interpret, in the above formula, the predictions $\hat{Y}(x)$ as constant and equal to the true mean of the stratum, so that $\mathbb{V}_x(\hat{Y}(x)) = \sum_{j=1}^{L} p_j(\bar{Y}_j - \bar{Y})^2$. Asymptotically, one can do as if the predictions were exact when calculating the variance (but not for the point estimate because we must use the residuals in order to ensure unbiasedness). Of course, this requires that the prediction model yields asymptotically the true means in each stratum, a condition that will play an important role when calculating the anticipated variance.

5.3 Two-phase one-stage cluster random sampling

This is a straightforward generalization of the previous sections. The first phase draws a large sample of n_1 clusters, whose origins $x \in s_1$ are uniformly and independently distributed in $A \supset F$. The geometry of the clusters is determined as usual by the M vectors $e_l \in \Re^2$. At each point $x_l = x +$ of a given cluster one collects the auxiliary information required to make a prediction $\hat{Y}(x_l)$. The second phase draws a sub-sample of n_2 clusters out of the n_1, with origin $x \in s_2 \subset s_1$ according to equal probability sampling without replacement. For any given cluster with $x \in s_2$ the local density $Y($ is determined at each point x_l of the cluster. The easiest way to calculate the predictions is by estimating β via multiple linear regression of the $Y(x_l)$ on the $\mathbf{Z}(x_l)$ in s_2, while ignoring the cluster structure, and setting $\hat{Y}(x_l) = \mathbf{Z}(x_l)^t \hat{\boldsymbol{\beta}}_{s_2}$.

The two-phase one-stage estimate for cluster random sampling defined

$$\hat{Y}_{c,reg} = \frac{\sum_{x \in s_1} M(x)\hat{Y}_c(x)}{\sum_{x \in s_1} M(x)} + \frac{\sum_{x \in s_2} M(x)R_c(x)}{\sum_{x \in s_2} M(x)} \qquad (5.13)$$

where the residuals at the cluster level are defined according to

$$R_c(x) = \frac{\sum_{l=1}^{M} I_F(x_l)(Y(x_l) - \hat{Y}(x_l))}{\sum_{l=1}^{M} I_F(x_l)}$$

Here, we directly provide the main result; further details are given by Mandallaz (1991).

$$\widehat{Y}_{c,reg} = \frac{\sum_{x \in s_1} M(x)\widehat{Y}_c(x)}{\sum_{x \in s_1} M(x)} + \frac{\sum_{x \in s_2} M(x)R_c(x)}{\sum_{x \in s_2} M(x)}$$

$$\mathbb{V}(\widehat{Y}_{c,reg}) = \left(1 - \frac{n_2}{n_1}\right)\frac{1}{n_2}\frac{\mathbb{E}_x M^2(x)(R_c(x) - \bar{R})^2}{\mathbb{E}_x^2 M(x)}$$
$$+ \frac{1}{n_1}\frac{\mathbb{E}_x M^2(x)(Y_c(x) - \bar{Y})^2}{\mathbb{E}_x^2 M(x)}$$

$$\widehat{\mathbb{V}}(\widehat{Y}_{c,reg}) = \left(1 - \frac{n_2}{n_1}\right)\frac{1}{n_2}\frac{1}{n_2 - 1}\sum_{x \in s_2}\left(\frac{M(x)}{\bar{M}_2}\right)^2 (R_c(x) - \widehat{R}_2)^2$$
$$+ \frac{1}{n_1}\frac{1}{n_2 - 1}\sum_{x \in s_2}\left(\frac{M(x)}{\bar{M}_2}\right)^2 (Y_c(x) - \widehat{Y}_2)^2$$

The estimation of a ratio can be done by combining, in an obvious way, Theorems 5.1.2 and 5.3.1.

5.4 Two-phase two-stage cluster random sampling

This is a straightforward combination of Sections 5.2 and 5.3. The first phase yields the predictions $\widehat{Y}(x_l), \widehat{Y}_c(x)$ and the second phase the generalized densities $Y^*(x_l), Y_c^*(x)$. The easiest way to obtain these predictions is by estimating β via multiple linear regression of the $Y^*(x_l)$ on the $Z(x_l)$ in s_2, while ignoring the cluster structure, and setting $\widehat{Y}(x_l) = Z(x_l)^t\widehat{\beta}_{s_2}$.

The two-phase two-stage estimate for cluster random sampling described by

$$\widehat{Y}_{c,reg}^* = \frac{\sum_{x \in s_1} M(x)\widehat{Y}_c(x)}{\sum_{x \in s_1} M(x)} + \frac{\sum_{x \in s_2} M(x)R_c^*(x)}{\sum_{x \in s_2} M(x)} \qquad (5.14)$$

where the generalized residuals at the cluster level are given by

$$R_c^*(x) = \frac{\sum_{l=1}^M I_F(x_l)(Y^*(x_l) - \widehat{Y}(x_l))}{\sum_{l=1}^M I_F(x_l)}$$

We can state the main result as follows:

$$\widehat{Y}^*_{c,reg} = \frac{\sum_{x \in s_1} M(x)\widehat{Y}_c(x)}{\sum_{x \in s_1} M(x)} + \frac{\sum_{x \in s_2} M(x)R^*_c(x)}{\sum_{x \in s_2} M(x)}$$

$$\mathbb{V}(\widehat{Y}^*_{c,reg}) = \left(1 - \frac{n_2}{n_1}\right) \frac{1}{n_2} \frac{\mathbb{E}_{x \in A} M^2(x)(R_c(x) - \bar{R})^2}{\mathbb{E}^2_{x \in A} M(x)}$$

$$+ \frac{1}{n_1} \frac{\mathbb{E}_{x \in A} M^2(x)(Y_c(x) - \bar{Y})^2}{\mathbb{E}^2_{x \in A} M(x)} + \frac{1}{n_2} \frac{\mathbb{E}_{x \in F} V(x)}{\mathbb{E}_{x \in A} M(x)}$$

$$\widehat{\mathbb{V}}(\widehat{Y}^*_{c,reg}) = \left(1 - \frac{n_2}{n_1}\right) \frac{1}{n_2} \frac{1}{n_2 - 1} \sum_{x \in s_2} \left(\frac{M(x)}{\bar{M}_2}\right)^2 (R^*_c(x) - \widehat{R}^*_2)^2$$

$$+ \frac{1}{n_1} \frac{1}{n_2 - 1} \sum_{x \in s_2} \left(\frac{M(x)}{\bar{M}_2}\right)^2 (Y^*_c(x) - \widehat{Y}^*_2)^2$$

If the model is unbiased and if the predictions and residuals are not correlated then, as with simple random sampling, it is possible to rewrite the theoretical variance as

$$\mathbb{V}(\widehat{Y}^*_{c,reg}) = \frac{1}{n_2} \frac{\mathbb{E}_{x \in A} M^2(x)(Y_c(x) - \bar{Y})^2}{\mathbb{E}^2_{x \in A} M(x)}$$

$$- \frac{1}{n_2}\left(1 - \frac{n_2}{n_1}\right) \frac{\mathbb{E}_{x \in A} M^2(x)(\widehat{Y}_c(x) - \bar{Y})^2}{\mathbb{E}^2_{x \in A} M(x)}$$

$$+ \frac{1}{n_2} \frac{\mathbb{E}_{x \in F} V(x)}{\mathbb{E}_{x \in A} M(x)} \tag{5.15}$$

This formula will be useful for calculating the anticipated variance. We have mentioned many times that the regression coefficients and the predictions can be calculated by ordinary least squares. As an example, let us generalize the double-sampling procedure for stratification described in Subsection 5.2.1. For any $x + e_l \in F_k$ the prediction $\widehat{Y}(x_l)$ is the ordinary mean $\bar{Y}_{2,k}$, i.e. ignoring the cluster structure, of all the estimated local densities $Y^*(x_m)$ at points within the same stratum, $x_m \in F_k$. It is clear that the residuals again sum up to zero. Let $M_k(x) = \sum_{l=1}^{M} I_{F_k}(x_l)$ and $M_{1,k} = \sum_{x \in s_1} M_k(x)$. That is, $M_{1,k}$ is the total number of points from the large sample that fall into the k stratum. The estimated proportion of the surface area of the kth stratum is given by $\widehat{p}_{1,k} = \frac{M_{1,k}}{\sum_{x \in s_1} M(x)}$. It is not difficult to show here that in this case the regression estimator is again given by

$$\widehat{Y}^*_{strat} = \sum_{k=1}^{L} \widehat{p}_{1,k} \bar{Y}^*_{2,k}$$

However, there is no shortcut for calculating the variance and so one must use Theorem 5.4.1. Note that a cluster can be spread over different strata.

Because stratification is widely practiced one could ask whether more general linear models are really necessary. A major practical argument in favor of a broad theory is that a few qualitative variables can generate a huge number of strata (e.g. 3 variables, each with 5 levels, will produce 125 different strata). In contrast, one could use instead, after classical ANOVA testing for interactions, a simplified additive model with few parameters, thus improving substantially the precision of the estimates. The mathematical advantage is obvious. Moreover, in some instances the auxiliary information is given by continuous variables, such that a reduction to a discrete variable will decrease efficiency. Finally, let us note that if one must provide estimates for various sub-populations (say timber volume per tree species), the resulting estimates will be additive if one uses the same auxiliary vector $Z(x)$ throughout.

The reader can now have a look at the simulation example given in Appendix A.4.

The next section gives some more technical insight into parameter estimations.

5.5 Internal linear models in two-phase sampling

We shall investigate three techniques for estimating a regression coefficient vector β: design-based and weighted least squares (both relying on the cluster structure if necessary), and ordinary least squares at the plot level (which ignores that cluster structure). These three approaches are equivalent in simple two-phase sampling. Let us consider at the point level the **theoretical predictions** that can be obtained via a design-based linear model of the form

$$\widehat{Y}_{\beta}(x) = \bar{Y} + \beta^t(Z(x) - \bar{Z}), \beta \in \Re^p, Z(x) \in \Re^p$$

The $R(x) = Y(x) - \hat{Y}_{\beta}(x)$ have by construction a zero expected value, that is $\mathbb{E}_{x \in F} R(x) = 0$, which is not necessarily true, in the design-based sense, for the standard model $Y(x) = \beta^t Z(x) + R(x)$. At the cluster level we set consequently

$$\widehat{Y}_{\beta,c}(x) = \bar{Y} + \beta^t(Z_c(x) - \bar{Z}), \beta \in \Re^p, Z(x) \in \Re^p$$

These predictions and residuals are purely theoretical because $\bar{Y}$ and $\bar{Z}$ unknown and β is, for the time being, arbitrary. These definitions ensure by Eq. 4.15 and 4.16 that

$$\frac{\mathbb{E}_{x \in A} M(x) \widehat{Y}_{\beta,c}(x)}{\mathbb{E}_{x \in A} M(x)} = \bar{Y} \ \forall \beta$$

The theoretical residuals are defined as

$$R_{\beta,c}(x) = Y_c(x) - \widehat{Y}_{\beta,c}(x)$$

precisely

$$\bar{R}_\beta = \frac{\mathbb{E}_{x \in A} M(x) R_c(x)}{\mathbb{E}_{x \in A} M(x)} = 0 \ \forall \beta$$

We can verify that the theoretical two-phase estimator, which is defined as the mean of the predictions plus the mean of the residuals, might be rewritten as

$$\hat{Y}_{reg,\beta} = \hat{Y}_2 + \beta^t (\hat{Z}_1 - \hat{Z}_2) \ \forall \beta$$

where $\hat{U}_i = \frac{\sum_{x \in s_i} M(x) U_c(x)}{\sum_{x \in s_i} M(x)}$ for any variable U and $i = 1, 2$. **We shall assume that the model has an intercept term**, which is nearly always the case in practice. This means that one component of $\boldsymbol{Z}(x)$, for instance the first $\boldsymbol{Z}_1($ is always equal to 1. We then partition the vectors as $\boldsymbol{Z}(x)^t = (1, \boldsymbol{Z}^*(x)^t)$ and $\beta^t = (\beta_1, \beta^{*t})$. **In a design-based approach** the optimal choice for the unknown parameter vector β is determined by minimizing the variance of the regression estimator. Because of Theorem 5.3.1 and the property of a zero mean residual, this is equivalent to minimizing the term:

$$\min_{\beta} \mathbb{E}_x M^2(x)(Y_c(x) - \bar{Y} - \beta^t (\boldsymbol{Z}_c(x) - \bar{\boldsymbol{Z}}))^2$$

Differentiating with respect to β leads to the normal equations for an optimal theoretical choice β_0

$$\mathbb{E}_x M^2(x)(\boldsymbol{Z}_c(x) - \bar{\boldsymbol{Z}})(\boldsymbol{Z}_c(x) - \bar{\boldsymbol{Z}})^t \beta_0 = \mathbb{E}_x M^2(x)(Y_c(x) - \bar{Y})(\boldsymbol{Z}_c(x) - \bar{\boldsymbol{Z}}) \quad (5.16)$$

It follows by simple algebra from those normal equations that the optimal theoretical residuals and predictions are uncorrelated, such that

$$\mathbb{E}_{x \in A} M^2(x)(R_{\beta_0,c}(x) - \bar{R})(\hat{Y}_{\beta_o,c}(x) - \bar{Y}) = 0$$

From this we gain the following decomposition, which plays a key role in calculating the anticipated variance:

$$\mathbb{E}_{x \in A} M^2(x)(R_c(x) - \bar{R})^2 = \mathbb{E}_{x \in A} M^2(x)(Y_c(x) - \bar{Y})^2 - \mathbb{E}_{x \in A} M^2(x)(\hat{Y}_c(x) - \bar{Y} \quad (5.17)$$

The (p, p) matrix on the left hand side of Eq. 5.16 is singular for models with an intercept term because, in this case, the first row and the first column are both zero. An elegant solution is to use generalized inverses (Mandallaz, 1991), but this is not always feasible with standard statistical software packages. Therefore, we adopt an alternative approach. Rewriting the normal equations in terms of the reduced vectors $\boldsymbol{Z}_c^*(x), \beta^*$ we see that the general solution of the original normal equations is

$$\beta_0^t = (\beta_1, \beta_0^{*t})$$
$$\mathbb{E}_x M^2(x)(\boldsymbol{Z}_c^*(x) - \bar{\boldsymbol{Z}}^*)(\boldsymbol{Z}_c^*(x) - \bar{\boldsymbol{Z}}^*)^t \beta_0^* = \mathbb{E}_x M^2(x)(Y_c(x) - \bar{Y})(\boldsymbol{Z}_c^*(x) -$$

where β_1 is arbitrary.

In practice those theoretical normal equations obviously are not available, so

$$\sum_{x \in s_2} M^2(x)(\boldsymbol{Z}_c^*(x) - \widehat{\boldsymbol{Z}}_2^*)(\boldsymbol{Z}_c^*(x) - \widehat{\boldsymbol{Z}}_2^*)^t \widehat{\boldsymbol{\beta}}_0^* = \sum_{x \in s_2} M^2(x)(Y_c(x) - \widehat{Y}_2)(\boldsymbol{Z}_c^*(x) -$$

In other words, $\widehat{\boldsymbol{\beta}}_0^*$ is obtained by linear regression with weights $M^2(x)$ of the centered response variable $Y_c(x) - \widehat{Y}_2$ on the centered explanatory variables but without the intercept term, that is on $\boldsymbol{Z}_c^*(x) - \widehat{\boldsymbol{Z}}_2^*$.

Empirical predictions are given by

$$\widehat{Y}_c(x) = \widehat{Y}_2 + \widehat{\boldsymbol{\beta}}_0^{*t}(\boldsymbol{Z}_c^*(x) - \widehat{\boldsymbol{Z}}_2^*)$$

Most software packages will enable direct predictions, for example $P_c(x$ $\widehat{Y}_c(x) - \widehat{Y}_2$, so that one can also write $\widehat{Y}_c(x) = \widehat{Y}_2 + P_c(x)$.

The empirical residuals are then stated as

$$R_c(x) = Y_c(x) - \widehat{Y}_c(x) = (Y_c(x) - \widehat{Y}_2) - \widehat{\boldsymbol{\beta}}^{*t}(\boldsymbol{Z}_c^*(x) - \widehat{\boldsymbol{Z}}_2^*)$$

which by construction satisfy

$$\widehat{R}_2 = \frac{\sum_{x \in s_2} M(x) R_c(x)}{\sum_{x \in s_2} M(x)} = 0$$

Hence, the point estimate is finally given by

$$\widehat{Y}_{c,reg} = \widehat{Y}_2 + \widehat{\boldsymbol{\beta}}_0^{*t}(\widehat{\boldsymbol{Z}}_1^* - \widehat{\boldsymbol{Z}}_2^*) = \widehat{Y}_2 + \widehat{\boldsymbol{\beta}}_0^t(\widehat{\boldsymbol{Z}}_1 - \widehat{\boldsymbol{Z}}_2)$$

where

$$\widehat{\boldsymbol{\beta}}_0^t = (\beta_1, \widehat{\boldsymbol{\beta}}_0^{*t})$$

and $\boldsymbol{\beta}_1$ is arbitrary. Note that the first component of $\widehat{\boldsymbol{Z}}_1 - \widehat{\boldsymbol{Z}}_2$ is zero and that all statistically relevant quantities are independent of the arbitrary choice of β_1.

In short, an optimal design-based regression estimate can be obtained through standard regression procedures.

The weighted least squares procedure finds $\boldsymbol{\beta}$ in order to minimize

$$\mathbb{E}_{x \in A} M(x)(Y_c(x) - \boldsymbol{\beta}^t \boldsymbol{Z_c}(x))^2$$

The choice of $M(x)$ rather than $M^2(x)$ as weights is suggested by the model-dependent approach. When $Y_c(x)$ is the mean of the $M(x)$ observations, its variance can be expected to be inversely proportional to $M(x)$. We will use this concept later with the g- weights technique. The corresponding theoretical normal equations then read

$$\mathbb{E}_{x \in A} M(x) \boldsymbol{Z_c}(x) \boldsymbol{Z_c}(x)^t \boldsymbol{\beta} = \mathbb{E}_{x \in A} M(x) Y_c(x) \boldsymbol{Z_c}(x)$$

An estimate for $\boldsymbol{\beta}$ can be obtained from almost any software package by taking a sample copy of the above equation. This procedure also leads to $\mathbb{E}_{x \in A} M(x) R_c(x) = 0$ and $\bar{R} = 0$. This follows from the orthogonality properties of the residuals with weighted least squares. Partitioning as before the

$$= (1 \quad)$$

weighted least squares β^* components coincide if $M(x) \equiv M$. This can be valid only with simple random sampling, i.e. $M(x) \equiv 1$. However, it will be approximately correct if boundary effects can be neglected so that most clusters will be entirely contained within the forest (although they could be across different strata).

The ordinary least squares technique at the plot level estimates β simple multiple linear regression of all $Y(x_l)$ on the $Z(x_l)$. That is, we ignore completely the cluster structure of the data as far as β is concerned. One then defines, in the usual way, the predictions and residuals at the cluster level. Because the model has an intercept the sum of the residuals is again zero. Clearly, all three procedures will produce the same regression estimates in simple random sampling. Empirical evidence from case studies and simulations suggest that, in cluster-sampling, the differences between the resulting regression estimates and their variances (always calculated, of course, according to the formulae for cluster-sampling) are negligible from a practical point of view. It is also evident that all the previous results are valid in two-phase two-stage sampling, such that it suffices to replace $Y_c(x)$ with $Y_c^*(x)$ everywhere. Because $\mathbb{E}_{3|1,2}(Y^*(x_l)) = Y(x_l)$ we are able to get unbiased estimates of the regression coefficients. The fact that the resulting $\widehat{\beta}$ may not be optimal is of secondary importance in the context of two-phase two-stage estimation. A great advantage of computing ordinary least squares at the plot level is that the resulting point estimates often have an intuitive form, as shown, for example, with double cluster-sampling for stratification.

5.6 Remarks on systematic sampling

Here, we introduce a simple formalism to describe systematic sampling in the plane as applied to a forest inventory. A grid is defined by two non-collinear vectors in the plane $e_1, e_2 \in \Re^2$ plus an infinite number of fundamental cells:

$$c_{(i,j)} = \{x \in \Re^2 \mid x = (i + \mu)e_1 + (j + \nu)e_2, \ \mu, \nu \in [0,1]\} \tag{5.18}$$

where $i, j = 0, \pm 1, \pm 2, \pm 3 \ldots$ run over the integers. The grid points are identified by $z_{ij} = ie_1 + je_2$. Obviously, one has $\Re^2 = \cup_{i,j} c_{(i,j)}$ and $F = \cup_{i,j} F \cap c_{(i,j)}$ Point z_0 is uniformly distributed in cell $c_{(0,0)}$ and the sample points are $x_{ij} = z_0 + z_{ij}$. x_{ij} is uniformly distributed in $c_{(i,j)}$. We say that sample points x_{ij} are on a systematic grid with a fixed orientation and a random start. The number of points for the systematic sample falling within a particular forest area F is the random variable $n = \sum_{i,j} I_F(x_{ij}) < \infty$. Its expected value is given by

$$\mathbb{E}_{z_0}(n) = \sum_{i,j} \mathbb{P}(x_{ij} \in F) = \frac{\sum_{i,j} \lambda(F \cap c_{(i,j)})}{\lambda(c_{(i,j)})} = \frac{\lambda(F \cap (\cup_{i,j} c_{(i,j)}))}{\lambda(c_{(0,0)})} = \frac{\lambda(F}{\lambda(c_{(0}}$$

$(_{(0,0)})$ is the surface area of the fundamental cell (all cells are congruent).

Let us consider a usual local density $Y(x_{ij})$ at point $x_{ij} \in F$, if necessary being adjusted for boundary effects at the forest edge. Then, by definition, we set $Y(x_{ij}) = 0$ whenever $x_{ij} \notin F$. The random variable is defined as

$$\hat{T}_Y = \lambda(c_{(0,0)}) \sum_{i,j} Y(x_{ij}) < \infty \tag{5.19}$$

and we calculate its expected value as

$$
\begin{aligned}
\mathbb{E}_{z_0}(\hat{T}_Y) &= \lambda(c_{(0,0)}) \sum_{i,j} \mathbb{E}_{z_0} Y(x_{ij}) \\
&= \lambda(c_{(0,0)}) \sum_{i,j} \frac{\int_{c_{(i,j)}} Y(u)du}{\lambda(c_{(ij)})} \\
&= \sum_{i,j} \int_{c_{(i,j)}} Y(u)du = \int_F Y(u)du \\
&= \sum_{i=1}^{N} Y_i
\end{aligned}
$$

Hence, the random variable $\hat{T}_Y$ is an unbiased estimate of the total. Because $\lambda(F) = \lambda(c_{(0,0)})n$ is obviously an unbiased estimate of the surface area, one can take the ratio

$$\tilde{Y} = \frac{\sum_{i,j} Y(x_{ij})}{n}$$

as an estimate of the density. However, because this a ratio of two random variables, it is not, in fact, an unbiased estimate of the density. Therefore, systematic samples yield an unbiased estimate of totals but not of densities, a fact sometimes overlooked. In practice, and with large samples, such bias is negligible because the variance of n is very small as compared to its expected value. The same holds true for $Y^*(x_{ij})$ in two-stage sampling as well as in two-phase sampling with the residuals $R(x_{ij})$, $R^*(x_{ij})$. To calculate the inclusion probabilities we use the following elegant argument: we superimpose all the cells $c_{(i,j)}$ as well as the forest segments $F \cap c_{(i,j)}$, together with the trees. Thus, we end up with only one "super-cell" containing all the trees. Drawing a systematic sample is equivalent to drawing a single point in the super cell. If, furthermore, we identify the opposite borders of that super-cell (topologically, this is then a torus), we see with some geometric insight that the inclusion probabilities are now given by

$$\tilde{\pi}_i = \frac{\lambda(K_i \cap F)}{\lambda(c_{(0,0)})} = \frac{\lambda(K_i)}{\mathbb{E}(n)\lambda(F)} \qquad \tilde{\pi}_{ij} = \frac{\lambda(K_i \cap K_j \cap F)}{\mathbb{E}(n)\lambda(F)}$$

This is the case because a tree near the border of the super-cell must have its circle reflected at those identified boundaries. It is clear that for most pairs

$_{ij} = 0$ because their circles do not overlap (on the torus) and we have a single sample point in the super-cell. Therefore, once again, we cannot estimate the design based variance with one single systematic sample. One verifies that the Horvitz-Thompson point estimate and its variance, based on the $\tilde{\pi}_i$ and $\tilde{\pi}_{ij}$, coincide with Eq. 4.9 and 4.10 if one sets $n_2 = \mathbb{E}_{z_0}($ which holds approximately in large samples. Superpositioning the original cells into the super-cell is likely to destroy any existing pattern in the original forest, which a posteriori partially justifies the usual practice of treating large systematic samples as random samples. We shall come back to this point in chapter 9, see also Problem 4.3. In our previous arguments the orientation of the grid was fixed. By conditioning on the orientation (i.e. on the realization of the random angle), it becomes obvious that the same results hold for a systematic grid with random start and random orientation.

5.7 Exercises

Problem 5.1. *The following table displays terrestrial inventory data obtained from a cluster sampling scheme (with nominally $M = 5$ plots) in the Zürichberg forest. This data set refers to the small area from a case study discussed at great length in chapter 8. The surface areas for the entire forest and*

Cluster no	Plot nr	Stand	$\bar{N}$	$\bar{G}\frac{m^2}{ha}$
1	1	6	233.33	34.04
2	1	4	1266.66	32.33
2	2	6	200.00	32.37
3	1	4	333.33	38.52
3	2	4	366.66	14.76
4	1	5	66.67	8.65
5	1	6	100.00	16.86
5	2	6	266.67	34.59
5	3	5	400.00	47.57
5	4	6	33.33	6.55
6	1	5	100.00	19.91
6	2	5	100.00	5.14
6	3	6	133.33	17.83
6	4	5	100.00	13.20
6	5	5	300.00	29.14
7	1	5	200.00	33.11
7	2	5	200.00	32.17
7	3	5	166.67	47.26
8	1	6	100.00	2.37

the three different Development stages (Stands) 4, 5, 6 are $\lambda(F) = 17.07ha$ $\lambda(F_4) = 4.26ha$, $\lambda(F_5) = 7.32ha$ and $\lambda(F_6) = 5.49$ respectively. Calculate the *(number of stems per ha) and basal area density $\bar{G}$ (in $\frac{m^2}{ha}$): for a) for the one-phase estimate based only on the above terrestrial data, and b) for the two-phase estimate with $n_1 = \infty$ if these predictions are based on the least squares estimates at the plot level. The prediction model is a simple Analysis of Variance (ANOVA), with stand type (three levels) as the only factor. The two-phase estimates will differ from those given in chapter 8, first because the prediction model is restricted in this exercise to development stage only, second, because the regression co-efficients used for the two-phase estimates given in chapter 8 are based on the full data rather than only the small area; and, third, because Plot 1 of Cluster 2, originally in Development stage 3 has been re-allocated to Stand 4 in order to have at least two plots per stand. Needless to say, this exercise represents the inferior limit, in terms of sample sizes, of what is feasible in practice.*

Problem 5.2. *Prove the result given in Eq. 5.6.*

CHAPTER 6

Forest Inventory: advanced topics

6.1 The model-dependent approach

In a design-based approach statistical inferences rest upon the randomization principle and confidence intervals are interpreted under the hypothetical repetitions of samples. Even when models are used, such as in two-phase sampling schemes, the validity of an inference is insured by the scheme and not by the validity of the model in the classical statistical sense (i.e. with assumptions such as normally distributed, independent random variables). As mentioned for local density, $Y(x)$ is random because point x is the realization of a random variable. This contrasts with a model-dependent approach, in which point fixed and $Y(x)$ is interpreted as the realization of a spatial stochastic process that depicts the actual forest under investigation. Inferences are then based on the hypothetical realization of all forests that can be generated by a given stochastic process. The actual forest is viewed as one out of an infinite number of "similar" forests. Thus, in principle, one must build a highly complex process that determines the positioning of the trees as well as all variables Y associated with N trees, N also being a random variable under that model. Obviously, such models are complex mathematical objects (e.g. marked point processes) so that estimating the parameters is a highly non-trivial task. This approach could be viewed as the microscopic approach, analogous to thermodynamics, where one attempts to model pressure, temperature, entropy and other physical quantities at the molecular level. For practical problems encountered when conducting a forest inventory, it suffices to take a more pragmatic approach. This might be then considered macroscopic, in which we directly model the locally observed density $Y(x)$ as a stochastic process or random function. There are of course hidden difficulties with this philosophy. First, microscopic models at the tree level and macroscopic models at the point level are generally incompatible in general, in the sense that the relationship

$$\mathbb{E}_{M_1} \frac{1}{\lambda(F)} \sum_{i=1}^{N} Y_i = \frac{1}{\lambda(F)} \int_F \mathbb{E}_{M_2} Y(x) dx$$

can be violated (with $\mathbb{E}_{M_1}$ and $\mathbb{E}_{M_2}$ denoting expectations under their respective models). Second, we must define the stochastic integral, i.e. the integral of a random function.

$$\bar{Y} = \frac{1}{\lambda(F)} \int_F Y(x)dx$$

To appreciate this difficulty, let us assume we have a random function, such that $Y(u)$ and $Y(v)$ are independent random variables for all points $u \neq$ If we interpret the above integral as the limit of approximating sums, we can then conclude that the random variable $\bar{Y}$ has zero variance (because the variance of the mean of k independent random variables $Y(x_i)$, the obvious approximation of $\bar{Y}$, goes to zero as k^{-1}), unless $V(x) = \mathbb{V}(Y(x)) = \infty$, which is absurd. Such a process is called "white noise" and its covariance function can be viewed as "a Dirac function" $\delta(u-v)$ with $\delta(u) = 0$ for all $u \neq 0$, $\delta(0) =$ and $\int_{\Re^2} \delta(u) = 1$. The mathematics required for such processes (labeled as generalized random functions or distributions) are beyond the scope of this book and, fortunately, not necessary for our purposes. It suffices to assume that the spatial covariance function $C(u,v) = \mathbb{COV}(Y(u), Y(v))$ is continuous and that $\int_F V(x)dx < \infty$ in order to define stochastic integrals $\int_G Y(x)$ for an arbitrary $G \subset \Re^2$ as the limit, in the quadratic mean, of sums I_k $\sum_{i=1}^k Y(x_i)\Delta_k$. There, $\Delta_k = \frac{\lambda(G)}{k}$, and points x_i are on a regular grid covering G. We say that a sequence I_k of random variables converges in the quadratic mean towards the random variable I if and only if $\lim_{k\to\infty} \mathbb{E}(I_k - I)^2 = 0$.

Spatial covariance $C(u,v)$ is the most difficult aspect in this model-dependent approach. Here, we shall assume that the correlation between two points u and v is a rapidly decreasing function of their distance $|u-v|$. This assumption can be removed, at least partially, in the geostatistical approach. Let us now set the stage.

We assume that we have a column vector of auxiliary variables $\mathbf{Z}(x) \in \Re$ that is known for all $x \in \Re^p$ and in particular at the sample points $x \in s_2$ (we denote as usual the terrestrial sample with s_2). This is the case when thematic maps, are available (e.g. stand maps for Switzerland). The spatial mean of the auxiliary vector is also known:

$$\bar{\mathbf{Z}} = \frac{1}{\lambda(F)} \int_F \mathbf{Z}(x)dx \tag{6.1}$$

The vector $\mathbf{Z}(x)$ is assumed to be fixed and not random. **In our model-dependent approach the sample of points $x \in s_2$ can be selected in a fully arbitrary way, e.g. via systematic or simple random sampling or cluster sampling, or even through purposive sampling**. Although this choice is totally irrelevant for model-dependent statistical inferences, it is highly recommended that the sample be "representative". For example, if you have a stand map with 10 different type of stands, you should ensure that all 10 stands are sufficiently represented in the sample.

We assume the following linear model:

$$
\begin{aligned}
Y(x) &= \mathbf{Z}(x)^t \boldsymbol{\beta} + R(x) \\
\mathbb{E}_M Y(x) &= \mathbf{Z}(x)^t \boldsymbol{\beta}
\end{aligned}
\tag{6.2}
$$

unknown but fixed parameter. The residuals $R(x)$, with $\mathbb{E}_M R(x) \equiv 0$, are viewed as random fluctuations about the "drift" $Z(x)^t \beta$. Therefore, the model-dependent approach relies on a super-population model, first introduced into forestry in 1960 by the statistician Matérn (see Matérn, 1986).

The notion of drift is not clear cut: it depends of the particular scale considered as well as on the available auxiliary information $Z(x)$. For example, we may have a process $Y(x)$ that appears stationary on a large scale, such as a globally homogenous forest, so that the auxiliary information can be summarized as $Z(x) \equiv 1$. Then, the covariance function $\mathbb{COV}(Y(u), Y(v)) = \mathbb{COV}(R(u), R(v)) = C(u, v)$ will have a rather large range. However, if we consult an adequate stand map, $Z(x)$ is a vector of 0/1 defining the stratification, the forest is only locally homogenous, and the covariance function of the residuals will have a short range.

We require the following notation:

$$ s_2 = \{x_1, x_2 \dots, x_{n_2}\} $$

being the sample of the n_2 points in which the local densities $Y(x_i)$ are observed.

$$ Y_{s_2} = (Y(x_1), Y(x_2), \dots, Y(x_{n_2}))^t $$

is then the $n_2 \times 1$ column vector of observations.

$$ Z_{s_2} = (Z(x_1), Z(x_2) \dots, Z(x_{n_2}))^t $$

is the $n_2 \times p$ design matrix. The ith row of Z_{s_2} is the vector $Z^t(x_i)$.

$$ V_{s_2} = diag(V(x_1), V(x_2), \dots, V(x_{n_2})) $$

is the diagonal $n_2 \times n_2$ matrix of the true variances under the model, i.e. $V(x) = \mathbb{V}_M(Y(x))$.

$$ W_{s_2} = diag(W(x_1), W(x_2), \dots, W(x_{n_2})) $$

is an $n_2 \times n_2$ matrix of weights. Those weights are considered correct if $V(x$ $\sigma^2 W(x)$ for some σ^2. The ordinary weighted least squares estimate of the regression parameter is then

$$ \hat{\beta}_{s_2} = (Z_{s_2} W_{s_2}^{-1} Z_{s_2})^{-1} Z_{s_2}^t W_{s_2}^{-1} Y_{s_2} \qquad (6.3) $$

This estimate $\hat{\beta}_{s_2}$ is model-unbiased, i.e. $\mathbb{E}_M(\hat{\beta}_{s_2}) = \beta$. Nevertheless, in most instances it is not the best linear-unbiased estimate of the parameter β, unless we have the correct weights and the observations reveal zero covariance.

The model-dependent predictor $\hat{Y}_{pred}$ of $\bar{Y}$ is given by

$$ \hat{Y}_{pred} = \frac{1}{\lambda(F)} \int_F Z(x)^t \hat{\beta}_{s_2} dx = \bar{Z}^t \hat{\beta}_{s_2} \qquad (6.4) $$

We can then demonstrate that $\hat{Y}_{pred}$ is the best linear-unbiased predictor

$\mathbb{E}_M(\hat{Y}_{pred} - \bar{Y})^2$, is minimized, provided that the weights are correct and the observations uncorrelated. This remains valid when the correlation is constant if $\sqrt{V(x)}$ is an element of $Z(x)$. It is true when $V(x) \equiv \sigma^2$ and $\mathbf{Z}(x)$ contains the intercept term, (Mandallaz, 1991). However, the assumption of constant correlation is rarely if ever fulfilled in a forest inventory. Although Royall (see Royall (1970) and related papers) is a strong advocate of the pure model-dependent approach, such extreme views are not shared by the majority of sample surveyors. **It must be emphasized that if a sample is randomly selected, then the model-dependent estimator $\hat{Y}_{pred}$ is usually not design-unbiased**. Rather, it might even be severely biased if the model is incorrect, e.g. if unsuspected important variables are not included in $\mathbf{Z}($ This means that, in general, $\mathbb{E}_p(\hat{Y}_{pred}) \neq \bar{Y}$, for a given realization of Y and $\bar{Y}$ of the forest and for a given design $p(\cdot)$ generating the points $x \in s$

As in Section 3.3 we again incorporate g-weights (N.B., because we are estimating spatial means and not totals here, the definition of the g-weights used in forest inventories is slightly different)

$$g_{s_2}(x) = \bar{\mathbf{Z}}^t \left(\frac{1}{n_2} \sum_{x \in s_2} \frac{\mathbf{Z}(x)\mathbf{Z}(x)^t}{W(x)} \right)^{-1} \frac{\mathbf{Z}(x)}{W(x)} \qquad (6.5)$$

Direct calculations lead to these important properties:

$$\frac{1}{n_2} \sum_{x \in s_2} g_{s_2}(x)\mathbf{Z}(x) = \bar{\mathbf{Z}}$$

$$\hat{Y}_{pred} = \frac{1}{n_2} \sum_{x \in s_2} g_{s_2}(x)Y(x)$$

$$\hat{Y}_{pred} - \bar{Y} = \frac{1}{n_2} \sum_{x \in s_2} g_{s_2}(x)R(x) - \frac{1}{\lambda(F)} \int_F R(x)dx \qquad (6.6)$$

The third equation follows directly from the first by setting $Y(x) = \mathbf{Z}(x)^t\boldsymbol{\beta}$ $R(x)$. The first relationship in Eq. 6.6 suggests that for sufficiently regular samples one will have $g_{s_2}(x) \approx 1$. We shall see later in the proof for Theorem 6.2.2 **that one can ensure without loss of generality that** $g_{s_2}(x) = 1 + O(n_2^{-\frac{1}{2}})$. This will be the case for systematic rectangular grids, the most important case in practice, as well as for points generated by simple or stratified random sampling. We are now able to calculate the **MSE**. First, let us note that because $\hat{Y}_{pred}$ is model-unbiased we have $\mathbb{E}_M(\hat{Y}_{pred} - \bar{Y})^2$

($pred$). Using the third relationship from Eq. 6.6, we arrive at

$$\mathbb{V}_M(\hat{Y}_{pred} - \bar{Y}) = \frac{1}{n_2^2} \sum_{u \in s_2} g_{s_2}^2(u) V(u)$$

$$+ \frac{1}{n_2^2} \sum_{u \neq v \in s_2} g_{s_2}(u) g_{s_2}(v) C(u,v)$$

$$- \frac{2}{n_2} \sum_{u \in s_2} g_{s_2}(u) \frac{1}{\lambda(F)} \int_F C(u,v) dv$$

$$+ \mathbb{V}_M \left(\frac{1}{\lambda(F)} \int_F R(v) dv \right) \tag{6.7}$$

Although our particular scheme is irrelevant at the estimation stage within the model-dependent approach it is useful, when conducting a pre-sampling evaluation of the efficiency of strategies (i.e. the choice of the estimators and of the sampling schemes), that we consider the so-called **anticipated mean square error**, **AMSE**, defined as the average of the **MSE** under all possible samples generated by the design. Then

$$\mathbf{AMSE} = \mathbb{E}_M \mathbb{E}_p(\hat{Y}_{reg} - \bar{Y})^2 = \mathbb{E}_p \mathbb{E}_M(\hat{Y}_{reg} - \bar{Y})^2 \tag{6.8}$$

The order of expectation can be interchanged provided the design is uninformative, which is here the case because the sampling density is $\frac{1 dx}{\lambda(F)}$ and this does not depend explicitly on $Y(x)$. The joint expectation $\mathbb{E}_M \mathbb{E}_p = \mathbb{E}_p$ with respect to design and model will be simply denoted by $\mathbb{E}$.

To get a better insight into the cumbersome expression found in Eq. 6.7, we use the following heuristic argument to derive the **AMSE** of $\hat{Y}_{pred}$ under simple random sampling. Replacing the sums with integrals, setting $g_{s_2}(u) = 1$ and noting that

$$\mathbb{V}_M \left(\frac{1}{\lambda(F)} \int_F R(v) dv \right) = \frac{1}{\lambda(F)^2} \int_F \int_F C(u,v) du dv$$

we see that the last two terms add up to $-\frac{1}{\lambda(F)^2} \int_F \int_F C(u,v) du dv$ as $n_2 \to \infty$ The second term is the sum of $n_2(n_2 - 1)$ terms in the form $C(u,v)$ with $u, v \in s_2$. The discrete sum can be approximated with $n_2(n_2 - 1)$ times the spatial average of $C(u,v)$, which is precisely $\frac{1}{\lambda(F)^2} \int_F \int_F C(u,v) du dv$, to finally obtain the following result:

Theorem 6.1.1. *The AMSE under simple random sampling of the model-dependent estimator $\hat{Y}_{pred}$ is asymptotically given by*

$$\mathbb{E}_p \mathbb{E}_M(\hat{Y}_{pred} - \bar{Y})^2 = \frac{1}{n_2} \left(\frac{1}{\lambda(F)} \int_F V(u) du - \frac{1}{\lambda(F)^2} \int_F \int_F C(u,v) du dv \right)$$

1. A word of caution is required for asymptotic arguments. If one naively replaces directly in Eq. 6.7 the g-weights with 1 and the sum with integrals, then some of the neglected terms are of order $n_2^{-\frac{1}{2}}$ whereas the leading term is of order n_2^{-1}. Nevertheless, the above heuristic derivation of **AMSE** produces the correct result. Asymptotically, we see that the drift is filtered out because the **AMSE** depends only on the covariance function of the residuals. In other words, one can neglect the impact of estimating β. In the next section, we shall provide a more general version of Theorem 6.1.1, which can be proven rigorously.

2. The term $\frac{1}{\lambda(F)^2} \int_F \int_F C(u,v)dudv$ is a double integral in the plane (i.e. a four-fold integral) which can be interpreted as the design-based expected value $\mathbb{E}_p C(U,V)$ of the covariance between two random points U,V independently and uniformly distributed in F.

To continue, we consider stationary stochastic processes in the plane, with a constant mean of $\mathbb{E}_M Y(x) = \mu$, constant variance $V(x) = \sigma^2$, and a positive isotropic correlation function $\rho(h)$. That is, we assume that $C(u,v) = \sigma^2$ $u - v \mid)$. The correlation between two points $u,v \in \Re^2$ depends only on their relative distance $\mid u - v \mid = h$. We shall also write $\rho((x,y))$ for $\rho(\sqrt{x^2 +}$ $(x,y \in \Re)$. Let us give three important examples:

1. **Spherical correlation with range r.**

$$\rho(h) = 1 - \frac{3h}{r} + \frac{h^3}{2r^3}$$

   for $h \le r$ and $\rho(h) = 0$ for $h > r$.

2. **Circular correlation with range r.**

$$\rho(h) = \frac{2}{\pi}\left(arccos(\frac{h}{r}) - \frac{h}{r}\sqrt{1 - (\frac{h}{r})^2} \right)$$

   for $h \le r$ and $\rho(h) = 0$ for $h > r$.

3. **Exponential correlation with range r.**

$$\rho(h) = exp(-\frac{3h}{r})$$

   for all $h \in \Re$. Though the range is infinite the correlation is, in practice, set to zero beyond the range r.

We now consider, as a preliminary calculation for systematic sampling, the **AMSE** under stratified random sampling. Without loss of generality we can examine the rectangular grid defined in Section 5.6 and a rectangular forest F which is the finite union of fundamental cells $c_{(i,j)}$. Sample points x_{ij} independently and uniformly distributed in c_{ij}. The x_{ij} display a pattern much closer to a systematic grid. Neglecting asymptotically the impact of estimating the drift, it is easy to obtain the following result (for a proof see Ripley, 1981):

Theorem 6.1.2. *The **AMSE** under stratified random sampling of the model-dependent estimator $\hat{Y}_{pred}$, is, assuming stationary and isotropic covariance of the residual process, asymptotically given by*

$$\mathbb{E}_p \mathbb{E}_M (\hat{Y}_{pred} - \bar{Y})^2 = \frac{1}{n_2} \left(\sigma^2 - \frac{1}{\lambda^2(c_{(0,0)})} \int_{c_{(0,0)}} \int_{c_{(0,0)}} C(u,v) du dv \right)$$

In a forest inventory the correlation is usually positive in natural forests, while plantations may display periodic positive and negative correlation. Comparing the positive second terms in Theorems 6.1.1 and 6.1.2 we see that stratified random sampling will lead to smaller **AMSE** than simple random sampling. Let us now evaluate the far more difficult case of systematic sampling as outlined for rectangular grids in Section 5.6. To simplify the notation we restrict ourselves to a square network, assuming again that $F = \cup_{i,j} c_{(i,j)}$ and that, asymptotically, one can neglect the impact of removing the drift. This is equivalent to considering the **AMSE** of a sample mean for the residual process, which is thought to be stationary with an $\mathbb{E}_M R(x) = 0$, $\mathbb{V}_M R(x) = \sigma^2$, and an isotropic correlation function $\sigma^2 \rho((x,y)) = \mathbb{COV}(Y((x,y)), Y((0,0)))$ (i.e. between the point in the plane with co-ordinates (x,y) and the origin $(0,$ The sample points are on the infinite grid $(m\Delta, n\Delta)$ with integer numbers m, n ranging from $-\infty$ to ∞. The asymptotic is the following: Δ is fixed, the forest F is square with $\lambda(F) \to \infty$, the density of sample point per unit area is a constant equal to $\frac{1}{\Delta^2}$ and the number of sample points is $n_2 = \frac{\lambda(F)}{\Delta^2} \to \infty$ **AMSE** is calculated with respect to $\mathbb{E}_{z_o} \mathbb{E}_M = \mathbb{E}_{z_o, M}$, that is with respect to a random start and fixed orientation as defined in Section 5.6. Using the proofs and discussions of Matérn (1986) and Ripley (1981) we can conclude that the following asymptotic result holds:

Theorem 6.1.3. *Under systematic sampling with square network (mesh and a stationary isotropic covariance the **AMSE** for our model-dependent estimator $\hat{Y}_{pred}$ is asymptotically (i.e. for $\lambda(F)$ and $n_2 \to \infty$ with $\frac{\lambda(F)}{n_2} = \Delta$ given by*

$$n_2 \mathbb{E}_{z_o, M} (\hat{Y}_{pred} - \bar{Y})^2 = \sum_{m,n=-\infty}^{\infty} \rho((m\Delta, n\Delta)) - \frac{1}{\Delta^2} \int_{-\infty}^{\infty} \int_{-\infty}^{\infty} \rho((x,y)) dx dy$$

The **spectral density function** is defined as the Fourier transform of the stationary covariance $c(x,y) = \sigma^2 \rho((x,y))$

$$f(\omega_1, \omega_2) = \frac{1}{4\pi^2} \int_{-\infty}^{\infty} \int_{-\infty}^{\infty} exp(-i(\omega_1 x + \omega_2 y)) c(x,y) dx dy \qquad (6.9)$$

Under systematic sampling with a square network (mesh Δ) and a stationary isotropic covariance the **AMSE** for the model-dependent estimator $\hat{Y}_{pred}$ is asymptotically (i.e. for $\lambda(F)$ and $n_2 \to \infty$ with $\frac{\lambda(F)}{n_2} = \Delta$) given by

$$n_2 \mathbb{E}_{z_o, M} (\hat{Y}_{pred} - \bar{Y})^2 = \frac{4\pi^2}{\Delta^2} \left(\sum_{m,n=-\infty}^{\infty} f(\frac{2\pi m}{\Delta}, \frac{2\pi n}{\Delta}) - f(0,0) \right)$$

In Theorems 6.1.3 and 6.1.4 one must require $\int_{-\infty}^{\infty} \int_{-\infty}^{\infty} \rho((x,y)) dx dy <$ and $f(0,0) < \infty$. Therefore, the correlation must decay sufficiently rapidly with distance and the spectral density function must be continuous at the origin. This excludes any process with a sharp peak at low frequencies or, equivalently, very long trends. Such processes do exist (e.g. the fractional Brownian motion) and have been used in hydrology. They are also related to fractal objects and self-similarity properties.

Thus, if low frequencies are dominant (corresponding to strong local positive correlations), $f(0,0)$ is large and systematic sampling is very efficient, unless there is a sharp peak in the spectral density at one of the frequencies summed in Theorem 6.1.4. This would occur if the process has a strong periodicity with a wavelength equal to the sampling interval Δ along either axis or with a wavelength $\sqrt{2}\Delta$ along a diagonal. As far as the spatial variation encountered in forest surveys is concerned, it may be said that no clear case of periodicity has ever been reported. However, if such a periodicity could be a priori suspected along a certain direction it would be wise to re-orient the grid in a different direction.

To understand the numerical relative efficiencies under random and systematic sampling we consider the exponential isotropic correlation function with a range r given by

$$\rho(x, y) = exp \left(-\frac{3\sqrt{x^2 + y^2}}{r} \right)$$

for a stationary residual process with unit variance $\sigma^2 = 1$. The spectral density function is given by

$$\frac{\nu}{2\pi} \frac{1}{(\nu^2 + \omega_1^2 + \omega_2^2)^{\frac{3}{2}}}$$

where $\nu = \frac{3}{r}$. We evaluate a square network under the asymptotic assumptions stated in Theorem 6.1.3. Setting $\beta = \frac{r}{3\Delta}$, one obtains the following expression for Theorem 6.1.3:

$$n_2 \mathbb{E}_{z_o, M} (\hat{Y}_{pred} - \bar{Y})^2 = \sum_{m,n=-\infty}^{\infty} exp \left(-\frac{\sqrt{m^2 + n^2}}{\beta} \right) - 2\pi\beta^2 \qquad (6.10)$$

$$n_2\mathbb{E}_{z_o,M}(\hat{Y}_{pred} - \bar{Y})^2 = 2\pi\beta^2 \left(\sum_{m,n=-\infty}^{\infty} \left(1 + 4\pi^2\beta^2(m^2 + n^2)\right)^{-\frac{3}{2}} - 1 \right)$$

$$(6.11)$$

The first series converges much faster than the second making it better for numerical work.

The second term in Theorem 6.1.1 will tend to zero as $\lambda(F) \to \infty$ for any correlation function with finite range and also for the exponential correlation (this is easy to prove by using polar coordinates). The second term in Theorem 6.1.2 is extremely difficult to calculate (see Matérn, 1986). Table 6.1 displays the **AMSE** for systematic sampling as a function of the correlation range.

Table 6.1 *Exponential correlation: AMSE for systematic sampling as a percentage of random sampling*

r	0.50	0.75	1.00	1.50	3.00
$\rho(\Delta)$	0.00	0.02	0.05	0.14	0.37
β	0.17	0.25	0.33	0.5	1.0
AMSE	83.6	69.7	58.4	42.2	22.4

Legend: Δ: distance between adjacent grid points; $r\Delta$, correlation range, $\beta = \frac{r}{3\Delta}$; $\rho(\Delta)$: correlation between two adjacent grid points

Remarks:

1. The asymptotic chosen $\lambda(F) \to \infty$ favors systematic sampling. For a finite forest with an irregular shape the advantage of systematic sampling over random sampling is likely to be reduced.

2. The loss of efficiency is important as soon as $r > \Delta$. Alternatively, treating a systematic sample as a random sample will substantially overestimate the error when the correlation of the residuals between adjacent grid points is larger than, say 0.10.

3. Triangular grid (or hexagonal network) are nearly equivalent to square network.

4. Rectangular grids with mesh Δ_x and Δ_y along the co-ordinates axes lead to much higher **MSE**, particularly if the ratio $\frac{\Delta_x}{\Delta_y}$ is larger than 2 or smaller than $\frac{1}{2}$. They should be avoided.

$$\bar{\rho} = \frac{1}{\lambda(D_r)} \int_{D_r} \rho((x,y)) dx dy = \frac{1}{\pi r^2} \int_0^r \rho(h) 2\pi h \, dh \qquad (6.12)$$

where D_r is the disk of radius r in the origin. The second expression is obtained by changing from Cartesian to polar co-ordinates. As an example, one might obtain the values of $\bar{\rho} = 0.2,\ 0.25$ and 0.18 for the spherical, circular and exponential correlations, respectively.

To better understand the behavior of the term

$$\mathbb{V}_M \bar{Y} = \frac{1}{\lambda(F)^2} \int_F \int_F C(u,v) du dv$$

we assume a short-range covariance structure of the following form:

$$C(u,v) = \begin{cases} 0 & \text{if } v \notin D_r(u) \\ \rho(u-v)V(u) & \text{if } v \in D_r(u) \end{cases} \qquad (6.13)$$

where $D_r(u)$ is the disk of radius r centered in u. Then one has

$$\frac{1}{\lambda(F)^2} \int_F \int_F C(u,v) du dv = \frac{1}{\lambda(F)^2} \int_F du \int_{D_R(u)} \rho(u-v)V(u)$$

which can be approximated if $V(u)$ is slowly varying in the neighborhood of u, as $\frac{\bar{\rho}\lambda(D_r)}{\lambda(F)^2} \int_F V(u) du = \bar{\rho}\bar{V}\frac{\lambda(D_r)}{\lambda(F)}$, where $\bar{V} = \frac{1}{\lambda(F)} \int_F V(u) du$ is the spatial mean of the variance.

The density of sample points per unit area is $\frac{n_2}{\lambda(F)}$ so that on average we have $k = n_2 \frac{\lambda(D_r)}{\lambda(F)}$ points within the correlation range. Hence, we can write $\bar{\rho}\bar{V}\frac{\lambda(D_r)}{\lambda(F)} = \frac{1}{n_2} k \bar{\rho}\bar{V}$ and approximate the **AMSE** according to

$$\mathbb{E}_p \mathbb{E}_M (\hat{Y}_{pred} - \bar{Y})^2 \approx \frac{1}{n_2}\bar{V} + \frac{1}{n_2^2} k \bar{\rho}\bar{V} \qquad (6.14)$$

The relative contribution of the correlation term to the **AMSE** is $\frac{k\bar{\rho}}{n_2}$ and is in practice negligible if $\frac{k}{n_2} < 0.2$. Under this assumption of **short-range correlation** we can say that the **AMSE** from simple random sampling is given by

$$\mathbb{E}_p \mathbb{E}_M (\hat{Y}_{pred} - \bar{Y})^2 \approx \frac{1}{n_2} \frac{1}{\lambda(F)} \int_F V(x) \qquad (6.15)$$

The right-hand side can be viewed as the variance of $\mathbb{V}(\hat{Y}_{pred})$ under the idealized model of uncorrelated residuals. From now on we shall work under this model to derive an estimate of $\mathbb{V}(\hat{Y}_{pred})$. In standard weighted least squares, assuming correct weights, the estimated covariance matrix of $\hat{\boldsymbol{\beta}}_{s_2}$ is given by

$$\hat{\boldsymbol{\Sigma}}_{\hat{\boldsymbol{\beta}}_{s_2}} = \hat{\sigma}^2 (\boldsymbol{Z}_{s_2} \boldsymbol{W}_{s_2}^{-1} \boldsymbol{Z}_{s_2})^{-1}$$

$$\hat{\sigma}^2 = \frac{(\boldsymbol{Y}_{s_2} - \boldsymbol{Z}_{s_2}\hat{\boldsymbol{\beta}}_{s_2})^t \boldsymbol{W}_{s_2}^{-1}(\boldsymbol{Y}_{s_2} - \boldsymbol{Z}_{s_2}\hat{\boldsymbol{\beta}}_{s_2})}{n_2 - p}$$

However, if the weights are incorrect, it is known that this estimate can be severely biased, even asymptotically. In the absence of prior knowledge it is natural to choose unit weights, i.e. $W(x) \equiv 1$. A **robust estimate** of the covariance matrix can be obtained with

$$\hat{\boldsymbol{\Sigma}}_{\hat{\boldsymbol{\beta}}_{s_2},R} = (\boldsymbol{Z}_{s_2}^t \boldsymbol{Z}_{s_2})^{-1} (\boldsymbol{Z}_{s_2}^t diag(\hat{r}^2(x)) \boldsymbol{Z}_{s_2}) (\boldsymbol{Z}_{s_2}^t \boldsymbol{Z}_{s_2})^{-1} \qquad (6.16)$$

where $\hat{R}(x) = Y(x) - \boldsymbol{Z}^t(x)\boldsymbol{\beta}_{s_2}$ are the empirical residuals (Huber (1967), Gregoire and Dyer (1989) and Mandallaz (1991)). This variance estimate is robust in the sense that it provides asymptotically the correct variance even if the unit weights are incorrect, that is, even if $V(x)$ is not constant. Because $\hat{Y}_{pred} = \bar{\boldsymbol{Z}}^t \hat{\boldsymbol{\beta}}_{s_2}$ one obtains a robust estimate of variance (under the model $\tilde{M}$ of uncorrelated observations) as $\bar{\boldsymbol{Z}}^t \hat{\boldsymbol{\Sigma}}_{\hat{\boldsymbol{\beta}}_{s_2},R} \bar{\boldsymbol{Z}}$. Direct calculations show that this is precisely equal to $n_2^{-2} \sum_{x \in s_2} g_{s_2}^2(x) \hat{R}(x)^2$, where $W(x) = 1$ is used in Eq. 6.5. We can summarize these findings in the following theorem:

Theorem 6.1.5. *The g-weights are defined according to*

$$g_{s_2}(x) = \bar{\boldsymbol{Z}}^t \left(\frac{1}{n_2} \sum_{x \in s_2} \boldsymbol{Z}(x)\boldsymbol{Z}(x)^t \right)^{-1} \boldsymbol{Z}(x)$$

The model-dependent predictor of the spatial mean $\bar{Y}$ is given by

$$\hat{Y}_{pred} = \frac{1}{n_2} \sum_{x \in s_2} g_{s_2}(x) Y(x)$$

Assuming short-range correlation the variance and the mean square error have the asymptotically unbiased and robust estimate of

$$\frac{1}{n_2^2} \sum_{x \in s_2} g_{s_2}^2(x) \hat{R}(x)^2$$

The empirical residuals $\hat{R}(x)$ are obtained through multiple linear regression with units weights of the $Y(x)$ on the $Z(x)$.

For illustration, we use an important example of stratification. Forest F partitioned in L strata $F = \cup_{k=1}^L F_k$ and likewise a sample of n_2 points $s_2 = \cup_{k=1}^L s_{2,k}$, $n_2 = \sum_{k=1}^L n_{2,k}$. The auxiliary vector is defined by $\boldsymbol{Z}(x$ $(Z_1(x), Z_2(x), \ldots, Z_L(x))^t$ with $Z_k(x) = 1$ if $x \in F_k$ or $Z_k(x) = 0$ otherwise. We assume that $n_{2,k} \geq 2$ $\forall k$ (obviously one would require in practice $n_{2,k}$ at least). Then, lacking prior knowledge, we choose $W(x) \equiv 1$. The model then

$$
\begin{aligned}
Y(x) &= \beta_k + R(x) \quad \text{if } x \in F_k \\
\mathbb{E}_M R(x) &\equiv 0 \\
\mathbb{V}(R(x)) &= \sigma_k^2 \quad \text{if } x \in F_k
\end{aligned}
$$

The true mean of the auxiliary vector is obviously given by $\bar{\boldsymbol{Z}} = (p_1, p_2, \ldots p$ with $p_k = \frac{\lambda(F_k)}{\lambda(F)}$. One immediately obtains $\frac{1}{n_2} \sum_{x \in s_2} Z(x) Z(x)^t = diag(\frac{n_2}{n}$ and consequently $g_{s_2}(x) = p_k \frac{n_2}{n_{2,k}}$ whenever $x \in F_k$. Straightforward calculations yield the least square estimate $\hat{\beta}_k = \bar{Y}_{s_2,k} = \frac{1}{n_{2,k}} \sum_{x \in s_{2,k}} Y(x)$, i.e. the observed strata means. Let us define in the usual way the strata empirical variances as $\hat{V}_k = \frac{1}{n_{2,k}-1} \sum_{x \in s_{2,k}} (Y(x) - \bar{Y}_{s_2,k})^2$. Simple calculations will show that Theorem 6.1.5 states:

$$
\hat{Y}_{pred} = \sum_{k=1}^{L} p_k \bar{Y}_{s_2,k}
$$

$$
\frac{1}{n_2^2} \sum_{x \in s_2} g_{s_2}^2(x) \hat{R}(x)^2 = \sum_{k=1}^{L} \frac{p_k^2}{n_{2,k}} \frac{n_{2,k}-1}{n_{2,k}} \hat{V}_k
$$

$$
\hat{\mathbb{V}}(\hat{Y}_{pred,R}) = \sum_{k=1}^{L} \frac{p_k^2}{n_{2,k}} \hat{V}_k + O(n_{2,k}^{-2})
$$

where we have assumed that $\frac{n_{2,k}-1}{n_{2,k}} = 1 + O(n_{2,k}^{-1})$, which is legitimate for n large.

For comparison let us calculate the variance according to standard least squares. One obtains $\boldsymbol{Z}_{s_2}^t \boldsymbol{Z}_{s_2} = diag(n_{2,k}^{-1})$, $\hat{\sigma}^2 = \frac{\sum_{k=1}^{L} (n_{2,k}-1) \hat{V}_k}{n_2 - L}$, $\hat{\boldsymbol{\Sigma}}_{\hat{\beta}_{s_2}} = \hat{\sigma}^2 diag(n_2^-$ With $\hat{\mathbb{V}}(\hat{Y}_{pred}) = \bar{\boldsymbol{Z}}^t \boldsymbol{\Sigma}_{\hat{\beta}_{s_2}} \bar{\boldsymbol{Z}}$ we finally arrive at

$$
\hat{\mathbb{V}}(\hat{Y}_{pred}) = \hat{\sigma}^2 \sum_{k=1}^{L} \frac{p_k^2}{n_{2,k}}
$$

which intuitively is not as appealing as $\hat{\mathbb{V}}(\hat{Y}_{pred,R})$. Under proportional allocation and for large sample sizes one has $n_{2,k} \approx n_2 p_k$ and, because $\sum_{k=1}^{L} p_k = 1$, it is easy to check that both variance estimates will be roughly equal. This is not true, however, under other allocation schemes, e.g. Neymann optimal allocation rule with constant costs, where according to Eq. 2.46 $n_{2,k} = \frac{n_2 p_k \sigma_k}{\sum_{j=1}^{L} p_j}$ In that case the estimate of standard least squares variance is asymptotically wrong (Mandallaz, 1991).

One could formally apply the previous results to generalized local densities $Y^*(x)$. There is, however, a mathematical difficulty, when one combines a model-dependent approach for the unobserved density $Y(x)$ with a design-based approach for $Y^*(x)$, with $\mathbb{E}_p Y^*(x) = Y(x)$. The design $p(\cdot)$ at point introduces a so called "nugget effect" for the spatial covariance structure of the $Y^*(x)$. This is due to the design-based sampling variances at neighboring points u and v: $Y(u)$ and $Y(v)$ may be close to each other but not $Y^*(u)$ and $Y^*(v)$. The nugget effect causes a discontinuity at zero in the correlation function: formally one has $\rho(0) = 1$ but $\rho(0^+) < 1$, making this precisely a jump of $\rho(0) - \rho(0^+)$. One can interpret the process $Y^*(x)$ as the sum of a regular, continuous component $Y(x)$, and of an erratic, "white noise" component e with an extremely short-range correlation function. This term was coined in the gold mining industry, reflecting the experience that gold nuggets are rather unpredictable phenomena. We shall return to this problem in our chapter on geostatistical methods, which bare, of course, also model-dependent. The nuance is in the spatial correlation: short-range (ideally smaller than the distance between adjacent points or clusters) in the model-dependent approach given here versus long-range in geostatistics. However, regularity conditions are also required for consistent estimations of the covariance function. Roughly speaking, that correlation range must be smaller than the dimension of the forest. Moreover, some kind of stationarity or ergodicity assumptions are always necessary in order to make inference based on a single realization (Mandallaz, 1993). One mathematical advantage of model-dependent estimation and geostatistics is that they provide a coherent framework for systematic samples.

6.2 Model-assisted approach

In this section we shall study, in the context of a forest inventory, the relationships between the model-assisted procedures (which rely on models for efficiency but on the randomization principle for validity) and pure model-dependent procedures. The two-phase sampling schemes and regression estimators presented in Chapter 5 are of course model-assisted. We shall derive for such schemes appropriate g-weights and compare the design-based variance with the model-dependent expected mean square error. Because the latter requires complete knowledge of the auxiliary information we first consider two-phase sampling schemes in which the first phase is exhaustive (census). In practice, this simply means that a map is available. We first examine a terrestrial inventory based on simple random sampling; as usual we shall assume that the same calculations will be performed with systematic samples.

6.2.1 Model-assisted estimation with census in the first phase and simple random sampling

We use the same notation as in Section 6.1. A sample s_2 of the n_2 points is generated by simple random sampling. The regression estimator is given by

$$\hat{Y}_{reg} = \bar{Y}_{s_2} + \hat{\boldsymbol{\beta}}_{s_2}^t (\bar{\boldsymbol{Z}} - \bar{\boldsymbol{Z}}_{s_2}) \tag{6.17}$$

$\frac{2}{n_2} \sum_{x \in s_2} (\quad)$ and $\frac{2}{n_2} \sum_{x \in s_2} (\quad)$. Requirements for the regression parameter β_{s_2} are rather vague but always fulfilled in practice, being ordinary, weighted or generalized least squares, or even a constant if an external model is used.

To illustrate this issue let us consider two simple examples:

1. **Simple linear regression through the origin with weighted least squares:**

   the model is

   $$Y(x) = Z(x)\beta + R(x)$$

   We assume, in our model-dependent approach, that $\mathbb{E}_M R(x) = 0$ and $\mathbb{V}_M R(x) = \sigma^2 Z(x)$, i.e. the weights are $W(x) = Z(x)$. Then, one obtains

   $$\hat{\beta}_{s_2} = \frac{\sum_{x \in s_2} Y(x)}{\sum_{x \in s_2} Z(x)} = \frac{\bar{Y}_{s_2}}{\bar{Z}_{s_2}}$$

   and

   $$\hat{Y}_{pred} = \bar{Z}\hat{\beta}_{s_2} = \left(\frac{\bar{Z}}{\bar{Z}_{s_2}}\right) \bar{Y}_{s_2}$$

   which is the ratio estimator.

   $$\hat{Y}_{reg} = \bar{Y}_{s_2} + \hat{\beta}_{s_2}(\bar{Z} - \bar{Z}_{s_2}) = \bar{Y}_{s_2} + \frac{\bar{Y}_{s_2}}{\bar{Z}_{s_2}}(\bar{Z} - \bar{Z}_{s_2}) = \frac{\bar{Z}}{\bar{Z}_{s_2}}\bar{Y}_{s_2}$$

   Hence, in this case, $\hat{Y}_{pred} = \hat{Y}_{reg}$

2. **Simple linear regression through the origin with ordinary least squares:**

   the model is

   $$Y(x) = Z(x)\beta + R(x)$$

   We assume, in this model-dependent approach, that $\mathbb{E}_M R(x) = 0$ and $\mathbb{V}_M R(x) \equiv \sigma^2$, i.e. with a constant variance and weights of $W(x) \equiv 1$. This produces

   $$\hat{\beta}_{s_2} = \frac{\sum_{x \in s_2} Z(x)Y(x)}{\sum_{x \in s_2} Z^2(x)}$$

   and

   $$\hat{Y}_{pred} = \bar{Z}\hat{\beta}_{s_2} = \bar{Z}\frac{\sum_{x \in s_2} Z(x)Y(x)}{\sum_{x \in s_2} Z^2(x)}$$

   $$\hat{Y}_{reg} = \bar{Y}_{s_2} + \frac{\sum_{x \in s_2} Z(x)Y(x)}{\sum_{x \in s_2} Z^2(x)}(\bar{Z} - \bar{Z}_{s_2})$$

   and, consequently,

   $$\hat{Y}_{pred} \neq \hat{Y}_{reg}$$

   Furthermore, we know that $\hat{Y}_{reg}$ is always asymptotically design-unbiased,

$$\mathbb{E}_p(\hat{Y}_{pred}) \approx \bar{Z} \left(\frac{\mathbb{E}_p(Z(x)Y(x))}{\mathbb{E}_p Z^2(x)} \right) \neq \bar{Y}$$

Therefore, $\hat{Y}_{pred}$ generally is not asymptotically design-unbiased.

We can show, under weak regularity assumptions, that the regression estimator $\hat{Y}_{reg}$ is asymptotically design-unbiased and consistent. That is

$$\lim_{n_2 \to \infty} \mathbb{E}_p(\hat{Y}_{reg}) = \bar{Y}$$

$$\lim_{n_2 \to \infty} \mathbb{P}_p\{| \hat{Y}_{reg} - \bar{Y} |> \varepsilon\} = 0 \quad \forall \varepsilon > 0$$

where $\mathbb{E}_p$ and $\mathbb{P}_p$ denote expectation and probability with respect to the sampling scheme (see (Mandallaz, 1991) for a proof). The above counter example demonstrates that this usually is not applicable for the model-dependent estimator $\hat{Y}_{pred}$. That, of course, is the main advantage in taking a design-based rather a model-dependent approach. From a more philosophical perspective, one could argue that both result in only a single realization, either of the sample or of the forest. Under the weak assumptions required for consistency and the following condition for the convergence of the regression parameter

$$\mathbb{E}_M(\hat{\beta}_j - \beta_j)^2 = O(n_2^{-1}) \quad \text{for } j = 1, \ldots p \tag{6.18}$$

one has the following result for the **AMSE** (the rather technical proof is presented by Mandallaz (1991))

$$\lim_{n_2 \to \infty} n_2 \mathbb{E}(\hat{Y}_{reg} - \bar{Y})^2 = \frac{1}{\lambda(F)} \int_F V(x) dx$$
$$- \frac{1}{\lambda^2(F)} \int_F \int_F \mathbb{COV}_M(Y(u), Y(v)) du dv$$
$$+ \mathbb{V}_p \left(\mu(x) - \boldsymbol{\beta}^t \boldsymbol{Z}(x) \right) \tag{6.19}$$

where $\mu(x) = \mathbb{E}_M Y(x)$ and $V(x) = \mathbb{V}_M Y(x)$. Condition 6.18 will be satisfied under the assumption of short-range correlation, in which case the second term in Eq. 6.19 will be negligible. This leads to the following theorem which, in some sense, summarizes all the previous findings:

Theorem 6.2.1.

$$\lim_{n_2 \to \infty} n_2 \mathbb{E}(\hat{Y}_{reg} - \bar{Y})^2 = \frac{1}{\lambda(F)} \int_F V(x) dx + \mathbb{V}_p \left(\mu(x) - \boldsymbol{\beta}^t \boldsymbol{Z}(x) \right)$$

Remarks:

1. In a pure design-based approach we have no super-population model, thereby being interpreted as $V(x) \equiv 0$ and $\mu(x) = Y(x)$. Setting $R(x) = Y(x \boldsymbol{\beta}^t \boldsymbol{Z}(x)$ we see that Theorem 6.2.1 precisely states the result given in Theorem 5.1.1 for $n_1 \to \infty$.

$\mathbb{E}_M Y(x) = \mu(x) = \boldsymbol{Z}(x)^t \boldsymbol{\beta}$, then both Eq. 6.19 and, under a small-range correlation, Theorem 6.2.1 are equivalent to Theorem 6.1.1.

3. The impact from estimating the drift is asymptotically negligible under short-range correlations.

4. As a first crude approximation the intuitive estimate of both components on the right-hand side of Theorem 6.2.1 is given by the mean sum of squares of the empirical residuals, or

$$\frac{1}{n_2} \sum_{x \in s_2} \hat{R}^2(x)$$

This implies that both approaches yield roughly the same estimated error under the assumption of a short-range correlation and a correct model, whether design-based or model-dependent. Indeed, this is rather reassuring.

We have seen that, in general, $\hat{Y}_{pred} \neq \hat{Y}_{reg}$. We now examine a condition under which both point estimates are equal. From now on, we will assume that this condition always holds.

Assumption I

$$\exists \lambda \in \Re^p \quad W(x) = \boldsymbol{\lambda}^t \boldsymbol{Z}(x) \tag{6.20}$$

Remarks:

1. If $\boldsymbol{Z}(x)$ contains the intercept and $W(x) \equiv 1$ (or any other constant) the condition is satisfied with $\boldsymbol{\lambda}^t = (1, 0, \ldots 0)$.

2. In the example "simple linear regression through the origin with weighted least squares" we have $W(x) = Z(x)$, so that Assumption I is trivially satisfied with $\lambda = 1$ and we find that $\hat{Y}_{pred} = \hat{Y}_{reg}$.

3. In the example "simple linear regression through the origin with ordinary least squares" we have $W(x) \equiv 1$. There is no λ with $1 \equiv \lambda Z(x)$ and we end up with $\hat{Y}_{pred} \neq \hat{Y}_{reg}$.

4. In the stratification example given at the end of Section 6.1 we have $W(x)$ 1 and $W(x) = \boldsymbol{\lambda}^t \boldsymbol{Z}(x)$ with $\boldsymbol{\lambda}^t = (1, 1, \ldots, 1) \in \Re^p$, where the components of $\boldsymbol{Z}(x)$ are the indicator variables of the L strata. Note that in this case the sum of the residuals is zero, although $\boldsymbol{Z}(x)$ does not contain the intercept term. We have $\hat{Y}_{pred} = \hat{Y}_{reg}$.

5. By extending, if necessary, the auxiliary vector with the component W one can always insure that Eq. 6.20 holds.

We arrive then at the following:

Theorem 6.2.2. *Under **Assumption I** we have*

$$\hat{Y}_{pred} = \hat{Y}_{reg} \tag{6.21}$$

One can rewrite the regression estimator $\hat{Y}_{reg}$ as

$$\hat{Y}_{reg} = \bar{\boldsymbol{Z}}^t \hat{\boldsymbol{\beta}}_{s_2} + \frac{1}{n_2} \sum_{x \in s_2} \hat{R}(x)$$

with the empirical residuals $\hat{R}(x) = Y(x) - \hat{\boldsymbol{\beta}}_{s_2}^t \boldsymbol{Z}(x)$. Hence, $\hat{Y}_{pred} = \hat{Y}_{reg}$ and only if the sum of the residuals is zero, or $\sum_{x \in s_2} \hat{R}(x) = 0$. Suppose that $\hat{\boldsymbol{\beta}}_{s_2}$ is obtained by ordinary weighted least squares, which is almost always the case in practice. That is, according to Eq. 6.3, we will have

$$\hat{\boldsymbol{\beta}}_{s_2} = (\boldsymbol{Z}_{s_2} \boldsymbol{W}_{s_2}^{-1} \boldsymbol{Z}_{s_2})^{-1} \boldsymbol{Z}_{s_2}^t \boldsymbol{W}_{s_2}^{-1} \boldsymbol{Y}_{s_2}$$

The residuals are by construction orthogonal to the components of $\boldsymbol{Z}$, i.e.,

$$\sum_{x \in s_2} \frac{\hat{R}(x) Z_i(x)}{W(x)} = 0 \quad \forall i = 1, 2 \ldots p$$

Multiplying this equality with the components λ_i of λ and summing over $i = 1, 2, \ldots p$ we obtain under condition 6.20 $\sum_{x \in s_2} \hat{R}(x) = 0$ because the $W(x)$ cancel out.

Recall that the g-weights are defined as

$$g_{s_2}(x) = \bar{\boldsymbol{Z}}^t \left(\frac{1}{n_2} \sum_{x \in s_2} \frac{\boldsymbol{Z}(x) \boldsymbol{Z}(x)^t}{W(x)} \right)^{-1} \frac{\boldsymbol{Z}(x)}{W(x)}$$

and that the model-dependent estimator can be stated as

$$\hat{Y}_{pred} = \frac{1}{n_2} \sum_{x \in s_2} g_{s_2}(x) Y(x)$$

It is straightforward to rewrite the regression estimator as

$$\hat{Y}_{reg} = \frac{1}{n_2} \sum_{x \in s_2} \tilde{g}_{s_2}(x)$$

with the $\tilde{g}$-weights

$$\tilde{g}_{s_2}(x) = 1 + (\bar{\boldsymbol{Z}} - \bar{\boldsymbol{Z}}_{s_2})^t \left(\frac{1}{n_2} \sum_{x \in s_2} \frac{\boldsymbol{Z}(x) \boldsymbol{Z}(x)^t}{W(x)} \right)^{-1} \frac{\boldsymbol{Z}(x)}{W(x)}$$

Both the g and the $\tilde{g}$-weights are independent of $Y(x)$. Therefore, the equality $\hat{Y}_{pred} = \hat{Y}_{reg}$ for all functions $Y(x)$ implies that $\tilde{g}_{s_2}(x) = g_{s_2}(x)$. Because the sample mean $\bar{\boldsymbol{Z}}_{s_2}$ converges towards the true mean $\bar{\boldsymbol{Z}}$ at the rate $O(n_2^{-\frac{1}{2}})$ we have under condition 6.20 and simple random sampling the asymptotic result $g_{s_2}(x) = 1 + O(n_2^{-\frac{1}{2}})$.

following important properties:

$$\frac{1}{n_2} \sum_{x \in s_2} g_{s_2}(x) \mathbf{Z}(x) = \bar{\mathbf{Z}}$$

$$\frac{1}{n_2} \sum_{x \in s_2} g_{s_2}(x) W(x) = \bar{W} = \frac{1}{\lambda(F)} \int_F W(x) dx$$

$$\frac{1}{n_2} \sum_{x \in s_2} g_{s_2}^2(x) W(x) = \frac{1}{\lambda(F)} \int_F g_{s_2}(x) W(x) dx \qquad (6.22)$$

In the previous section we derived a g-weight model-dependent, robust variance estimate of $\hat{Y}_{pred}$ with the assumption of a short-range correlation.

The classical estimate for the design-based variance of $\hat{Y}_{reg}$ is given by

$$\hat{\mathbb{V}}_p(\hat{Y}_{reg}) = \frac{1}{n_2} \frac{1}{n_2 - 1} \sum_{x \in s_2} \hat{R}^2(x) \qquad (6.23)$$

where the

$$\hat{R}(x) = Y(x) - \hat{\boldsymbol{\beta}}_{s_2}^t \mathbf{Z}(x)$$

are the empirical residuals. For $n_1 \to \infty$, this is a direct consequence of Theorem 5.1.1 because the sample mean of the empirical residuals is zero. We shall now derive a g-weight design-based variance estimate, under condition 6.20, of $\hat{Y}_{reg} = \hat{Y}_{pred}$. First, we consider the theoretical, unobservable, least squares estimate

$$\hat{\boldsymbol{\beta}} = \left(\frac{1}{\lambda(F)} \int_F \frac{\mathbf{Z}(x)\mathbf{Z}(x)^t}{W(x)} dx \right)^{-1} \left(\frac{1}{\lambda(F)} \int_F \frac{Y(x)\mathbf{Z}(x)}{W(x)} \right)$$

and the theoretical, unobservable, residuals

$$R(x) = Y(x) - \hat{\boldsymbol{\beta}}^t \mathbf{Z}(x)$$

Applying the same arguments as in the proof for Theorem 6.2.2 one first obtains $\bar{R} = \frac{1}{\lambda(F)} \int_F R(x) dx = 0$. Then, substituting $Y(x) = R(x) + \hat{\boldsymbol{\beta}}^t \mathbf{Z}$ in $\hat{Y}_{reg} = \frac{1}{n_2} \sum_{x \in s_2} g_{s_2}(x) Y(x)$, using $\bar{R} = 0$ and the first property from Eq. 6.22 one arrives at the important relationship of

$$\hat{Y}_{reg} - \bar{Y} = \frac{1}{n_2} \sum_{x \in s_2} g_{s_2}(x) R(x)$$

Adopting the same trick with the empirical residuals, we get $0 = \hat{Y}_{reg} - \hat{Y}_{pred}$ $\frac{1}{n_2} \sum_{x \in s_2} g_{s_2}(x) Y(x) - \hat{\boldsymbol{\beta}}_{s_2}^t \bar{\mathbf{Z}}$, which by Eq. 6.22 is equal to $\frac{1}{n_2} \sum_{x \in s_2} g_{s_2}(x) Y$ ($\frac{1}{n_2} \sum_{x \in s_2} g_{s_2}(x) \mathbf{Z}(x)^t \hat{\boldsymbol{\beta}}_{s_2}$). Finally, we obtain

$$\frac{1}{n_2} \sum_{x \in s_2} g_{s_2}(x) \hat{R}(x) = 0$$

variance as

$$\hat{\mathbb{V}}_{p,g}(\hat{Y}_{reg}) = \frac{1}{n_2^2} \sum_{x \in s_2} g_{s_2}^2(x)\hat{R}^2(x) \qquad (6.24)$$

Because $g_{s_2}(x) = 1 + O(n_2^{-\frac{1}{2}})$, it follows in design probability that $\hat{\mathbb{V}}_{p,g}(\hat{Y}$ is a consistent estimate of design-based variance. We make the following assumption:

Assumption II

$$\begin{aligned}
\mathbb{E}_M(Y(x)) &= \mathbf{Z}(x)^t\boldsymbol{\beta} \\
\mathbb{COV}_M(Y(u), Y(v)) &= \sigma^2(u) = \sigma^2 W(u) \quad \text{if } u = v \\
&= 0 \quad \text{if } u \neq v \qquad (6.25)
\end{aligned}$$

The main advantage of a g-weight variance estimate is characterized by the following result

Theorem 6.2.3. *Under **Assumptions I and II** we have these asymptotic properties:*

$$\begin{aligned}
\mathbb{V}_M(\hat{Y}_{reg}) &= \frac{1}{n_2}\frac{\sigma^2}{\lambda(F)} \int_F g_{s_2}(x) W(x) dx \\
\mathbb{E}_M \hat{\mathbb{V}}_p(\hat{Y}_{reg}) &= \frac{1}{n_2}\frac{\sigma^2}{\lambda(F)} \int_F W(x) dx + O(n^{-\frac{3}{2}}) \\
\mathbb{E}_M \hat{\mathbb{V}}_{p,g}(\hat{Y}_{reg}) &= \frac{1}{n_2}\frac{\sigma^2}{\lambda(F)} \int_F g_{s_2}(x) W(x) dx + O(n_2^{-2})
\end{aligned}$$

In other words, **the g-weight estimate of the design-based variance is closer to the model-dependent variance estimate if indeed the model is correct** (recall that $\hat{Y}_{reg} = \hat{Y}_{pred}$ under Assumption I). The theorem will remain approximately valid if Assumption II is weakened to a short-range correlation. Its proof is described by Mandallaz (1991). Furthermore, the g-weight estimate has better conditional properties and is often intuitively more appealing than the classical variance estimate. The stratification example given in Section 6.1 can be used mutatis mutandis to illustrate this concept because Assumption I holds here, so that the calculations in the model-dependent and model-assisted approaches are exactly the same. Let us examine the ratio estimator in this context.

Example: the ratio estimator

The model reads:

$$\begin{aligned}
Y(x) &= \beta Z(x) + R(x) \\
\mathbb{E}_M R(x) &= 0 \\
\mathbb{V}_M R(x) &= \sigma^2 Z(x)
\end{aligned}$$

$$\hat{\beta}_{s_2} = \frac{\bar{Y}_{s_2}}{\bar{Z}_{s_2}}, \quad \hat{Y}_{pred} = \hat{Y}_{reg} = \frac{\bar{Z}}{\bar{Z}_{s_2}}\bar{Y}_{s_2}, \quad g_{s_2}(x) = \frac{\bar{Z}}{\bar{Z}_{s_2}}$$

Then

$$\hat{\mathbb{V}}_p(\hat{Y}_{reg}) = \frac{1}{n_2(n_2-1)}\sum_{x\in s_2}(Y(x) - \hat{\beta}_{s_2}Z(x))^2$$

$$\hat{\mathbb{V}}_{p,g}(\hat{Y}_{reg}) = \left(\frac{\bar{Z}}{\bar{Z}_{s_2}}\right)^2\frac{1}{n_2^2}\sum_{x\in s_2}(Y(x) - \hat{\beta}_{s_2}Z(x))^2$$

until we finally obtain

$$\mathbb{V}_M(\hat{Y}_{reg}) = \frac{\sigma^2}{n_2}\frac{\bar{Z}^2}{\bar{Z}_{s_2}}$$

$$\mathbb{E}_M\hat{\mathbb{V}}_p(\hat{Y}_{reg}) = \frac{\sigma^2\bar{Z}_{s_2}}{n_2} + O(n_2^{-2}) = \frac{\sigma^2}{n_2}\frac{\bar{Z}^2}{\bar{Z}_{s_2}} + O(n_2^{-\frac{3}{2}})$$

$$\mathbb{E}_M\hat{\mathbb{V}}_{p,g}(\hat{Y}_{reg}) = \frac{\sigma^2}{n_2}\frac{\bar{Z}^2}{\bar{Z}_{s_2}} + O(n_2^{-2})$$

as stated in the general result from Theorem 6.2.3.

It is clear from Theorem 5.2.1, and can be shown by using B.3, that all the results given in this section can be generalized immediately to two-stage sampling. Moreover, it suffices to replace $Y(x)$ with $Y^*(x)$ in the formulae for the regression estimate, the point estimates and the variance estimate (the theoretical variance will of course entail the second stage variance $\mathbb{V}_{3|1,2}Y^*(x)$).

6.2.2 Model-assisted estimation with two-phase simple random sampling

Because of the results obtained in Subsection 6.2.1 we can restrict ourselves to heuristic arguments (formal proofs do not entail new arguments). The only difference is that, in two-phase sampling, the true spatial mean $\bar{Z}$ is of course unknown, and so must be estimated. The obvious choice is $\bar{Z}_{s_1} = \frac{1}{n_1}\sum_{x\in s_1}Z($
i.e. the mean of the auxiliary vector in the large sample. Our two-phase estimator is defined as

$$\hat{Y}_{reg} = \frac{1}{n_1}\sum_{x\in s_1}\hat{Y}(x) + \frac{1}{n_2}\sum_{x\in s_2}\hat{R}(x) \tag{6.26}$$

with empirical residuals $\hat{R}(x) = Y(x) - \hat{Y}(x)$ and predictions $\hat{Y}(x) = \mathbf{Z}(x)^t\hat{\boldsymbol{\beta}}$
The estimate $\hat{\boldsymbol{\beta}}_{s_2}$ of the regression parameter is given by Eq. 6.3. Under **Assumption I** the mean residual in Eq. 6.26 is zero (thus the residuals do not enter explicitly into calculations of the point estimate, although they are

essential for the variance). The

$$g_{s_2}(x) = \bar{Z}_{s_1}^t \left(\frac{1}{n_2} \sum_{x \in s_2} \frac{Z(x)Z(x)^t}{W(x)} \right)^{-1} \frac{Z(x)}{W(x)} \tag{6.27}$$

Again we have $g_{s_2}(x) = 1 + O(n_2^{-\frac{1}{2}})$ and

$$\hat{Y}_{reg} = \frac{1}{n_2} \sum_{x \in s_2} g_{s_2}(x)Y(x) = \bar{Y}_{s_2} + (\bar{Z}_{s_1} - \bar{Z}_{s_2})^t \hat{\beta}_{s_2} \tag{6.28}$$

Using the conditional arguments from B.3 we obtain the *g*-**weight estimate of the design-based variance in two-phase sampling** as

$$\hat{\mathbb{V}}_{p,g}(\hat{Y}_{reg}) = \frac{1}{n_1} \frac{1}{n_2 - 1} \sum_{x \in s_2} (Y(x) - \bar{Y}_{s_2})^2$$

$$+ \left(1 - \frac{n_2}{n_1} \right) \frac{1}{n_2^2} \sum_{x \in s_2} g_{s_2}^2(x)\hat{R}(x)^2 \tag{6.29}$$

Compared with the classical estimate given in Theorem 5.2.1, the *g*-weight variance estimate from Eq. 6.29 is better (with respect to the observed sample sizes, as found via post-stratification, and also with respect to the model if it is correct).

The generalization to two-stage sampling in the second phase is straightforward, one simply replacing $Y(x)$ with $Y^*(x)$.

6.2.3 Model-assisted estimation with two-phase cluster random sampling

Again, we restrict ourselves to heuristic arguments (formal proofs are presented by Mandallaz (1991)). The notation used for cluster sampling is defined in Sections 4.3 and 5.3. Model variance of the cluster mean $Y_c(x)$ is expected to be inversely proportional to the number $M(x)$ of plots in the cluster. That is, we set $W(x) = M(x)^{-1}$. We assume that the first component of $Z(x$ the intercept term such that the first component of $Z_c(x)$ is also always equal to 1. The weighted least square estimate of β is

$$\hat{\beta}_{c,s_2} = \left(\frac{1}{n_2} \sum_{x \in s_2} M(x)Z_c(x)Z_c(x)^t \right)^{-1} \left(\frac{1}{n_2} \sum_{x \in s_2} M(x)Y_c(x)Z_c(x) \right) \tag{6.30}$$

At the cluster level, the predictions are $\hat{Y}_c(x) = Z_c(x)^t \hat{\beta}_{c,s_2}$ and the residuals are $\hat{R}_c(x) = Y_c(x) - \hat{Y}_c(x)$. Because this model contains the intercept we have

$$\sum_{x \in s_2} M(x)\hat{R}_c(x) = 0$$

The regression estimate under two-phase cluster-sampling is then

$$\hat{Y}_{c,reg} = \hat{Y}_2 + (\hat{Z}_1 - \hat{Z}_2)^t \hat{\beta}_{c,s_2} \tag{6.31}$$

$$\hat{\boldsymbol{Z}}_k = \frac{\sum_{x \in s_k} M(x)\boldsymbol{Z}(x)}{\sum_{x \in s_k} M(x)}, \quad k = 1,2$$

are the estimated spatial means of the auxiliary vectors in the large and small samples. Likewise,

$$\hat{Y}_2 = \frac{\sum_{x \in s_2} M(x)Y(x)}{\sum_{x \in s_2} M(x)}$$

We define the g-weights at the cluster level as

$$g_{s_2,c}(x) = \hat{\boldsymbol{Z}}_1^t \left(\frac{1}{n_2} \sum_{x \in s_2} M(x)\boldsymbol{Z}_c(x)\boldsymbol{Z}_c(x)^t \right)^{-1} M(x)\boldsymbol{Z}_c(x) \qquad (6.32)$$

Then one can rewrite the classical estimator as

$$\hat{Y}_{c,reg} = \frac{1}{n_2} \sum_{x \in s_2} g_{s_2,c}(x)Y_c(x) \qquad (6.33)$$

By analogy with simple random sampling, one obtains the g-weight estimate of the design-based variance for the two-phase regression estimate under cluster-sampling as

$$\begin{aligned}
\hat{\mathbb{V}}_{p,g}(\hat{Y}_{c,reg}) &= \frac{1}{n_1} \frac{1}{n_2 - 1} \sum_{x \in s_2} \left(\frac{M(x)}{\bar{M}_2} \right)^2 (Y_c(x) - \hat{Y}_2)^2 \\
&+ \left(1 - \frac{n_2}{n_1} \right) \frac{1}{n_2^2} \sum_{x \in s_2} g_{s_2,c}^2(x)\hat{R}_c(x)^2 \qquad (6.34)
\end{aligned}$$

We can verify that, that asymptotically, $g_{c,s_2}^2(x) \approx \frac{M^2(x)}{\bar{M}_2^2}$ in agreement with Theorem 5.3.1. As before, one generalizes to two-stage sampling by replacing $Y_c(x)$ with $Y_c^*(x)$.

Remarks:

So far, in two-phase sampling, the probability that a point $x \in s_1$ will be included in the terrestrial sample has always been $\frac{n_2}{n_1}$. However, it is possible to have unequal inclusion probabilities in the second phase: a point $x \in$ is included in s_2 with probability $p(x)$, independently of the other points. Most of our previous results can be generalized to this procedure and optimal sampling probabilities $p(x)$ can be constructed for a class of cost functions. It turns out that the optimal choice satisfies $p(x) \propto |R(x)|$. Very often one has $R(x) \propto Y(x)$ so that the optimal scheme is roughly a **PPS** (Mandallaz, 1991). However, this additional complexity is worth the effort only if the sampling costs and variances vary greatly over the forested area F. We shall return to this topic in the chapter about optimal sampling schemes.

An overall sample size is usually designed to achieve specific accuracy for a global estimation over the entire domain F and frequently also for a few large sub-domains of F. For example, the 1984 Swiss National Forest Inventory stipulated an accuracy of 1 percent (i.e. $\frac{\sqrt{\mathbb{V}(\hat{Y})}}{\hat{Y}} = 0.01$) for Switzerland as a whole, with respect to timber volume in m^3 per ha. Clearly, the relative error achieved for the five main regions (Jura, Swiss Plateau, Pre-Alps, Alps and Southern Alps) was higher (roughly 1.8, 1.7, 1.9, 1.9 and 3.2 percent respectively). Accurate estimates for many less expansive regions (such as small Cantons) would require overall sample sizes much larger than could be accommodated by budgetary constraints. A related problem is the growing demand to reduce the costs of forest inventories at the enterprize level by using data from regional inventories, which themselves are frequently an extension of national inventories, with higher sampling densities. This is called **the small-area estimation problem** (obviously, "small" is a relative concept). It is intuitively clear that a satisfactory answer can be provided only if one uses auxiliary information, ideally thematic maps (e.g. if the first-phase is exhaustive). Let us set the stage for simple random sampling:

1. We have a forest area F with small areas $G_k \subset F$ and an auxiliary information vector $\boldsymbol{Z}(x)$. $\bar{\boldsymbol{Z}} = \frac{1}{\lambda(F)} \int_F \boldsymbol{Z}(x) dx$ and $\bar{\boldsymbol{Z}}_k = \frac{1}{\lambda(G_k)} \int_{G_k} \boldsymbol{Z}(x) dx$ known.

2. The estimate of the regression parameter, based on our entire data set and assuming units weights of $W(x) = 1$, is according to Eq. 6.3

$$\hat{\boldsymbol{\beta}}_{s2} = (\boldsymbol{Z}_{s2}\boldsymbol{Z}_{s2})^{-1}\boldsymbol{Z}_{s2}^t \boldsymbol{Y}_{s2}$$

and its robust estimate of variance is by Eq. 6.16

$$\hat{\boldsymbol{\Sigma}}_{\hat{\beta}_{s2},R} = (\boldsymbol{Z}_{s2}^t\boldsymbol{Z}_{s2})^{-1}(\boldsymbol{Z}_{s2}^t diag(\hat{r}^2(x))\boldsymbol{Z}_{s2})(\boldsymbol{Z}_{s2}^t\boldsymbol{Z}_{s2})^{-1}$$

3. Our objective is to estimate

$$\bar{Y}_k = \frac{\sum_{i \in G_k} Y_i}{\lambda(G_k)} = \frac{1}{\lambda(G_k)}\int_{G_k} Y(x) dx$$

We neglect any boundary effect with respect to G_k but not with respect to F.

The pure model-dependent estimator, also called the **synthetic estimator** is obviously given by

$$\hat{Y}_{k,syn} = \bar{\boldsymbol{Z}}_k^t \hat{\boldsymbol{\beta}}_{s2} \tag{6.35}$$

Its estimated model-dependent variance is, therefore, determined by

$$\hat{\mathbb{V}}(\hat{Y}_{k,syn}) = \bar{\boldsymbol{Z}}_k^t \hat{\boldsymbol{\Sigma}}_{\hat{\beta}_{s2},R} \bar{\boldsymbol{Z}}_k \tag{6.36}$$

biased. The variance is valid under the assumption of short-range correlations, which can be faulty in small areas. Recall that the sum of the residuals is zero only for the entire sample. Whereas the model is globally valid (zero mean residual over F), it could underestimate or overestimate locally (i.e. over G) Correcting the synthetic estimator with the mean residual over a small area leads to the so-called **small-area estimator**

$$\hat{Y}_{k,small} = \hat{Y}_{k,syn} + \frac{1}{n_{2,k}} \sum_{x \in s_{2,k}} \hat{R}(x) \tag{6.37}$$

where $s_{2,k} = s_2 \cap G_k$ and n_2 is the number of points of s_2 falling within G_k. Derivation of the design-based variance of $\hat{Y}_{k,small}$ is not trivial but relies on conditional arguments given in Mandallaz (1991). Intuitively, one gets the correct answer via Theorem 5.1.1 (as with an external model) by restricting samples s_1 and s_2 to $s_1 \cap G_k$, $s_2 \cap G_k$, and then letting n_1 tends to infinity. One obtains the design-based estimated variance as

$$\hat{\mathbb{V}}_p(\hat{Y}_{k,small}) = \frac{1}{n_{2,k}} \frac{1}{n_{2,k} - 1} \sum_{x \in s_{2,k}} (\hat{R}(x) - \bar{\hat{R}}_{2,k})^2 \tag{6.38}$$

where

$$\bar{\hat{R}}_{2,k} = \frac{1}{n_{2,k}} \sum_{x \in s_{2,k}} \hat{R}(x)$$

is the mean residual over a small area.

Remarks:

1. The empirical residual $\hat{R}(x)$ are always calculated with respect to the global model.

2. In general $\bar{\hat{R}}_{2,k} \neq 0$ because the model can locally overestimate or under-estimate. This is precisely the rule of the correction term in $\hat{\mathbb{V}}_p(\hat{Y}_{k,small}$ as compared with the synthetic estimator. Thus, it would be completely erroneous to use $\sum_{x \in s_{2,k}} \hat{R}^2(x)$ instead of $\sum_{x \in s_{2,k}} (\hat{R}(x) - \bar{\hat{R}}_{2,k})^2$ in Eq. 6.38.

3. Although somewhat surprising, it is absolutely correct that the synthetic component in $\hat{Y}_{k,small}$ does not contribute to design-based variance. In other words, the model is fitted globally and can be considered locally as an external model.

4. The result is valid, strictly speaking, only asymptotically. In practice this means that $n_{2,k}$ cannot be too small.

5. In cluster-sampling, the synthetic component remains the same because the sampling scheme is irrelevant in a model-dependent approach. The second term in Eq. 6.38 must be adjusted in the usual way for cluster-sampling by

$$\frac{1}{n_{2,k}} \frac{1}{n_{2,k}-1} \sum_{x \in s_{2,k}} \left(\frac{M(x)}{\bar{M}_{2,k}}\right)^2 (\hat{R}_c(x) - \bar{\hat{R}}_{c,2,k})^2$$

after restricting all the cluster points to G_k. A non-void cluster (with respect to F) with all its point outside G_k will be considered as a void cluster in the small area context. Likewise, $n_{2,k}$ is the number of non-void clusters with respect to G_k.

6. The formulae for two-stage sampling can be obtained by replacing, as usual, $Y(x)$ with $Y^*(x)$.

7. The confidence intervals for numerous G_k should be interpreted in the context of simultaneous inference, i.e. by adjusting the confidence levels. The probability that, for example, 1 out of 10 confidence intervals at the 95 percent confidence does not include the true value is of course larger than 5 percent.

8. When multiplying the mean spatial densities with the surface areas $\lambda($ we obtain estimates of totals that are only approximately additive.

Our generalization to genuine two-phase sampling is straightforward:

1. Calculate the estimate of the regression coefficient based on the entire data set (using ordinary least squares in simple random sampling or weighted least squares in cluster-sampling).

2. Determine predictions and residuals for all points in F. For each small area G_k restrict the samples to points or non-void clusters in G_k. If necessary, also calculate those predictions and residuals at the cluster level.

3. For each small area, use the formulae given in Theorems 5.1.1, 5.2.1, 5.3.1, 5.4.1 by considering the model to be an external one. That is, the non-zero mean residual over each small area must be subtracted from each residual when calculating the residual component of the variance.

Other small-area estimates and further references have been reported by K¨ et al. (2006).

6.4 Modeling relationships

So far our focus has been on estimating totals, means and ratios. We have seen that by using regression models one can increase the efficiency of sampling procedures. Estimating the regression parameter $\boldsymbol{\beta}$ is an important issue. However, the point estimate of $\boldsymbol{\beta}$ enters only into the construction of the various regression estimates, its variance is of secondary importance (in Chapter 8 this will be confirmed by practical examples). As a matter of fact, we can provide only an estimate of the model-dependent variance (see Eq. 6.16) and we do not know what the design-based variance is. Similarly, one might be

ulation (such as variances or correlation coefficients). We conduct **descriptive study**, in the design-based sense, if the survey data are employed to make inferences about known functions for the response variables $Y_i^{(m)}$ considered fixed values for a given population. Often, however, we might want to view the $Y_i^{(m)}$ as realizations of random variables in a super-population setup and make inferences about parameters in that super-population model. This would be called an **analytical study**. The subject, though rather subtle, remains a very active field of research, primarily in the socio-economic context. The reader can consult Chaudhuri and Stenger (1992) and Särndal et al. (2003) for an introduction. The mathematics are particularly difficult for finite populations under elaborated sampling schemes (such as multi-stage with clustering). Fortunately, in forest inventories and because of the infinite population approach, these matters seem to be getting a little bit easier, though little had been done so far. To illustrate the main issues we shall examine linear models and contingency tables.

6.4.1 Linear models

Let us assume that the response variable of interest is Y_i for the i-th tree and that we have a vector of auxiliary variables $\mathbf{Z}_i \in \Re^p$. To present the main features, let us consider the simple model:

$$
\begin{aligned}
Y_i &= \mathbf{Z}_i^t \boldsymbol{\beta} + R_i \quad \boldsymbol{\beta} \in \Re^p \\
\mathbb{E}_M(R_i) &= 0 \\
\mathbb{V}_M(R_i) &= \sigma^2 \quad \text{with } R_i \text{ independently and normally distributed.}
\end{aligned}
$$

$$(6.39)$$

If we examine this super-population model for fixed $\mathbf{Z}_i$ the likelihood for the entire population of N trees is

$$
L(\boldsymbol{\beta}) = \prod_{i=1}^{N} \frac{1}{\sqrt{2\pi}} exp[-\frac{(Y_i - \mathbf{Z}_i^t \boldsymbol{\beta})^2}{2\sigma^2}]
$$

The log likelihood is up to a constant

$$
l(\boldsymbol{\beta}) = \sum_{i=1}^{N}(Y_i - \mathbf{Z}_i^t \boldsymbol{\beta})^2
$$

Maximization of $l(\boldsymbol{\beta})$ yields a **model-dependent least-squares-census estimate**

$$
\hat{\boldsymbol{\beta}} = \left(\sum_{i=1}^{N} \mathbf{Z}_i \mathbf{Z}_i^t \right)^{-1} \sum_{i=1}^{N} Y_i \mathbf{Z}_i
$$

$$(6.40)$$

which is of course not available because the data are limited to all trees included at all points x in the sample $s = \cup_{x \in s_2} s_2(x)$. In a **model-dependent**

inclusion probabilities and achieve the classical estimate

$$\hat{\beta}_{s,mod} = \left(\sum_{i\in s} Z_i Z_i^t\right)^{-1} \sum_{i\in s} Y_i Z_i \qquad (6.41)$$

This is the approach advocated by Royall (1970). The resulting estimate is model-unbiased if the model is correct, i.e. $\mathbb{E}_M Y_i = Z_i^t\beta$. In the design-based sense, however, it can be severely biased. Its major drawback in forestry applications is that the spatial covariance is ignored between neighboring trees. Consequently, the estimated variance covariance matrix for β can be totally misleading when testing hypothesis. This difficulty can be partially surmounted by incorporating explanatory variables that describe the plot at x, and therefore are common to all trees $i \in s_2(x)$. Doing so reduces the correlation of the residuals at the tree level. Such an approach is useful with more general likelihood functions, in particular a logistic model for binary data with $Y_i = 0, 1$. The key issue is that the random response variables of neighboring trees are conditionally independent given all the explanatory variables Z_i the tree and plot levels). This can be assessed by inspecting the goodness-of-fit and the covariance structure at the plot level. Mandallaz et al. (1986) gives an example of logistic regression to forest health. Another approach is to introduce random plot effects and to implement the theory of full or restricted maximum likelihood in mixed models (Christensen, 1987 and Christensen, 1990). The correlation matrix of the Y_i is then block-diagonal with constant correlation in each block. Using a **design-based approach**, one can work, in principle, at either the tree or the plot level. However, the former is extremely difficult because the variance depends on pair-wise inclusion probabilities, which are usually unknown in practice. Because the statistical unit is point x it is natural to consider the local densities for all the variables in Eq. 6.39, which, because of the linearity, leads to the following design-based model at the point level

$$Y(x) = Z(x)^t\beta + R(x) \qquad (6.42)$$

Some of the components of $Z(x)$ can be based on explanatory variable that describe the plot, thus being constant for all trees in $s_2(x)$.

The theoretical least squares estimate β minimizes

$$\int_F R^2(x)dx = \int_F (Y(x) - Z^t(x)\beta)^2$$

It satisfies the normal equation

$$\left(\int_F Z(x)Z(x)^t\right)\beta = \int_F Y(x)Z(x)dx$$

and the orthogonality relationship

$$\int_F R(x)Z(x) = 0$$

$( )\ ( )\quad ( ) = \quad ( )\ ( )$. The normal equation then reads

$$\boldsymbol{A}\boldsymbol{\beta} = \mathbb{E}_x \boldsymbol{U}(x) := \boldsymbol{U}$$

Of course, only a sample-based normal equation is available, i.e.,

$$\boldsymbol{A}_{s_2}\hat{\boldsymbol{\beta}}_{s_2} = \frac{1}{n_2}\sum_{x \in s_2} \boldsymbol{U}(x) = \boldsymbol{U}_{s_2}$$

where we have set

$$\boldsymbol{A}_{s_2} = \frac{1}{n_2}\sum_{x \in s_2} \boldsymbol{Z}(x)\boldsymbol{Z}(x)^t$$

The theoretical and empirical regression vector parameters are

$$\begin{aligned} \boldsymbol{\beta} &= \boldsymbol{A}^{-1}\boldsymbol{U} \\ \hat{\boldsymbol{\beta}}_{s_2} &= \boldsymbol{A}_{s_2}^{-1}\boldsymbol{U}_{s_2} \end{aligned} \tag{6.43}$$

$\hat{\boldsymbol{\beta}}_{s_2}$ is asymptotically design-unbiased for $\boldsymbol{\beta}$. To calculate the design-based variance-covariance matrix

$$\mathbb{E}_p(\hat{\boldsymbol{\beta}}_{s_2} - \boldsymbol{\beta})(\hat{\boldsymbol{\beta}}_{s_2} - \boldsymbol{\beta})^t$$

we shall use the Taylor linearization technique (Rao and Scott (1984), Chaudhuri and Stenger (1992)). Let us consider the function $f(\cdot, \cdot)$ of an arbitrary (p, p) matrix $\boldsymbol{A}$ and an arbitrary $(p, 1)$ vector $\boldsymbol{U}$ defined by $f(\boldsymbol{A}, \boldsymbol{U}) = \boldsymbol{A}^{-1}$ We can write

$$\hat{\boldsymbol{\beta}}_{s_2} - \boldsymbol{\beta} = f(\boldsymbol{A}_{s_2}, \boldsymbol{U}_{s_2}) - f(\boldsymbol{A}, \boldsymbol{U})$$

which can be viewed as the differential of the function $f()$ at the point $\boldsymbol{P}_0$ $(\boldsymbol{A}, \boldsymbol{U})$. This is the expected value of the random point $\boldsymbol{P}_{s_2} = (\boldsymbol{A}_{s_2}, \boldsymbol{U}_s$ Distances between fixed and random point are of the order $n_2^{-\frac{1}{2}}$ in probability. The differential of $f(\cdot, \cdot)$ at $\boldsymbol{P}_0$ is, by the derivation rule for product

$$df = d(\boldsymbol{A}^{-1})\boldsymbol{U} + \boldsymbol{A}^{-1}d\boldsymbol{U}$$

For any matrix $\boldsymbol{A} = (a_{ij})$ let us differentiate with respect to a_{ij} the identity $\boldsymbol{A}^{-1}\boldsymbol{A} = \boldsymbol{I}$. We then have

$$\boldsymbol{0} = \frac{\partial \boldsymbol{A}^{-1}}{\partial a_{ij}}\boldsymbol{A} + \boldsymbol{A}^{-1}\frac{\partial \boldsymbol{A}}{\partial a_{ij}}$$

which leads to

$$d(\boldsymbol{A}^{-1}) = -\boldsymbol{A}^{-1}(d\boldsymbol{A})\boldsymbol{A}^{-1}$$

and the following first-order Taylor expansion:

$$\hat{\boldsymbol{\beta}}_{s_2} - \boldsymbol{\beta} = -\boldsymbol{A}^{-1}(\boldsymbol{A}_{s_2} - \boldsymbol{A})\boldsymbol{A}^{-1}\boldsymbol{U} + \boldsymbol{A}^{-1}(\boldsymbol{U}_{s_2} - \boldsymbol{U})$$

Expanding this expression and substituting $\boldsymbol{A}^{-1}\boldsymbol{U} = \boldsymbol{\beta}$ we obtain the Taylor linearization

$$\hat{\boldsymbol{\beta}}_{s_2} - \boldsymbol{\beta} = \boldsymbol{A}^{-1}\left(-\boldsymbol{A}_{s_2}\boldsymbol{\beta} + \frac{1}{n_2}\sum_{x \in s_2} \boldsymbol{U}(x)\right)$$

$$A^{-1}\left(-\frac{1}{n_2}\sum_{x\in s_2}Z(x)Z(x)^t\beta+\frac{1}{n_2}\sum_{x\in s_2}Y(x)Z(x)\right)$$

and consequently also to

$$A^{-1}\left(\frac{1}{n_2}\sum_{x\in s_2}\left(Y(x)-Z(x)^t\beta\right)Z(x)\right)=A^{-1}\left(\frac{1}{n_2}\sum_{x\in s_2}R(x)Z(x)\right)$$

Thus, we finally arrive at

$$\hat{\beta}_{s_2}-\beta=A^{-1}\left(\frac{1}{n_2}\sum_{x\in s_2}R(x)Z(x)\right) \tag{6.44}$$

with an error decreasing to zero faster than $n_2^{-\frac{1}{2}}$. It is interesting to note, that setting $Y(x)=\beta^t Z(x)+R(x)$ in Eq. 6.43, one gets exactly

$$\hat{\beta}_{s_2}-\beta=A_{s_2}^{-1}\left(\frac{1}{n_2}\sum_{x\in s_2}R(x)Z(x)\right) \tag{6.45}$$

Replacing in this expression A_{s_2} with A we achieve an error of order n_2^- probability. The Taylor expansion demonstrates that this error is of smaller order.

The variance-covariance matrix of $\hat{\beta}_{s_2}$ can be obtained immediately from Eq. 6.44

$$\Sigma_{\hat{\beta}_{s_2}}=A^{-1}\left(\frac{1}{n_2}\mathbb{E}_x R^2(x)Z(x)Z(x)^t\right)A^{-1}$$

which can be estimated by replacing the theoretical residual $R(x)$ with their empirical counterparts $\hat{R}(x)=Y(x)-\hat{\beta}_{s_2}^t Z(x)$ and A with A_{s_2}. We then get the **estimated design-based variance-covariance matrix** as

$$\hat{\Sigma}_{\hat{\beta}_{s_2}}=A_{s_2}^{-1}\left(\frac{1}{n_2^2}\sum_{x\in s_2}\hat{R}^2(x)Z(x)Z(x)^t\right)A_{s_2}^{-1} \tag{6.46}$$

which is verified as being equal to the model-dependent (g-weight-based) robust estimate given in Eq. 6.16. Alternatively, because the matrix inversion is a continuous operation and A_{s_2} converges in probability towards A we can apply Slutsky's theorem, e.g. see (Lehmann, 1999), to obtain the asymptotic equivalence of Eq. 6.44 and 6.45 and, therefore, of the model-dependent and the design-based approaches. This is a reassuring result indeed.

Using bootstrapping methods, one can also obtain an estimated variance-covariance matrix. This is done by drawing in s_2 n_2 points, with equal probability and with replacement, a large number B of samples $s_{2k}^*\subset s_2$, k $1,2\ldots B$. Note that in $s_{2,k}$ the same point can occur many times. For each $s_{2,k}^*$ calculate the corresponding estimate $\hat{\beta}_{s_{2,k}}^*$ according to Eq. 6.43 with $s_{2,k}^*$ in place of s_2. It can happen, albeit with small probability, that the matrix $A_{s_{2,k}}$ is singular, in which case the sample is rejected. The **bootstrap**

$$\hat{\boldsymbol{\beta}}^* = \frac{1}{B}\sum_{k=1}^{B}\hat{\boldsymbol{\beta}}^*_{s_2,k}$$

$$\hat{\mathbb{V}}(\hat{\boldsymbol{\beta}}^*) = \frac{1}{B-1}\sum_{k=1}^{B}(\hat{\boldsymbol{\beta}}^*_{s_2,k} - \hat{\boldsymbol{\beta}}^*)^2 \tag{6.47}$$

Alternatively one can take sub-grids of s_2.

The previous results are straightforward for generalizing to cluster-sampling: replace $Y(x)$, $\boldsymbol{Z}(x)$ and $\hat{R}(x)$ with respectively $M(x)Y_c(x)$, $M(x)\boldsymbol{Z}_c(x)$ and $M(x)\hat{R}_c(x)$, respectively. Likewise, for two-stage sampling substitute Y^* for $Y(x)$.

Classical independence tests for contingency tables obtained from survey data are not valid, particularly with forest inventories. We now demonstrate how one can adjust the classical theory.

6.4.2 Contingency tables

We consider two qualitative variables, $Y_i^{(1)}$, which takes values in $\{1, 2, 3 \ldots r\}$ and $Y_i^{(2)}$ with values in $\{1, 2, \ldots c\}$, where $i = 1, 2 \ldots N$. We also introduce a trivial variable $Y_i^{(0)} \equiv 1$. The indicator variable $Y_i^{(kl)}$ is equal to 1 if $Y_i^{(1)} =$ **and** $Y_i^{(2)} = l$ but is equal to 0 otherwise. We also need $Y_i^{(k+)} = \sum_{l=1}^{c}Y_i$ and $Y_i^{(+l)} = \sum_{k=1}^{r}Y_i^{(kl)}$. True cell and marginal frequencies in the population are defined as

$$p_{kl} = \frac{\sum_{i=1}^{N}Y_i^{(kl)}}{\sum_{i=1}^{N}Y_i^{(0)}} \quad \text{for } k = 1, 2 \ldots m \text{ and } l = 1, 2 \ldots c$$

$$p_{k+} = \sum_{l=1}^{c}p_{kl} \quad \text{for } k = 1, 2 \ldots r$$

$$p_{+l} = \sum_{k=1}^{r}p_{kl} \quad \text{for } l = 1, 2 \ldots c \tag{6.48}$$

The (r, c) matrix $\boldsymbol{P}$ with entries $P_{kl} = p_{kl}$ is a contingency table comprising rows and c columns. The rc variables $Y_i^{(kl)}$, the $r+c$ variables $Y_i^{(k+)}$, $Y_i^{(+l)}$ and $Y_i^{(0)}$ define the corresponding local densities $Y^{(kl)}(x)$, $Y^{(+l)}(x)$, $Y^{(k+)}(x)$ and $Y^{(0)}(x)$ depending on the sampling scheme with probabilities π_i. If we have a sample s_2 of n_2 points that are independently and uniformly distributed

$$\hat{p}_{kl} = \frac{n_2^{-1} \sum_{x \in s_2} Y^{(kl)}(x)}{n_2^{-1} \sum_{x \in s_2} Y^{(0)}(x)}$$

$$\hat{p}_{k+} = \frac{n_2^{-1} \sum_{x \in s_2} Y^{(k+)}(x)}{n_2^{-1} \sum_{x \in s_2} Y^{(0)}(x)} = \sum_{l=1}^{c} \hat{p}_{kl}$$

$$\hat{p}_{+l} = \frac{n_2^{-1} \sum_{x \in s_2} Y^{(+l)}(x)}{n_2^{-1} \sum_{x \in s_2} Y^{(0)}(x)} = \sum_{k=1}^{r} \hat{p}_{kl} \qquad (6.49)$$

The estimated variances $\hat{\mathbb{V}}(\hat{p}_{kl})$, $\hat{\mathbb{V}}(\hat{p}_{+l})$ and $\hat{\mathbb{V}}(\hat{p}_{k+})$ are calculated with Theorem 4.2.2. To estimate the covariance between any two estimated cell frequencies we use

$$\hat{\mathbb{COV}}(\hat{p}_{kl}, \hat{p}_{mn}) = \frac{1}{2}\left(\hat{\mathbb{V}}(\hat{p}_{klmn}) - \hat{\mathbb{V}}(\hat{p}_{kl}) - \hat{\mathbb{V}}(\hat{p}_{mn}) \right) \qquad (6.50)$$

where $p_{klmn} = p_{kl} + p_{mn}$. We get $\hat{p}_{klmn}$ and $\hat{\mathbb{V}}(\hat{p}_{klmn})$ via the local densities of the variables $Y_i^{(kl)} + Y_i^{(mn)}$ and $Y_i^{(0)}$. We can, therefore, determine the variance-covariance matrix of the vector $\hat{\boldsymbol{p}}$, which we shall write as $\hat{\boldsymbol{\Sigma}}_{\hat{\boldsymbol{p}}}$.

The generalization to cluster-sampling is immediate if one uses the results given in Theorem 4.3.1 with the local densities defined above.

In the classical statistical approach to contingency tables, the random variables $Y_i^{(kl)}$ are, for the different units $i = 1, 2 \ldots N$, identically and independently distributed according to a multinomial distribution with parameter $\boldsymbol{p} = (p_{11}, p_{12}, \ldots p_{r,c-1})^t \in \Re^{rc-1}$ (we use a $rc - 1$ dimensional parameter because the cell probabilities must sum up to 1). The hypothesis of independence, for a given unit i, of the categorial response random variables is formulated as

$$p_{kl} = \mathbb{P}(\{Y_i^{(1)} = k\} \cap \{Y_i^{(2)} = l\}) = \mathbb{P}(\{Y_i^{(1)} = k\})\mathbb{P}(\{Y_i^{(2)} = l\}) = p_{k+}p$$

for $k = 1, 2 \ldots r - 1$, $l = 1, 2 \ldots c - 1$ and $i = 1, 2 \ldots N$. The probability is with respect to simple random sampling without replacement at the tree level and not with respect to a super-population model, in which the spatial correlation between neighboring trees should be taken into account. We face two difficulties: first, we almost never have simple random sampling at the tree level in forest inventory and, second, if we had a census, the exact relative frequencies p_{kl} will, in practice, never satisfy exactly the relation $p_{kl} = p_{k+}p$ We solve this contradiction in a very pragmatic way by testing the hypothesis H_0 on the basis of the inventory data, taking into account the special features of the sampling schemes (varying inclusion probabilities, clustering at the tree level and the plot level, boundary effects etc), and decide whether or not H_0 is compatible, at a given significance level, with the data or not. If is not rejected we have good reasons to believe that this would also be the case if we were sampling the trees by simple random sampling. The super-variable by means other explanatory variables, such as in the logistic regression (Mandallaz et al., 1986). There is almost always the implicit assumption, that the response variables of two different trees are stochastically independent (with respect to the model) given the values of their explanatory variables. Let us go back to design-based inference and define

$$h_{kl} = p_{kl} - p_{k+}p_{+l} \quad k = 1, 2 \ldots r, \quad l = 1, 2 \ldots c \tag{6.51}$$

The h_{kl} satisfy the identities $\sum_{k=1}^{r} h_{kl} = h_{+l} = 0$ and similarly $h_{k+} = 0$, $h_{++} = 0$. Hence, only $rc - (r + c - 1) = (r - 1)(c - 1)$ components are independent, which is the number of degrees of freedom. Therefore, we define the vector

$$\boldsymbol{h} = (h_{11}, h_{12}, \ldots h_{1,c-1}, h_{21} \ldots h_{2,c-1}, \ldots h_{r-1,c-1})^t \in \Re^{(r-1)(c-1)}$$

Our hypothesis of independence then reads

$$H_0 : \boldsymbol{h} = \boldsymbol{0}$$

The $(r - 1)(c - 1)$ dimensional vector $\boldsymbol{h}$ depends on the $rc - 1$ dimensional vector $\boldsymbol{p}$. Our objective is to construct a design-based confidence interval for the unknown $\boldsymbol{h}$. If it contains $\boldsymbol{0}$, we shall say, in a pragmatic way, that the independence hypothesis is compatible with the inventory data. We need the $((r - 1)(c - 1), (r - 1)(c - 1))$ variance-covariance matrix of $\boldsymbol{h}$. Again, we use the Taylor expansion technique, i.e., setting

$$\hat{\boldsymbol{h}} - \boldsymbol{h} = \boldsymbol{h}(\hat{\boldsymbol{p}}) - \boldsymbol{h}(\boldsymbol{p}) = \boldsymbol{H}(\hat{\boldsymbol{p}})(\hat{\boldsymbol{p}} - \boldsymbol{p})$$

where $\boldsymbol{H}$ is the $((r - 1)(c - 1), rc - 1)$ Hessian matrix of the partial derivatives $\frac{\partial h_{ij}}{\partial p_{kl}}$. The variance-covariance matrix $\hat{\Sigma}_{\hat{\boldsymbol{h}}}$ of the estimated vector $\hat{\boldsymbol{h}}$ is given by

$$\hat{\Sigma}_{\hat{\boldsymbol{h}}} = \hat{\boldsymbol{H}} \hat{\Sigma}_{\hat{\boldsymbol{p}}} \hat{\boldsymbol{H}}^t \tag{6.52}$$

where $\hat{\boldsymbol{H}} = \boldsymbol{H}(\hat{\boldsymbol{p}})$. Therefore, under the null hypothesis of independence, the statistic

$$\chi^2 = \hat{\boldsymbol{h}}^t (\hat{\Sigma}_{\hat{\boldsymbol{h}}})^{-1} \hat{\boldsymbol{h}} \tag{6.53}$$

is distributed asymptotically, i.e. for n_2 large, as a chi-square with $(r-1)(c-$ degrees of freedom.

To illustrate the notation, let us consider the simplest but most important case of a two-by-two contingency table. In this case $\boldsymbol{p} = (p_{11}, p_{12}, p_{21})^t$ and $\boldsymbol{h} = (h_{11})$. One checks that $h_{12} = -h_{11}$, $h_{21} = -h_{11}$ and $h_{22} = h_{11}$. Simple calculations lead to

$$h_{11} = p_{11} - p_{11}^2 - p_{12}p_{11} - p_{11}p_{21} - p_{12}p_{21}$$

and to a $(1, 3)$ Hessian matrix (in this case a row vector) $\boldsymbol{H} = (H_1, H_2, H$ with components $H_1 = \frac{\partial h_{11}}{\partial p_{11}} = p_{22} - p_{11}$, $H_2 = \frac{\partial h_{11}}{\partial p_{12}} = -p_{+1}$ and $H_3 = \frac{\partial h_{11}}{\partial p_{21}}$ $-p_{1+}$. Here, the matrix $\hat{\Sigma}_{\hat{\boldsymbol{h}}} = \hat{\boldsymbol{H}} \hat{\Sigma}_{\hat{\boldsymbol{p}}} \hat{\boldsymbol{H}}^t$ is in this case the value of a quadratic

$$\chi^2 = \frac{\hat{h}_{11}^2}{\hat{\mathbb{V}}(\hat{h}_{11})}$$

is, under H_0, asymptotically distributed (i.e. n_2 large) as a chi-square with one degree of freedom. The reader should record the formulae for a $(3, 3)$ table, for which one has $\boldsymbol{p} = (p_{11}, p_{12}, p_{13}, p_{21}, p_{22}, p_{23}, p_{31}, p_{32})^t$, $\boldsymbol{h} = (h_{11}, h_{12}, h_{21}, h$ and the Hessian $\boldsymbol{H}$ is a $(4, 8)$ matrix. The ith row, $i = 1, 2, 3, 4$, of $\boldsymbol{H}$ comprises the eight partial derivatives of the ith component of $\boldsymbol{h}$ with respect to the eight components of $\boldsymbol{p}$, in the corresponding order.

Testing for independence in two-dimensional contingency tables is relatively straightforward. However, practice conceals many pitfalls. To illustrate the problem let us consider a forest F with two strata F_1 and F_2, $\lambda(F_1) = \lambda(F_2$ $\frac{\lambda(F)}{2}$ and a two-by-two table with the following true frequencies in our two strata

$$\boldsymbol{P}_1 = \begin{pmatrix} 0.05 & 0.20 \\ 0.15 & 0.60 \end{pmatrix}$$

$$\boldsymbol{P}_2 = \begin{pmatrix} 0.60 & 0.15 \\ 0.20 & 0.05 \end{pmatrix}$$

Obviously, the hypothesis of independence holds in both strata (rows and columns are proportional but with different proportionality ratios). Hence, if one samples uniformly in F_1, the hypothesis of independence will be wrongly rejected in less than, say 5% of the inventories, and likewise in F_2. However, sampling in F leads to the average table

$$\boldsymbol{P}_{12} = \begin{pmatrix} 0.325 & 0.175 \\ 0.175 & 0.325 \end{pmatrix}$$

for which the hypothesis of independence clearly does not hold. Thus, pooling two-dimensional contingency tables over geographical strata or (as a mathematical equivalent) over categories of one or more further categorial variables can be very misleading. This is known as Simpson's paradox (Fienberg, 1980). In practice, one should, prior to formal testing, perform an exploratory data analysis, most importantly by stratifying according to geographical areas (e.g. stand map, elevation, soil characteristics etc), which may have an impact on the two response variables defining the table. Likewise, if one suspects that a further categorical variable may have an impact on the variables in the table, by examining the tables obtained by conditioning on that variable. In this case, one has to modify carefully the indicator variables previously defined, in particular the variable $Y_i^{(0)} = 1$ must be replaced by as many indicator variables as there are categories in the conditioning variables (clearly, one should conduct separate analysis for various sub-populations). It is also intuitively obvious that one cannot cope with multi-dimensional tables containing many categories because the cell frequencies cannot be reliably estimated. From a practical point of view $3 \times 3 \times 3$ tables are probably at the upper limit.

log-linear models in a slightly modified setup (Fienberg, 1980). Let us first reconsider the simple 2×2 table. The model reads

$$ln(p_{kl}) = u + u_{1(k)} + u_{2(l)} + u_{12(kl)} \quad k, l = 1, 2 \tag{6.54}$$

with the constraint $u_{1(2)} = -u_{1(1)}$, $u_{2(2)} = -u_{2(1)}$ and $u_{12(11)} = -u_{12(12)}$ $-u_{12(21)} = u_{12(22)}$. We can verify that the hypothesis of independence is equivalent to $u_{12(11)} = \frac{1}{4}ln(\frac{p_{11}p_{22}}{p_{12}p_{21}}) = 0$, i.e. a log-odds ratio equal to 1. This model can be rewritten matrix notation if we have the following:
the vector of the theoretical log-probabilities

$$l = (ln(p_{11}), ln(p_{12}), ln(p_{21}), ln(p_{22}))^t = l(p) \in \Re^4$$

the vector of ones for the grand mean parameter

$$\mathbf{1} = (1, 1, 1, 1)^t$$

the vector of the other parameters

$$\boldsymbol{\theta} = (u_{1(1)}, u_{2(1)}, u_{12(11)})^t$$

and the design matrix

$$\boldsymbol{X} = \begin{pmatrix} 1 & 1 & 1 \\ 1 & -1 & -1 \\ -1 & 1 & -1 \\ -1 & -1 & +1 \end{pmatrix}$$

We can then write

$$l = u\mathbf{1} + \boldsymbol{X}\boldsymbol{\theta} \tag{6.55}$$

Let us now examine an arbitrary three dimensional table and the model

$$ln(p_{klm}) = u + u_{1(k)} + u_{2(l)} + u_{3(m)} + u_{12(kl)} + u_{13(km)} + u_{23(lm)} + u_{123(klm)}$$

where we have, for the main effects, the constraints

$$\sum_k u_{2(k)} = \sum_l u_{2(l)} = \sum_m u_{2(m)} = 0$$

for the second-order interactions the constraints,

$$\sum_k u_{12(kl)} = \sum_l u_{12(kl)} = 0$$

similarly, for the terms $u_{13(km)}, u_{23(lm)}$, and, finally,

$$\sum_k u_{123(klm)} = \sum_l u_{123(klm)} = \sum_m u_{123(klm)} = 0$$

for the third-order interaction term. If only the main-effect terms are present then the three categorial variables are independent. Let us assume that $u_{12(kl}$ $u_{123(klm)} = 0$ for all k, l, m. We will have interactions only between the pairs of categorial variables $_i$, Y_i $_i$, Y_i. Simple calculations show, with the usual "+" notation, that we have

$$p_{klm} = \frac{p_{k+m}p_{+lm}}{p_{++m}} \quad \forall k, l, m \tag{6.56}$$

which is equivalent to saying that the two response variables $Y_i^{(1)}$ and Y are conditionally independent given the value of the third one $Y_i^{(3)}$. This follows from the definition of the conditional probabilities and the equalities ($\forall i = 1, 2 \dots N$):

$$
\begin{aligned}
p_{klm} &= \mathbb{P}(\{Y_i^{(1)} = k\} \cap \{Y_i^{(2)} = l\} \mid Y_i^{(3)} = m)\mathbb{P}(Y_i^{(3)} = m) \\
&= \mathbb{P}(Y_i^{(1)} = k \mid Y_i^{(3)} = m)\mathbb{P}(Y_i^{(2)} = l \mid Y_i^{(3)} = m)\mathbb{P}(Y_i^{(3)} = m)
\end{aligned}
$$

Therefore, testing for conditional independence is equivalent to testing the null hypothesis $u_{12(kl)} = u_{123(klm)} = 0$ for all k, l, m. To illustrate the notation let us examine the easiest case, the $2 \times 2 \times 2$ table. Set

$$l = (l_{111}, l_{112}, l_{121}, l_{122}, l_{211}, l_{212}, l_{221}, l_{222})^t$$

where $l_{klm} = ln(p_{klm})$. Because of constraints, the parameter vector has only seven components:

$$\theta = (u_{1(1)}, u_{2(1)}, u_{3(1)}, u_{12(11)}, u_{13(11)}, u_{23(11)}, u_{123(111)})^t \in \Re^7$$

As usual, we set $1 = (1, 1, 1, 1, 1, 1, 1, 1)^t$ for the grand mean. The design matrix now has eight rows and seven columns, and reads

$$
X = \begin{pmatrix}
1 & 1 & 1 & 1 & 1 & 1 & 1 \\
1 & 1 & -1 & 1 & -1 & -1 & -1 \\
1 & -1 & 1 & -1 & 1 & -1 & -1 \\
1 & -1 & -1 & -1 & -1 & 1 & 1 \\
-1 & 1 & 1 & -1 & -1 & -1 & -1 \\
-1 & 1 & -1 & -1 & 1 & 1 & -1 \\
-1 & -1 & 1 & 1 & -1 & -1 & 1 \\
-1 & -1 & -1 & 1 & 1 & 1 & -1
\end{pmatrix}
$$

The saturated model, having the maximum number of parameters, is given by

$$l = u1 + X\theta$$

We shall estimate the parameter θ by weighted least squares with the $l(\hat{p})$ as observations. This essentially follows the method described in (Rao and Scott (1984), appendix). Such a method is valid for arbitrary design provided that there are no "structural" zero cells, i.e. $p_{klm} = 0$ by definition. Because of the logarithm, we also need $\hat{p}_{klm} > 0$. If this is not true, then a coarser classification must be used.

We first calculate the variance-covariance matrix of $\hat{l}$, which is obtained from the variance-covariance matrix of $\hat{p}$ via first-order Taylor expansion. The generic indexes $s, t = 1, 2 \dots T$ represent arbitrary cells, e.g. $p_{123} = p_s, p_{215}$

$T = 8$ in this example. For an arbitrary, differentiable function $f(\cdot)$ and for a random variable X_n that tends in probability to a constant a, we have

$$f(X_n) \approx f(\mathbb{E}X_n) + (X_n - \mathbb{E}X_n)f'(\mathbb{E}X_n)$$

This leads to the well known approximations for the variances and covariances:

$$\mathbb{V}(f(X_n)) = \mathbb{V}(X_n)(f'(\mathbb{E}(X_n))^2$$
$$\mathbb{COV}(f(X_n), f(Y_n)) = \mathbb{COV}(X_n, Y_n)f'(\mathbb{E}(X_n))f'(\mathbb{E}(Y_n)) \qquad (6.57)$$

For the function $f(x) = ln(x)$ we obtain the approximations

$$\hat{\mathbb{V}}(ln(\hat{p}_t)) = \frac{1}{\hat{p}_t^2}\hat{\mathbb{V}}(\hat{p}_t)$$

and

$$\hat{\mathbb{COV}}(ln(\hat{p}_s), ln(\hat{p}_t)) = \frac{1}{\hat{p}_s\hat{p}_t}\hat{\mathbb{COV}}(\hat{p}_s, \hat{p}_t)$$

The variances and covariances of the $\hat{p}_s$, $\hat{p}_t$ are of order n_2^{-1} and the errors for the above approximations are of smaller order. Let $\boldsymbol{D_p}$ be the (T, T) diagonal matrix with diagonal elements p_t. Using the above expressions for the variances and covariances of $ln(\hat{p}_t)$ we have the compact versions

$$\begin{aligned}\boldsymbol{\Sigma_l} &= \boldsymbol{D_p^{-1}}\boldsymbol{\Sigma_p}\boldsymbol{D_p^{-1}} \\ \hat{\boldsymbol{\Sigma}}_{\hat{l}} &= \hat{\boldsymbol{D}}_{\hat{p}}^{-1}\hat{\boldsymbol{\Sigma}}_{\hat{p}}\hat{\boldsymbol{D}}_{\hat{p}}^{-1}\end{aligned} \qquad (6.58)$$

Because the cell probabilities add up to 1 the matrices $\boldsymbol{\Sigma_l}$ and $\boldsymbol{\Sigma_p}$ are singular which introduces a minor technical difficulty. We require a $(T-1, T)$ matrix $\boldsymbol{F}$ of full rank $T-1$, such that $\boldsymbol{F1} = \boldsymbol{0}$, where $\boldsymbol{1}$ is the T dimensional column vector of 1 and $\boldsymbol{0}$ is the $T-1$ dimensional vector of 0. In the two examples above, one could take $\boldsymbol{F} = \boldsymbol{X}^t$ (the columns of $\boldsymbol{X}$ are orthogonal and contain as many 1 as -1). By introducing the new $T-1$ column vectors of true values and observations:

$$f = Fl = FX\theta, \quad \hat{f} = F\hat{l} \qquad (6.59)$$

where $\hat{l} = l(\hat{p})$. We reduce the dimensionality of the problem by eliminating the grand mean and, by the same token, we also get rid of the singular covariance matrices. Indeed, the estimated variance-covariance matrix of the new observation vector $\hat{f}$ is given by

$$\hat{\boldsymbol{\Sigma}}_{\hat{f}} = F\hat{\boldsymbol{\Sigma}}_{\hat{l}}F^t = F\hat{\boldsymbol{D}}_{\hat{p}}^{-1}\hat{\boldsymbol{\Sigma}}_{\hat{p}}\hat{\boldsymbol{D}}_{\hat{p}}^{-1}F^t \qquad (6.60)$$

which is a regular $(T-1, T-1)$ matrix due to the full rank of $(T-1, T$ matrix $\boldsymbol{F}$. We now estimate $\boldsymbol{\theta}$ by considering the model

$$\hat{f} = FX\theta + \varepsilon$$

with a $T-1$ column vector of residuals $\boldsymbol{\varepsilon}$. A generalized weight least squares is used with the regular weight matrix $\hat{\boldsymbol{\Sigma}}_{\hat{f}}^{-1}$ (see Rao, 1967, chapter 4). This

$$\hat{\theta} = \left(\boldsymbol{X}^t \boldsymbol{F}^t \hat{\boldsymbol{\Sigma}}_f^{-1} \boldsymbol{F} \boldsymbol{X} \right)^{-1} \boldsymbol{X}^t \boldsymbol{F}^t \hat{\boldsymbol{\Sigma}}_f^{-1} \boldsymbol{F} \hat{\boldsymbol{l}} \tag{6.61}$$

from which we obtain after some algebra the estimated variance-covariance matrix

$$\hat{\boldsymbol{\Sigma}}_{\hat{\theta}} = \left(\boldsymbol{X}^t \boldsymbol{F}^t \hat{\boldsymbol{\Sigma}}_f^{-1} \boldsymbol{F} \boldsymbol{X} \right)^{-1} \tag{6.62}$$

One may think that a given solution depends on the choice of the matrix but this is not the case. Instead, with a matrix $\boldsymbol{F}^* = \boldsymbol{GF}$ and $\boldsymbol{f}^* = \boldsymbol{F}^* \boldsymbol{l}$, where $\boldsymbol{G}$ is a regular $(T-1, T-1)$ matrix, one eventually arrives at

$$\boldsymbol{F}^{*t} \hat{\boldsymbol{\Sigma}}_{f^*}^{-1} \boldsymbol{f}^* = \boldsymbol{F}^t \hat{\boldsymbol{\Sigma}}_f^{-1} \boldsymbol{f}$$

$$\boldsymbol{F}^t \hat{\boldsymbol{\Sigma}}_f^{-1} \boldsymbol{F} = \boldsymbol{F}^{*t} \hat{\boldsymbol{\Sigma}}_{f^*}^{-1} \boldsymbol{F}^*$$

and, therefore, at the equivalence of the two estimates. Alternatively, one could work from the onset with $\hat{\boldsymbol{l}}$, include the grand mean $u\boldsymbol{1}$ in $\boldsymbol{\theta}$ and use a generalized inverse for the variance-covariance matrix $\hat{\boldsymbol{\Sigma}}_{\hat{l}}$ (Rao, 1967).

To test the hypothesis let us split the parameter vector into disjoined components, so that we have

$$\boldsymbol{\theta}^t = (\boldsymbol{\theta}_1^t, \boldsymbol{\theta}_2^t)$$

For testing conditional independence in the $2 \times 2 \times 2$ table we would take $\boldsymbol{\theta}$ $(u_{1(1)}, u_{2(1)}, u_{3(1)}, u_{13(11)}, u_{23(11)})^t$ and $\boldsymbol{\theta}_2 = (u_{12(11)}, u_{123(111)})^t$. Rearranging, if necessary, the ordering of the components of $\boldsymbol{\theta}$ and the entries in our full variance-covariance matrix, we can write

$$\hat{\boldsymbol{\Sigma}}_{\hat{\theta}} = \begin{pmatrix} \hat{\boldsymbol{\Sigma}}_{11} & \hat{\boldsymbol{\Sigma}}_{12} \\ \hat{\boldsymbol{\Sigma}}_{21} & \hat{\boldsymbol{\Sigma}}_{22} \end{pmatrix}$$

Under the hypothesis $H_o : \boldsymbol{\theta}_2 = \boldsymbol{0}$ the test statistic

$$\chi^2 = \hat{\boldsymbol{\theta}}_2^t \hat{\boldsymbol{\Sigma}}_{22}^{-1} \hat{\boldsymbol{\theta}}_2 \tag{6.63}$$

is asymptotically distributed (i.e. n_2 large) as a chi-square, with a number of degrees of freedom equal to the number of components in the sub-vector $\boldsymbol{\theta}_2$. Problem 6.1 of this chapter gives an interesting real life example that illustrates the chi-square tests above and shows that the usual chi-square for contingency table is erroneous.

In principle the previous technique can deal with any contingency table without structural zeros. Nevertheless, recall that a $3 \times 3 \times 3$ table requires us to estimate 27 variances, $\frac{27 \times 26}{2} = 351$ covariances and a $(27, 26)$ design matrix, possibly with that analysis stratified by areas. This is a lot of work indeed.

In the next chapter we shall investigate the model-dependent geostatistical approach for estimating integrals $\int_F Y(x) dx$.

Problem 6.1. *The following table displays the counts n_{ij} of 2×2 contingency tables obtained from an inventory with 20 plots (independently and uniformly distributed in the small area F described for the case study in Chapter 8). Each plot comprises a single $500m^2$ circle. To simplify the calculations, boundary effects have been neglected so that the inclusion probabilities are constant in a first approximation. The interpretation is $i = 1$ for conifers, $i = 2$ broadleaf species, $j = 1$ for healthy trees and $j = 2$ for slightly damaged trees (based on a foliage loss criterion).*

Plot nr	n_{11}	n_{12}	n_{21}	n_{22}
1	2	1	0	0
2	2	3	2	0
3	6	4	0	1
4	6	2	40	27
5	3	5	28	26
6	7	2	1	1
7	5	2	6	2
8	2	5	2	2
9	13	5	0	0
10	5	2	14	32
11	0	1	0	0
12	0	2	4	1
13	5	5	0	0
14	3	1	1	3
15	2	2	0	0
16	5	2	26	37
17	0	0	1	0
18	4	4	0	3
19	3	0	0	0
20	5	5	1	0

Using the technique developed in Problem 4.7 and Eq. 6.50, test, applying the chi-squares tests from Eq. 6.53 and 6.63, the hypothesis that the state of health is independent from the species. Compare these results with the classic chi-square test that is valid under a multinomial distribution.

CHAPTER 7

Geostatistics

This chapter briefly introduces geostatistical estimation techniques, with emphasis on the context of forest inventory. The mathematical and numerical tools required for a good understanding of this subject are more sophisticated than those presented so far and formal proofs will not be given here. The theory was first described by Matheron (1965) and was developed primarily at the Ecole des Mines de Paris, to address practical problems in the mining and oil industries. Mathematically speaking, it dealt originally with stochastic processes defined in the euclidian plane $\Re^2$ or space $\Re^3$, but has since been partially extended to spatiotemporal processes. Its current range of applications is nowadays very large: mining, petrology, soil physics, meteorology, oceanography, etc, and forestry. Excellent general references are books by Christensen (1990), Cressie (1991), Wackernagel (2003), and Schabenberger and Gotway (2005). In addition, Mandallaz (1993, 2000) addresses some specific problems associated with forest inventories from an advanced mathematical perspective, further illustrating the theory with an extensive case study, particularly with respect to small area estimations. As mentioned above, geostatistical methods are purely model-dependent. A key issue is the modeling of the underlying spatial covariance function. Once that function or, rather, its variogram is available, the optimal linear estimation is determined via Kriging. This term was coined by Matheron in honor of Krige, a geologist in South Africa. The first section in this chapter will present the concept of variogram, which generalizes, in some sense, the covariance function.

7.1 Variograms

Basing our approach on the theory of Matheron (1965) we shall assume that stochastic processes describing the local density $Y(x)$, the prediction $\hat{Y}$ and the residual $R(x) = Y(x) - \hat{Y}(x)$ are **intrinsic**. That is,

$$
\begin{aligned}
\mathbb{E}(Y(x+h) - Y(x)) &= 0 \quad \forall x, h \in \Re^2 \\
\mathbb{E}(Y(x+h) - Y(x))^2 &= 2\gamma_Y(h) \quad \text{independent of } x \\
&= \mathbb{V}(Y(x+h) - Y(x)) \quad (7.1)
\end{aligned}
$$

and similarly for $\hat{Y}(x)$ and $R(x)$. Functions $\gamma_Y(h)$, $\gamma_{\hat{Y}}(h)$, $\gamma_R(h)$ are called **semi-variograms** (or **variograms**). We shall often drop the lower index

process being considered. Note that $\gamma(h) = \gamma(-h)$. The increment vector h known as the **lag** and the variogram is called **isotropic** if $\gamma(h) = \gamma(|h|)$.

Though the intrinsic hypothesis does not require finite expectation and variance for the process itself but only for its increments, we shall assume that this is the case. The main advantage of this intrinsic hypothesis is that it requires the stationarity in mean and variance of the increments only. For example, Brownian motion on the real line is intrinsic but not stationary (i.e. it has a nonconstant variance).

Variograms satisfy the following important conditions:

$$\gamma(0) = 0$$

$$\lim_{h \to \infty} \frac{\gamma(h)}{|h|^2} = 0$$

$$\sum_{i=1}^{n} \lambda_i \lambda_j \gamma(x_i - x_j) \leq 0 \quad \forall n \ \forall x_i, x_j \ \text{provided that} \ \sum_{i=1}^{n} \lambda_i = 0$$

That last property means that the variogram is a **conditionally negative definite function**.

An **authorized linear combination** is a quantity $\sum_{i=1}^{n} \lambda_i Y(x_i)$ where the weights λ_i satisfy the relationship $\sum_{i=1}^{n} \lambda_i = 0$. The variance of an authorized linear combination is well defined and is given by the formula

$$\mathbb{V}\left(\sum_{i=1}^{n} \lambda_i Y(x_i)\right) = -\sum_{i,j=1}^{n} \lambda_i \lambda_j \gamma_Y(x_i - x_j) \tag{7.2}$$

If the underlying process itself is second-order stationary, i.e.,

$$\mathbb{E}(Y(x)) = m \quad \forall x \in \Re^2$$
$$\mathbb{COV}(Y(x+h), Y(x)) = C_Y(h) \tag{7.3}$$

then we have an important relationship between the variogram and the correlation function

$$\gamma(h) = C(0) - C(h) \tag{7.4}$$

Many properties of these variograms can be easily checked in this special case by using Eq. 7.4. If $\lim_{|h| \to \infty} C(h) = 0$ then $C(0) = \gamma(\infty)$ is called the **sill** The smallest vector r_o for which $\gamma(r_o(1 + \varepsilon)) = C(0)$ for any $\varepsilon > 0$ is called the **range** in the direction r_o.

If $\gamma(h)$ is continuous at the origin, the process is continuous in the mean square sense (i.e. $\lim_{|h| \to 0} \mathbb{E}(Y(x+h) - Y(x))^2 = 0$). Otherwise, the discontinuity at the origin $c_o = \gamma(0^+) - \gamma(0) > 0$ is called the **nugget effect**. For a correct interpretation, it is useful to recall that averaging white noise over a disk, no matter how small, yields a continuous variogram, likewise for a Poisson-marked point process. Therefore, the only way for a nugget effect to occur is in the presence of measurement or sampling errors (e.g. for ( ) in two-stage sampling). Thus, in principle $Y(x)$ has a continuous variogram. However, in applications, ad hoc variograms with a nugget effect will be used occasionally, simply to reflect our numerical ignorance with respect to the behavior of the process at the micro-scale. This is because observations are available only for points lying at a minimal distance from each other, a distance determined by the sampling scheme. The value of this pseudo nugget effect is often inversely proportional to a power of the surface area for the largest circle utilized in the inventory scheme. The following variograms play an important role in applications.

1. **Linear model**, valid in $\Re^d \quad d \geq 1$:

$$\gamma(h) = \sigma^2 \mid h \mid \tag{7.5}$$

This is an unbounded variogram, in $\Re$ induced by the Brownian motion.

2. **Spherical model**, valid in $\Re^d \quad d \leq 3$:

$$\gamma(h) = \begin{cases} \sigma^2 \left(\frac{3|h|}{2r} - \frac{1}{2}(\frac{|h|}{r})^3 \right) & \text{if } \mid h \mid \leq r \\ \sigma^2 & \text{otherwise} \end{cases} \tag{7.6}$$

The sill is σ^2 and the range r. This variogram results from averaging a white noise over spheres of radius $\frac{r}{2}$.

3. **Exponential variogram**, valid in $\Re^d \quad d \leq 3$:

$$\gamma(h) = \sigma^2 \left(1 - exp(-\frac{\mid h \mid}{a}) \right) \tag{7.7}$$

The sill is σ^2. Although the range is infinite, in practice it is defined as 3

4. **Circular variogram**, valid in $\Re^d \quad d \leq 2$:

$$\gamma(h) = \begin{cases} \sigma^2 \left(1 - \frac{2}{\pi}(arcos(\frac{|h|}{r}) - \frac{|h|}{r} \sqrt{1 - (\frac{|h|}{r})^2} \right) & \text{if } \mid h \mid \leq r \\ \sigma^2 & \text{otherwise} \end{cases} \tag{7.8}$$

This variogram results from averaging a white noise over a circle of radius $\frac{r}{2}$.

5. **Pure nugget effect**:

$$\gamma(h) = \begin{cases} 0 & \text{if } \mid h \mid = 0 \\ \sigma^2 & \text{otherwise} \end{cases} \tag{7.9}$$

6. **Compound variograms**:

Given p variograms $\gamma_i(h) \quad i = 1, 2 \ldots p$ one can construct a further variogram by setting

$$\gamma(h) = \sum_{i=1}^{p} \gamma_i(h) \tag{7.10}$$

Let $\gamma_0(\cdot)$ be a valid one-dimensional variogram and $\boldsymbol{A}$ a (d, d) regular matrix. An anisotropic variogram $\gamma(h)$ is considered geometrically anisotropic if $\gamma(h) = \gamma_0(|\boldsymbol{A}h|)$.

In practice, the underlying variograms $\gamma(h)$ must be estimated from the data. Under the intrinsic assumption, a natural **non-parametric estimator** based on the methods of moments is the **empirical variogram**

$$\hat{\gamma}_Y(h) = \frac{\sum_{i,j \in N(h)} (Y(x_i) - Y(x_j))^2}{|N(h)|} \qquad (7.11)$$

where $N(h) = \{(x_i, x_j) \mid x_i - x_j = h\}$ and $|N(h)|$ is the number of pairs in $N(h)$. In most instances, because of the positions of the sample points, only smooth versions of Eq. 7.11 are available, namely

$$\hat{\gamma}_Y(h_l) = average\{(Y(x_i) - Y(x_j))^2 \mid (x_i, x_j) \in N(h), h \in T(h_l)\}$$

where $T(h_l)$ is some specified tolerance region around h_l. Such regions should be as small as possible to retain spatial resolution, yet large enough to ensure stability of the estimates (say 60 pairs, at least).

With this technique, it is better to estimate directly the variogram rather than the corresponding covariance function. After the empirical variogram $\hat{\gamma}$ has been obtained one must still fit a model to it (e.g. among the examples mentioned above). Again, several methods are possible and some software offers interactive "fitting-by-eye" procedures (ISATIS (1994)). There are also more advanced procedures that rely on restricted maximum likelihood and direct least squares estimation of the covariance matrix, Mandallaz (1993). Generally speaking, the behavior of the variogram near the origin is crucial and large lags h should not be retained unless enough pairs exist. Finally, an unbounded empirical variogram may indicate that the intrinsic hypothesis is violated (e.g. when a linear drift is found in the data). Then more advanced procedure such as Universal Kriging must be used (Mandallaz, 2000).

7.2 Ordinary Kriging

Assuming the underlying variogram has been adequately modeled we want to estimate quantities of the form

$$Y(V_i) = \frac{1}{\lambda(V_i)} \int_{V_i} Y(x) dx \qquad (7.12)$$

for arbitrary domains of the plane $V_i \subset F$. If the domain V_i is reduced to a single point x_0 we will be concerned with **Punctual Kriging**, but that scenario is hardly relevant in a forest inventory. Therefore, for our purposes V is usually a small area in the forest F, so that we can proceed with **Block** (a term originally applied in mining). We shall discuss this problem under the assumption that the uncertainties with respect to the variogram can be neglected. The following important result is known as the **extension variance of V_2 to V_1**:

$$\mathbb{V}\left(Y(V_1) - Y(V_2)\right) = -\frac{1}{\lambda^2(V_1)} \int_{V_1} \int_{V_1} \gamma_Y(x - y) dx dy \qquad (7.13)$$

$$-\frac{1}{\lambda^2(V_2)} \int_{V_2} \int_{V_2} \gamma_Y(x - y) dx dy$$

$$+2\frac{1}{\lambda(V_1)\lambda(V_2)} \int_{V_1} \int_{V_2} \gamma_Y(x - y) dx dy$$

$$=: \quad -\bar{\gamma}_Y(V_1, V_1) - \bar{\gamma}_Y(V_2, V_2) + 2\bar{\gamma}_Y(V_1, V_2)$$

Indeed, the difference is an authorized linear combination because both integrals can be approximated by discrete sums (with constant weights summing up to 1), for which 7.2 can be used and the limit then taken. Similarly one obtains this fundamental result:

Theorem 7.2.1. *If $\sum_i \lambda_i = 1$ then*

$$\mathbb{V}\left(\sum_i \lambda_i Y(x_i) - \frac{1}{\lambda(V)} \int_V Y(x) dx\right) = -\sum_{i,j} \lambda_i \lambda_j \gamma_Y(x_i - x_j)$$

$$-\frac{1}{\lambda^2(V)} \int_V \int_V \gamma_Y(x - y) dx dy$$

$$+2\sum_i \lambda_i \frac{1}{\lambda(V)} \int_V \gamma_Y(x_i - y)$$

which can be abbreviated to

$$-\sum_{i,j} \lambda_i \lambda_j \gamma_Y(x_i - x_j) - \bar{\gamma}_Y(V, V) + 2\sum_i \lambda_i \bar{\gamma}_Y(x_i, V)$$

Remarks:

1. The formula remains valid for **Punctual Kriging**. That is, $V = \{x_0\}$, x_i $\forall i$, $\frac{1}{\lambda(V)} \int_V Y(x) dx = Y(x_0)$, $\bar{\gamma}_Y(V, V) = 0$ and $\bar{\gamma}_Y(x_i, V) = \gamma(x_i -$

2. Calculating the terms $\bar{\gamma}_Y(x_i, V)$ and $\bar{\gamma}_Y(V, V)$ can be a difficult numerical task.

Ordinary Kriging yields the **Best Linear Unbiased Estimator (BLUE** of

$$Y(V_0) = \frac{1}{\lambda(V_0)} \int_{V_0} Y(x)$$

in the class

$$\hat{Y}_{OK}(V_0) = \sum_{i \in s_2} \lambda_i Y(x_i)$$

$$\mathbb{E}\left(\hat{Y}_{OK}(V_0) - Y(V_0)\right) = 0$$

$$\mathbb{E}\left(\hat{Y}_{OK}(V_0) - Y(V_0)\right)^2 = minimum \qquad (7.14)$$

Remarks:

1. As usual s_2 denotes the set of sample points for a terrestrial inventory, so we write $i \in s_2$ for $x_i \in s_2$.

2. The first constraint also requires that the error $\hat{Y}_K(V_0) - Y(V_0)$ is an authorized linear combination, which is equivalent to the condition $\sum_{i \in s_2} \lambda_i = 1$.

3. Minimization under constraint is performed according to Lagrange's multiplier technique. That is, one must minimize with respect to the λ_i and the Lagrange multiplier μ the Lagrange function $L(\lambda_i, \mu)$ given by

$$-\sum_{i,j}\lambda_i\lambda_j\gamma_Y(x_i - x_j) - \bar{\gamma}_Y(V,V) + 2\sum_i \lambda_i\bar{\gamma}_Y(x_i, V) - 2\mu\left(\sum_{i \in s_2}\lambda_i - 1\right)$$

The minus sign is a convenient convention.

Taking the partial derivatives of $L(\lambda_i, \mu)$ with respect to the λ_i and μ obtain the linear system of $n_2 + 1$ equations and $n_2 + 1$ unknowns

$$\sum_{i \in s_2}\lambda_i\gamma_Y(x_i - x_j) + \mu = \bar{\gamma}_Y(x_j, V_0)$$

$$\sum_{i \in s_2}\lambda_i = 1 \qquad (7.15)$$

The mean square error for the λ_i and μ solutions of the linear system 7.15 is then given by

$$\mathbb{E}\left(\hat{Y}_{OK}(V_0) - Y(V_0)\right)^2 = \sum_{i \in s_2}\lambda_i\bar{\gamma}_Y(x_i, V_0) - \bar{\gamma}_Y(V_0, V_0) + \mu \qquad (7.16)$$

Remarks:

1. We shall always assume that the Kriging system given by Eq. 7.15 is regular. This is not the case if for instance the same point occurs twice.

2. If $V_0 = \{x_0\}$ and $x_o \in s_2$ it is easy to check that $\hat{Y}_K(x_0) = Y(x_0)$. In this sense **Ordinary Kriging** is an exact interpolator for punctual estimation.

3. Variograms with a nugget effect lead to a mathematical inconsistency (frequently overlooked) because Punctual Kriging cannot be considered as the limit of genuine Block Kriging when the block V_0 shrinks to a point $x_0 \notin$ Indeed, with a pure nugget effect $\gamma(0) = 0$ and $\gamma(h) = \sigma^2$ for $h \neq 0$, it is easy to verify that Eq. 7.15 gives for the **MSE** $\sigma^2 + \frac{\sigma^2}{n_2}$ for Punctual Kriging

for any true domain, with interpretation of this paradox is, again, that a nugget effect is a coarse numerical approximation of the true underlying short-range behavior of the variogram. It is valid as long as this range is negligible with respect to the size of the domain.

4. In applications, s_2 is frequently reduced to a so-called Kriging neighborhood $U \cap s_2$ with $V_0 \subset U$. However, this numerical simplification should be avoided in the presence of a nugget effect (for a pure nugget effects the weights are constant and equal to n_2^{-1}).

5. The numerical aspects of Kriging, especially with very large data sets, are non trivial, such that the use of recognized software packages is recommended, (Mandallaz, 1993).

7.3 Kriging with sampling error

With two-stage sampling schemes one observes only $Y^*(x)$ and not the true local density $Y(x)$. We can write $Y^*(x) = Y(x) + \delta(x)$ with the sampling error $\delta(x)$. Recall that $\mathbb{E}_p \delta(x) = 0$ because $Y^*(x)$ is design-unbiased for $Y(x)$. In addition, the design-based variance and its estimate are

$$V(x) = \mathbb{V}(\delta(x)) = \frac{1}{\lambda^2(F)} \sum_{i \in s_2(x)} \frac{R_i^2(1 - p_i)}{\pi_i^2 p_i}$$

$$\hat{V}(x) = \frac{1}{\lambda^2(F)} \sum_{i \in s_3(x)} \frac{R_i^2(1 - p_i)}{\pi_i^2 p_i^2}$$

The Kriging estimate with sampling error is the **BLUE** in the class

$$\hat{Y}_{KE}^*(V_0) = \sum_{i \in s_2} \lambda_i Y^*(x_i) \tag{7.17}$$

If $\sum_{i \in s_2} \lambda_i = 1$ then $\mathbb{E}_{Y,\delta}(\hat{Y}_{KE}^*(V_0) - Y(V_0)) = 0$ and the overall **MSE** with respect to the underlying unobservable process $Y(x)$ and the second-stage sampling procedure δ) is

$$\mathbb{E}_{Y,\delta}(\hat{Y}_{KE}^*(V_0) - Y(V_0))^2 = -\sum_{i,j} \lambda_i \lambda_j \gamma_Y(x_i - x_j) - \bar{\gamma}_Y(V_0, V_0)$$

$$+ 2 \sum_i \lambda_i \bar{\gamma}_Y(x_i, V_0) + \sum_{i \in s_2} \lambda_i^2 V(x_i)$$

Minimizing this expression under the constraint yields the Kriging equations with sampling errors

$$\sum_{i \in s_2} \lambda_i \gamma_Y(x_i - x_j) - \sum_{i \in s_2} \lambda_i V(x_i) + \mu = \bar{\gamma}_Y(x_j, V_0)$$

$$\sum_{i \in s_2} \lambda_i = 1 \tag{7.18}$$

$$\mu$$

$$\mathbb{E}_{Y,\delta}\left(\hat{Y}_{KE}^*(V_0) - Y(V_0)\right)^2 = \sum_{i \in s_2} \lambda_i \bar{\gamma}_Y(x_i, V_0) - \bar{\gamma}_Y(V_0, V_0) + \mu \qquad (7.19)$$

To apply these equations in practice one can substitute $\hat{V}(x_i)$ for $V(x_i)$. The main difficulty is to obtain an estimate for the underlying variogram $\gamma_Y($ This is straightforward if we assume that 1) $Y(x)$ is second-order stationary and 2) the sampling variance is constant $V(x) \equiv \sigma^2$. Here, one can use the estimate $\hat{\sigma}^2 = \frac{1}{n_2} \sum_{i \in s_2} \hat{V}(x)$. Furthermore, one obtains the relationship

$$\gamma_{Y^*}(h) = \frac{1}{2} \mathbb{E}_{Y,\delta} (Y^*(x+h) - Y^*(x))^2 = \gamma_Y(h) + \sigma^2$$

Hence, it suffices to shift downwards the variogram of the observed process $Y^*(x)$ by σ^2, while retaining $\gamma_Y(0) = 0$, to obtain the variogram of the un-observed process $Y(x)$. The structure of the Kriging equations with sampling errors (Eq. 7.18) shows that it would be wrong, for Kriging purposes, to treat the second-stage sampling variance σ^2 as an ordinary nugget effect. As a matter of fact, we can verify that, for punctual estimation, the Kriging procedure with sampling errors does not yield an exact interpolator.

7.4 Double Kriging for two-phase sampling schemes

In this section we present a simple geostatistical generalization of the regression estimators derived in a model-assisted framework. Although a rigorous mathematical treatment is beyond the scope of this book, Mandallaz (1993, 2000) provides complete proofs and more sophisticated numerical procedures. Fortunately, one can give very convincing heuristic arguments here. Let us set the stage.

1. We have a large sample s_1 of n_1 points in which the auxiliary information vector $Z(x)$ is available. Because in a model-dependent geostatistical approach the sampling scheme is irrelevant for estimation it makes no difference whether s_1 has been generated by simple or cluster (systematic) sampling.

2. We have a small sub-sample $s_2 \subset s_1$ in which the true local density $Y(x$ available, or, for two-stage procedures, the generalized local density $Y^*($

3. We have a model linking the auxiliary information and the terrestrial inventory data, i.e.

$$Y(x) = Z(x)^t \beta + R(x)$$
$$Y^*(x) = Z(x)^t \beta + R^*(x)$$

as well as a model for the variogram $\gamma_R(h)$ or the covariance function C_R that depends on a parameter vector θ, usually including the sill, range and nugget effect.

of $\boldsymbol{\beta}$ and parameters $\boldsymbol{\theta}$ of the variogram. Nevertheless, it can be done, under some regularity assumptions, via Least Squares (LS) or Restricted Maximum Likelihood (RML). Once $\boldsymbol{\theta}$ is available, one can estimate $\boldsymbol{\beta}$ generalized least squares. However, especially with large data sets, this is a numerical challenge.

Double Kriging is a special case of **Co-Kriging** (see Cressie (1991) and Mandallaz (1993)). Recall that the model-assisted regression estimate is simply the mean of the predictions plus the mean of the residuals. The generalization is obvious: instead of using constant weights equal to n_1^{-1} and n_2^{-1} allow variable weights λ_i for the predictions $(\sum_{i \in s_1} \lambda_i = 1)$ and μ_j for the residuals $(\sum_{j \in s_2} \mu_j = 1)$. Optimal weights are determined via two separated Kriging procedures. The algorithm is straightforward to implement:

1. **Step 1**
   Estimate $\boldsymbol{\beta}$ by ordinary linear regression of $Y(x)$, or $Y^*(x)$ on $\boldsymbol{Z}(x)$. Although not optimal, the result is still unbiased.

2. **Step 2**
   Calculate the predictions $\hat{Y}(x) = \hat{\boldsymbol{\beta}}^t \boldsymbol{Z}(x)$ for all $x \in s_1$ and the residuals $R(x) = Y(x) - \hat{Y}(x)$, or $R^*(x) = Y^*(x) - \hat{Y}(x)$, for all $x \in s_2$. Then, one determines the empirical variograms of predictions and residuals, and proceed to fit adequate models, usually among the examples given in Section 7.1, to obtain $\gamma_{\hat{Y}}(h)$, $\gamma_{R^*}(h)$, and $\gamma_R(h)$. We assume that all involved processes are stationary and that the predictions and residuals are independent. Recall from Section 7.3 that in this case $\gamma_R(h)$ can be obtained from $\gamma_{R^*}(h)$ by subtracting the second-stage variance $\hat{\sigma}^2$. Note that one could also use an external model, i.e. with $\boldsymbol{\beta}$ obtained from a source other than the available inventory.

3. **Step 3**
   Perform Ordinary Kriging of the predictions to obtain for the domain of interest V_0 the estimate
   $$\hat{\hat{Y}}_{OK}(V_0)$$
   and its estimated mean square error
   $$MSE(\hat{\hat{Y}}_{OK}(V_0))$$
   This is accomplished by using Eq. 7.15 and 7.16 for the process $\hat{Y}(x)$ rather than $Y(x)$.

4. **Step 4**
   Perform Ordinary Kriging of the residuals $R(x)$ or Kriging with sampling errors of the $R^*(x)$ to obtain the estimate
   $$\hat{R}_{OK}(V_0)$$
   and its estimated **MSE**
   $$MSE(\hat{R}_{OK}(V_0))$$

$(\ )$ instead of $(\ )$. In two-stage sampling, one would apply Eq. 7.18 and 7.19 to obtain

$$\hat{R}_{KE}(V_0), \quad MSE(\hat{R}_{KE}(V_0))$$

5. **Step 5**

To arrive at the final **Double Kriging Estimator** $\hat{Y}_{DK}(V_0)$ simply add up the Kriging estimates of the predictions and of the residuals. The same is done for the mean square errors, that is

$$\hat{Y}_{DK}(V_0) = \hat{\tilde{Y}}_{OK}(V_0) + \hat{R}_{OK}(V_0)$$

$$MSE(\hat{Y}_{DK}(V_0)) = MSE(\hat{\tilde{Y}}_{OK}(V_0)) + MSE(\hat{R}_{OK}(V_0))$$

in one-stage sampling, or

$$\hat{Y}_{DK}(V_0) = \hat{\tilde{Y}}_{OK}(V_0) + \hat{R}_{KE}(V_0)$$

$$MSE(\hat{Y}_{DK}(V_0)) = MSE(\hat{\tilde{Y}}_{OK}(V_0)) + MSE(\hat{R}_{KE}(V_0))$$

in two-stage sampling.

Remarks:

1. The variogram of the predictions usually will be smooth, without the nugget effect and with a rather large range that corresponds roughly to the average diameter of the stands used for the prediction-model.

2. The variogram of the residuals normally will have a much shorter range. It is wise to retain only lags h with enough data points and to exclude from the calculations values at large lags h in order to reduce the bias in the resulting variogram. Furthermore, it is important that the fitted variogram represents well the behavior of the empirical $\gamma_R(h)$ near the origin.

3. With large data sets, especially for the predictions and for global estimation (such as $V_0 = F$) it might be necessary to divide F into smaller domains F_k. The final estimate can then be obtained by taking the weighted means and mean square errors

$$\sum_k \frac{\lambda(F_k)}{\lambda(F)} \hat{Y}_{DK}(F_k) , \quad \sum_k (\frac{\lambda(F_k)}{\lambda(F)})^2 MSE(\hat{Y}_{DK}(F_k))$$

(while neglecting the correlation between the estimates $\hat{Y}_{DK}(F_k)$).

4. Usually the forest F and the V_i are represented as polygons. The number of vertices is not crucial as long as the polygon approximately reflects the shape of the domain.

5. If a thematic map is available (i.e. $n_1 = \infty$) one can implement **Universal Kriging** (see Mandallaz (2000, 1993) and Cressie (1991)). This procedure is more efficient but also more difficult. Alternatively, one can use a very large n_1 in Double Kriging. The contribution of the predictions to the Kriging error will usually be much smaller than that of the residuals. In

$(\quad) = \quad(\quad)\qquad(\quad)$ (Drift + residual) is by definition non-stationary. In contrast, all of the processes involved in Double Kriging are assumed to be stationary. As already mentioned, the decomposition "drift + residual" and the stationarity concepts are not as clear cut as they may seem, very often depending on the scale and the availability of auxiliary information. Even if $Y(x)$ is stationary, Double Kriging is better than Ordinary Kriging because it does, in fact, utilize the auxiliary information $Z(x)$, as summarized into the predictions $\hat{Y}(x)$.

6. Kriging procedures are usually more efficient than model-assisted procedures for local estimation (small area estimation). The advantage for global estimation appears to be minimal. Of course, one drawback of Kriging procedures is that they are time-consuming, even when performed with sophisticated software. In addition, one must model a variogram for each variable of interest (e.g., timber volume for conifers, broadleaf, or all species combined). During routine work, it can also be disturbing to discover that the resulting tables are not exactly additive.

In the next chapter we shall see how what we have learned so far works in practice.

7.5 Exercises

Problem 7.1. *We consider a process $Y(x)$ on the real line with observations at the points $x_1 < x_2 < x_3 < x_4$ and a point x_0 with $x_2 < x_0 < x_3$. This process has a linear variogram: $\gamma(h) =\mid h \mid$ for $h \in \Re$ and $\sigma^2 = 1$. In other words, $Y(x)$ is the Brownian motion in one dimension. Show that the Punctual Ordinary Kriging estimate at point x_0 is given by*

$$\hat{Y}(x_0) = \lambda Y(x_2) + (1 - \lambda)Y(x_3)$$

with $\lambda = \frac{x_3 - x_0}{x_3 - x_2}$. That is, we will have Kriging weights of

$$\lambda_1 = \lambda_4 = 0, \ \lambda_2 = \lambda, \ \lambda_3 = 1 - \lambda$$

Give a geometrical interpretation of this result and examine whether it remains valid if one adds further observations to the left of x_2 and/or to the right of What happens if one inserts a nugget effect into the variogram $\gamma(h) =\mid$ What analogous findings would you expect in the plane?

Problem 7.2. *Suppose that you want to perform Block Kriging of a domain V_0 but that you have only Punctual Kriging software at your disposal. How would you proceed to get a point estimate $\hat{Y}(V_0)$? Justify your answer. What can you say about the mean square error?*

CHAPTER 8

Case Study

We can illustrate some theoretical results by summarizing the most important findings from an inventory conducted in the early 1990s in parts of the Zürichberg Forest belonging to the City and Canton of Zürich (see Mandallaz (1991, 1993) for details). The inventoried area covered 217.9 ha, of which 17.1 ha (thereafter referred to as "small area") served for a full census, in which the tree coordinates were recorded. Figure 8.1 displays the 16 stands, the perimeter of the small area with full census and the location of the cluster points in the terrestrial inventory.

Establishment of stand map was based on aerial photographs and "expert judgement", and encompassed three qualitative variables:

1. **Developmental stage**
   This entails four categories "pole stage=3," "young timber tree=4," "middle age timber tree=5," and "old timber tree=6." These were assigned according to the dominant diameter.

2. **Degree of mixture**
   This variable was simplified to the categories of "predominantly conifers=1" and "predominantly broadleaves=2."

3. **Crown closure**
   This variable was based on canopy density, defined as the proportion of the entire ground surface within the stand that was covered by the tree crowns. It was simplified to the categories of "dense=1" and "close=2."

These factors produced $4 \times 2 \times 2 = 16$ possible stands, all of which were found on the study site. Fig. 8.1 displays the sample plots $(+)$ as well as the perimeters of the stands and of the small area.

Fig. 8.2 shows the location for the trees in the small area.

This inventory utilized systematic cluster sampling. The cluster comprises five points: central point, two points each established 30 m east or west of the central point; two other points each established 40 m either north or south of the central point. Here, we will assess only two of the four sampling schemes.

1. **Scheme 1** $n_1 < \infty$
   The first phase sets the central cluster point on a 120 m W-E by 75

Figure 8.1 *Stand map of the Zürichberg Forest*

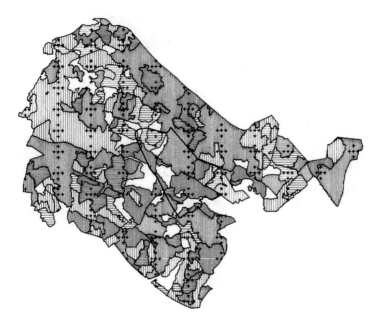

Figure 8.2 *Small area: 17.1 ha, 4784 trees*

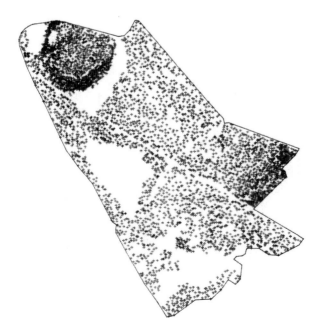

N-S rectangular grid (note that the clusters partially overlapped in the N-S direction). This resulted in a nominal density of 5.6 points per ha. The second, terrestrial phase, place the central point on a 1:4 sub-grid of the first phase, i.e. on a 240 m W-E by 150 m N-S systematic rectangular grid, thereby yielding a nominal density of 1.2 point per ha. The terrestrial inventory was purely one-stage with simple circular plots of $300m^2$ horizontal surface area, and an inventory threshold set at 12cm DBH.

2. **Scheme 2** $n_1 = \infty$

    The first phase was actually a census because it used the available thematic maps (this is required for the model dependent estimation procedures). The terrestrial phase was exactly the same as that described for Scheme 1.

Of the many variables recorded for each tree, only two are examined here. These include $Y_i \equiv 1$, which represented stand density (number of stems per ha) and $Y_i = G_i$, which represented the basal area density in m^2 per ha. Tables 8.1 and 8.2 list the observed number of clusters and plots as well as the mean $\bar{M}$ and variance V of the number of plots per cluster. Recall that the nominal value was $M = 5$.

Table 8.1 *Sample sizes over the entire inventoried area*

	n_2	$n_2 \bar{M}$	$\bar{M}$	V
1st phase	298	1203	4.04	1.92
2nd phase	73	298	4.08	1.99

Table 8.2 *Sample sizes for the small inventoried area*

	n_2	$n_2 \bar{M}$	$\bar{M}$	V
1st phase	29	92	3.17	2.37
2nd phase	8	19	2.38	2.27

A preliminary analysis of variance using the three qualitative variables that define the stand map revealed that these data could be adequately described with a simple ANOVA model containing the main effects only (all significant) and without interactions. This then led to the following linear model with the vector $\mathbf{Z}(x)$:

- $Z_1(x) \equiv 1$ intercept term

$(\ ) = 1$ if $\qquad\qquad\qquad\qquad\qquad\qquad$ $(\ ) = 0$ otherwise

$Z_3(x) = 1$ if x lies in Development Stage 4 and $Z_3(x) = 0$ otherwise

$Z_4(x) = 1$ if x lies in Development Stage 5 and $Z_4(x) = 0$ otherwise

$Z_2(x) = Z_3(x) = Z_4(x) = -1$ if x lies in Development Stage 6

- $Z_5(x) = 1$ if x lies in a coniferous stand and $Z_5(x) = -1$ otherwise
- $Z_6(x) = 1$ if x lies in a dense stand and $Z_6(x) = -1$ otherwise

Only 6 parameters were required rather than 16 for a model with full stratification, which resulted here in some difficulties due to the small number of observations in a few stands. The standard least square estimates (i.e. ignoring the cluster structure) were $\hat{\boldsymbol{\beta}}^t = (412.0, 291.6, 19.86, -126.3, 11.2, 32.5)$ for the stem density and $\hat{\boldsymbol{\beta}}^t = (30.4, -11-6, -0.2, 7.1, 4.3, 2.0)$ for the basal area density. The coefficients of determination in the classical least square sense were $R^2 = 0.5$ for the stem density, which was satisfactory, and $R^2 = 0.2$ for the basal area, which was somewhat disappointing. To estimate the intra-cluster correlation coefficient we use a sample copy of Eq. 4.20 to obtain $\hat{\rho} = 0$ for the stem density and $\hat{\rho} = 0.28$ for the basal area. In contrast, the corresponding values for the ordinary least squares residuals were, as expected, much smaller, namely $\hat{\rho} = 0.05$ and 0.15. Therefore, the model-dependent variance of the ordinary sample mean (which is also the point estimate under cluster-sampling and under a trivial model that consists of the intercept term only) would have been too small. Our results are summarized in Tables 8.3, 8.4, 8.5, 8.6, 8.7 and 8.8. There, the various estimation procedures have been abbreviated as

- **SM** ordinary sample mean in one-phase sampling, the variance being calculated according to cluster-sampling.
- **OK** Ordinary Kriging in one-phase sampling.
- **DB1** Design-based estimate in two-phase sampling (Schemes 1 and 2) with least squares estimates at the plot level.
- **DB2** Design-based estimate in two-phase sampling (Schemes 1 and 2) with least squares estimates at the cluster level.
- **MA** Model-assisted estimate in two-phase sampling (Schemes 1 and 2).
- **MD** Model-dependent estimate under Scheme 2 (exhaustive first-phase).
- **DK** Double Kriging estimate under Scheme 1.
- **UK** Universal Kriging estimate under Scheme 2.

Table 8.3 *One-phase estimates for the entire domain*

	SM	OK
Stem density	321.03	324.76
Error	18.44	15.14
Basal area m^2/ha	31.90	31.85
Error	1.08	0.67

Table 8.4 *One-phase estimates for the small area*

	True	SM	OK
Stem density	280.23	245.61	294.35
Error	-	65.51	40.56
Basal area m^2/ha	29.6	24.54	27.68
Error	-	3.59	2.08

Table 8.5 *Two-phase estimates for the entire domain, Scheme 1*

	DB1	DB2	MA	DK
Stem density	324.61	325.08	325.79	325.84
Error	12.15	12.12	12.39	11.15
Basal area m^2/ha	31.35	31.23	31.34	31.39
Error	0.91	0.89	0.91	0.71

Table 8.6 *Two-phase estimates for the entire domain, Scheme 2 $n_1 = \infty$*

	DB1	DB2	MA	MD	UK
Stem density	328.55	328.86	329.41	328.55	327.47
Error	9.22	9.18	9.65	9.00	8.45
Basal area m^2/ha	31.49	31.45	31.46	31.49	31.48
Error	0.85	0.83	0.83	0.69	0.71

Table 8.7 *Two-phase estimates for the small area, Scheme 1*

	True	DB1	DB2	MA	DK
Stem density	280.23	256.04	257.14	257.34	281.67
Error	-	48.52	48.59	48.29	26.91
Basal area m^2/ha	29.6	24.10	24.00	23.99	29.46
Error	-	3.63	3.70	3.68	1.39

Table 8.8 *Two-phase estimates for the small area, Scheme 2 $n_1 = \infty$*

	True	DB1	DB2	MA	MD	UK
Stem density	280.23	269.88	271.24	271.73	296.78	290.58
Error	-	40.20	40.33	39.83	9.54	17.04
Basal area m^2/ha	29.6	23.86	23.69	23.67	30.41	29.28
Error	-	3.65	3.74	3.72	1.03	1.42

1. For the entire domain the point estimates and their associated estimated errors were all very close to each other, the two-phase errors being smaller, as expected, than the one phase-errors.

2. For the small area the design-based and model-assisted point and error estimates were very similar but with a higher bias (though not significantly different from zero) than the model-dependent estimates **MD, DK UK**. This demonstrates that the Kriging estimates performed very well. Perhaps because the plots in the small area were not very "representative" and outside plots also were taken into account. Under the assumption of short-range correlation, the errors of synthetic model-dependent estimates were too small, as evident when they were compared with their geostatistical counterparts.

3. Figures 8.3 and 8.4 display the spherical variograms for our observed basal area and the residuals thereof, as obtained via the maximum likelihood procedure described by Mandallaz (2000, 1993). The correlation range of the basal area, $466\ m$, was of the same order as the average dimension of the stands. In contrast, the correlation range of the residuals, $45\ m$ was, as expected, much shorter and of the same magnitude as the distance between two adjacent points of the same cluster ($30\ m$ and $40\ m$). Note that the radius of the circular plots was roughly $10\ m$. The nugget effect for the basal area was needed so that we could get a good fit of the spherical variogram to the empirical variogram. This was not necessary for the variogram of the residuals. The corresponding ranges for the stem density and residuals were similar, namely $325\ m$ and $45\ m$ respectively. This is in perfect agreement with the small intra-cluster correlation coefficients of the residuals. The intuitive reason for this, is that the plots of the same cluster were often in the same stand, so that subtracting the stand mean reduced the correlation.

4. The Kriging procedures yielded good estimates even with areas as small as $1\ ha$ (Mandallaz, 1993).

Summarizing the findings (which have been confirmed by other empirical investigations) we can conclude that the various procedures were comparable for global estimation, with no apparent reason for one not to select the easiest to implement (in some sense **DB1**). For local estimations, however, the Kriging techniques were superior.

Figure 8.3 *Variogram of the basal area*

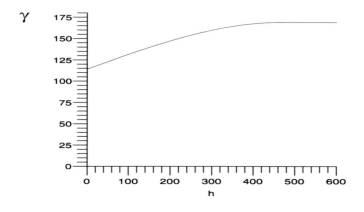

Spherical variogram: nugget=114, sill=169, range=466m

Figure 8.4 *Variogram of the basal area residuals*

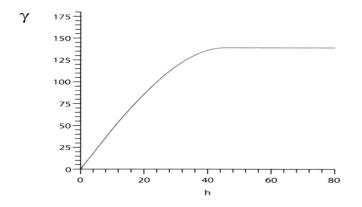

Spherical variogram: nugget=0, sill=139, range=45.5m

CHAPTER 9

Optimal sampling schemes for forest inventory

9.1 Preliminaries

In this chapter we shall design optimal sampling schemes. We saw in Section 3.2 that in sampling theory the concept of optimality is not as straightforward as it seems. Meaningful optimality criteria must rely somehow on a super-population model. The actual population to be surveyed is viewed as one realization of many similar ones. Design-based variance, i.e. under hypothetical repetition of the samples, is fixed for the given realization at hand. The average of that variance under the super-population model is called the **anticipated variance**. Optimal sampling schemes are those which minimize the anticipated variance for given costs or, conversely, minimize the costs for a given anticipated variance. This concept has been used successfully for many standard problems when sampling finite populations. The Y_i are assumed to be random variables, usually described by a linear multiple regression model with uncorrelated errors (Särndal et al. 2003). However, that approach is not well suited for forest inventory. Instead, we shall assume that the Y_i (e.g. tree volume) are fixed but that the location of those trees has been determined stochastically. We have seen that the troublesome term for the variance of the Horvitz-Thompson estimator is due to the pair-wise inclusion probabilities π_{ij} of any two trees in a forest. The solution is very simple: we assume that those locations are generated by a so-called **local Poisson model**: The forest is partitioned into **Poisson strata**, in which the trees are independently and uniformly distributed. Then, by calculating the anticipated variance under this model we can eliminate the π_{ij}. This is, of course, a very crude approximation because we do not account for repulsion or aggregation mechanisms. To do so would require the use of marked stochastic point processes for which the calculation of the anticipated variance is an extremely difficult problem and, so far, unsolved. On the other hand, the local Poisson model leads to simple and very intuitive results, that have been validated already by simulations. Though the basic ideas are very simple there are some technical difficulties with boundary effects. Here, we shall provide the main results and sketch the proofs for simple random sampling. Likewise, only some of the results will be stated for cluster-sampling, which are technically speaking more difficult to

dallaz (1997, 2001, 2002), Mandallaz and Ye (1999) and Mandallaz and Lanz (2001).

9.2 Anticipated variance under the local Poisson model

First, we consider a global Poisson forest, one with a single large Poisson stratum. This means that the coordinates $u_i(\omega) \in \Re^2$ of N trees are uniformly and independently distributed within F. The ω emphasizes the random element generated by the the global Poisson process. Likewise, we redefine the inclusion indicator variable as

$$I_i(x, \omega) = \begin{cases} 1 \text{ if } x \in K_i(\omega) \\ 0 \text{ if } x \notin K_i(\omega) \end{cases} \tag{9.1}$$

where $K_i(\omega)$ is the inclusion circle associated with the ith tree, with its center at $u_i(\omega)$. For a given realization ω the inclusion probabilities with respect to random point x, uniformly distributed in F, will therefore also depend on ω. We have $\pi_i(\omega) = \mathbb{E}_{x|\omega} I_i(x, \omega)$ and $\pi_{ij}(\omega) = \mathbb{E}_{x|\omega} I_i(x, \omega) I_j(x, \omega)$. The local density is defined, as usual, by $Y(x, \omega) = \frac{1}{\lambda(F)} \sum_{i=1}^{N} \frac{Y_i(x,\omega)}{\pi_i(\omega)}$. For ease of notation we shall continue to identify it as $Y(x)$, knowing that it depends on ω, the spatial pattern of the trees. An assumption of **negligible boundary effects** means that we set, for mathematical arguments and not for field work, $\pi_i(\omega) = \frac{\lambda(K_i)}{\lambda(F)} = \pi_i \ \forall \omega$. A general treatment of boundary effects is beyond our analytical treatment. For straight-line boundaries one can use techniques borrowed from integral geometry (Mandallaz, 1997) to show that the effects are indeed negligible if the sizes of the circles are small in comparison to the size of the forest F. By Theorem 4.2.1, the variance of the local density for a given ω is given by

$$\mathbb{V}_{x|\omega}(Y(x)) = \frac{1}{\lambda(F)^2} \left\{ \sum_{i=1}^{N} \frac{Y_i^2(1 - \pi_i(\omega))}{\pi_i(\omega)} + \sum_{i \neq j}^{N} \frac{Y_i Y_j (\pi_{ij}(\omega) - \pi_i(\omega)\pi_j(\omega))}{\pi_i(\omega)\pi_j(\omega)} \right.$$

The anticipated variance under the global Poisson model is

$$\mathbb{E}_\omega \mathbb{V}_{x|\omega}(Y(x))$$

Under the assumption of negligible boundaries, the expectation of the first term is trivial (Here, we replace $\pi_i(\omega)$ with π_i, likewise in the denominator of the second term). For the numerator of the second term we see by interchanging the order of expectation (which is legitimate because the Y_i are fixed and the $\pi_i(\omega)$ is considered as non-random under the assumption of negligible boundary effects) that

$$\mathbb{E}_\omega \mathbb{E}_{x|\omega} \pi_{ij}(\omega) = \mathbb{E}_\omega \mathbb{E}_{x|\omega} I_i(x, \omega) I_j(x, \omega) = \mathbb{E}_x \mathbb{E}_{\omega|x} I_i(x, \omega) I_j(x, \omega)$$

dependent conditionally on x given. Hence,

$$\mathbb{E}_{\omega|x}I_i(x,\omega)I_j(x,\omega) = \mathbb{E}_{\omega|x}I_i(x,\omega)\mathbb{E}_{\omega|x}I_j(x,\omega)$$

which, under negligible boundary, is equal to $\pi_i\pi_j$. Therefore, the expected value over ω of the numerator in the second term is zero and we finally obtain

$$\mathbb{E}_{\omega}\mathbb{V}_{x|\omega}(Y(x)) = \frac{1}{\lambda^2(F)}\sum_{I=1}^{N}\frac{Y_i^2}{\pi_i} - \frac{1}{\lambda^2(F)}\sum_{i=1}^{N}Y_i^2$$

The first term can be rewritten as $\frac{1}{\lambda(F)}\sum_{i=1}^{N}\frac{Y_i^2}{\lambda(K_i)}$, which is the spatial density of the "response" variable $\frac{Y_i^2}{\lambda(K_i)}$. The second term is usually much smaller than the first and is asymptotically zero when $\lambda(F) \to \infty$. Note also that the second term does not depend on the inclusion probabilities and is, therefore, irrelevant for the optimization. The anticipated variance under the global Poisson model is then given by

$$\mathbb{E}_{\omega}\mathbb{V}(Y(x)) = \frac{1}{\lambda^2(F)}\sum_{i=1}^{N}\frac{Y_i^2}{\pi_i} = \frac{1}{\lambda(F)}\sum_{i=1}^{N}\frac{Y_i^2}{\lambda(K_i)} \quad (9.2)$$

To calculate the anticipated variance under the local Poisson model, we assume that the previous assumptions hold within each of the Poisson strata F $k = 1,\ldots P_1$, $F = \cup_{k=1}^{P_1}F_{1,k}$ (the reason for introducing the further index 1 will be explained later). Again, by Theorem 4.2.1 the design-based variance is

$$\mathbb{V}(Y(x)) = \frac{1}{\lambda(F)}\int_F (Y(x) - \bar{Y})^2 dx = \frac{1}{\lambda(F)}\sum_{k=1}^{P_1}\int_{F_{1,k}}(Y(x) - \bar{Y})^2$$

The true spatial mean in the Poisson stratum F_{1k} is denoted by $\bar{Y}_{1,k}$. Writing in $F_{1,k}$ the difference as $(Y(x) - \bar{Y}) = (Y(x) - \bar{Y}_{1,k}) + (\bar{Y}_{1,k} - \bar{Y})$ and expanding the square, the double product vanishes and we get with $p_{1,k} = \frac{\lambda(F_{1,k})}{\lambda(F)}$

$$\mathbb{V}(Y(x)) = \sum_{k=1}^{P_1}p_{1,k}\frac{1}{\lambda(F_{1,k})}\int_{F_{1,k}}(Y(x) - \bar{Y}_{1,k})^2 + \sum_{k=1}^{P_1}p_{1,k}(\bar{Y}_{1,k} - \bar{Y})^2$$

The first term is the weighted sum of the conditional variances of $Y(x)$ given that x is in the Poisson stratum $F_{1,k}$, that is $\mathbb{V}(Y(x) \mid x \in F_{1,k})$. In contrast, the second term, without random component, is the variance among Poisson strata, defined as

$$\beta_1^2 = \sum_{k=1}^{P_1}p_{1,k}(\bar{Y}_{1,k} - \bar{Y})^2 \quad (9.3)$$

Applying Eq. 9.2 within each $F_{1,k}$ we can write

$$\mathbb{E}_{\omega}\mathbb{V}(Y(x) \mid x \in F_{1,k}) = \frac{1}{\lambda(F_{1,k})}\sum_{i\in F_{1,k}}\frac{Y_i^2}{\lambda(K_i)}$$

abilities via $\lambda(F)\lambda(K_i) = \lambda^2(F)\pi_i$, and recall that for the estimator $\hat{Y}$ $\frac{1}{n_2}\sum_{x \in s_2} Y(x)$ we have $\mathbb{V}(\hat{Y}) = \frac{1}{n_2}\mathbb{V}(Y(x))$. Finally, we obtain a fundamental result:

Theorem 9.2.1. *Assuming negligible boundary effects the anticipated variance under simple random sampling of the one-phase one-stage estimator is approximately given by*

$$\mathbb{E}_\omega \mathbb{V}(\hat{Y}) = \frac{1}{n_2\lambda^2(F)}\sum_{i=1}^{N}\frac{Y_i^2}{\pi_i} + \frac{1}{n_2}\beta_1^2$$

The anticipated variance under cluster random sampling is given by

$$\mathbb{E}_\omega \mathbb{V}(\hat{Y}_c) = \frac{1}{n_2\mathbb{E}M(x)\lambda^2(F)}\sum_{i=1}^{N}\frac{Y_i^2}{\pi_i} + \frac{1}{n_2\mathbb{E}M(x)}(1+\theta_1)\beta_1^2$$

where

$$(1+\theta_1)\beta_1^2 = \frac{\mathbb{V}_x\left(\sum_{k=1}^{P_1} M_{1,k}(x)(\bar{Y}_{1,k} - \bar{Y})\right)}{\mathbb{E}M(x)}$$

is the inflation factor for cluster-sampling and $M_{1,k}(x) = \sum_{l=1}^{M} I_{F_k}(x_l)$ is the number of points in a cluster with its origin at x falling into the Poisson stratum $F_{1,k}$.

Remarks:

1. When comparing simple with cluster-sampling, one must note that the number of points and the number of clusters are both represented by Then, the expected total number of points in cluster-sampling is $n_2\mathbb{E}M($ We see that for a global Poisson forest and the same overall number of points the estimators for simple and cluster random sampling have the same anticipated variance, which is intuitively obvious.

2. The anticipated variance under the local Poisson model and simple random sampling is therefore equal to the variance under the global Poisson model plus the variance among Poisson strata. This is a very nice and intuitive result.

3. Under cluster-sampling we have an inflation factor due to the intra-cluster correlation. If we assume, purely formally, that given $M(x)$ the $M_{1,k}$ have a multinomial distribution with probabilities $p_{1,k}$, then one obtains $1+\theta_1 = 1$ and cluster-sampling is again equivalent to simple sampling. Of course, the $M_{1,k}(x)$ cannot be independent. Our interpretation is that if the points of the cluster are spread over several strata the intra-cluster correlation and the inflation factors will be small. The other extreme case is that most of the points of a cluster will be in the same stratum. Mathematically, this can be modeled approximately via $M_{1,k}(x) = M(x)$ with probability $p_{1,k}$ and via $M_{1,k}(x) = 0$ with probability $1 - p_{1,k}$. In contrast, $M_{1,k}(x)M_{1,l}(x) = 0$

$$1 + \theta_1 = \mathbb{E}M(x) + \frac{\mathbb{V}M(x)}{\mathbb{E}M(x)}$$

which can be expected to be the maximum possible value. Here, by using Eq. 4.20, the intra-cluster correlation coefficient can be interpreted as

$$\rho \approx \frac{\beta_1^2}{\frac{1}{\lambda^2(F)} \sum_{i=1}^{N} \frac{Y_i^2}{\pi_i} + \beta_1^2}$$

4. In practice, this intra-cluster correlation is expected to be positive and $1 + \theta_1 \geq 0$. To minimize the inflation factor, cluster geometry should be chosen in such a way as to maximize the probability that cluster points are spread over different strata. Hence, under practical constraints, the points of the cluster should be as far away from each other as possible.

5. The quantity $\frac{1}{\lambda^2(F)} \sum_{i=1}^{N} \frac{Y_i^2}{\pi_i}$ is called the **pure error term**. It is the variance of the local density under simple random sampling and the global Poisson model, and can be written as $\frac{1}{\lambda(F)} \sum_{i=1}^{N} \frac{Y_i^2}{\lambda(K_i)}$, which is the spatial mean of the new response variable $Z_i = \frac{Y_i^2}{\lambda(K_i)}$. This pure error term can be estimated from a previous or pilot inventory for any set of π_i. By equating the empirical variance from that inventory with its anticipated variance, one immediately obtains, under simple random sampling, an estimate of the β_1^2 term (which is independent of the sampling scheme!), **even if the Poisson strata are unknown.** We only have to assume that they do exist! It is also possible to estimate the pure error term with the empirical distribution of the Y_i, which is generally a standard output of any inventory. The same is true under cluster-sampling, where the inflation factor can be obtained by equating anticipated and empirical variances. Therefore, one can achieve an estimate of the anticipated variance for any cluster-sampling schemes, provided it has the same cluster geometry as the pilot inventory. If not, one can at least consider some possible values for that inflation factor $1 + \theta_1$, especially in the worst case.

9.3 Optimal one-phase one-stage sampling schemes

To derive optimal one-phase one-stage sampling schemes we introduce a cost function

$$\phi(n_2) + n_2 c_0 + n_2 c_2 \sum_{i=1}^{N} \pi_i = \phi(n_2) + n_2 c_0 + n_2 c_2 m_1 = C \qquad (9.4)$$

C is the overall available budget and c_0 is the installation cost per point, the latter possibly including the time required to locate a point exactly, but not to access it, to assess a sample plot (e.g. stratum, slope, boundary or other relevant characteristics), or to delimit the circles. In addition, cost per first-stage tree to determine the exact response variable Y_i. Then, $m_1 = \sum_{i=1}^{N} \pi_i$ is the expected number of first-stage trees (by 4.7). $\phi(n$ represents the travel costs to access n_2 sample points. This is usually the largest cost component. For geometric reasons; it can be expected to grow like $\sqrt{n_2}$. However, an exact modeling is usually extremely difficult, so one must rely on a linear approximation valid for a sufficiently large range of values. Note that, due to the constraint, n_2 and m_1 are functions of each other, i.e. $m_1 = m_1(n_2)$ or $n_2 = n_2(m_1)$.

According to the Cauchy-Schwarz inequality 3.21 with $a_i = \frac{Y_i}{\sqrt{\pi_i}}$ and $b_i = \sqrt{}$ we have

$$\sum_{i=1}^{N} \frac{Y_i^2}{\pi_i} \geq \frac{(\sum_{i=1}^{n} Y_i)^2}{\sum_{i=1}^{N} \pi_i} \tag{9.5}$$

with equality if and only if $a_i \propto b_i$, i.e. $\pi_i = \alpha Y_i$ for some α. We implicitly assume that $Y_i \geq 0$, which is almost always the case in practice (if not, then we replace Y_i by $|Y_i|$).

For given n_2 the lower bound in Theorem 9.2.1 is achieved when its first term is minimized, which, using the previous argument, occurs whenever $\pi_i = \alpha Y$ By the inequality 9.5 it is then equal to

$$\frac{1}{n_2 \lambda^2(F) \alpha} \sum_{i=1}^{N} Y_i$$

Replacing π_i with αY_i in the constraint, we get

$$\alpha = \frac{C - n_2 c_0 - \phi(n_2)}{n_2 c_2} \frac{1}{\sum_{I=1}^{N} Y_i}$$

As the optimal solution for given sample sizes m_1, n_2, we finally arrive at

$$\lambda(F)\pi_i = \frac{C - n_2 c_0 - \phi(n_2)}{n_2 c_2} \frac{Y_i}{\bar{Y}} = m_1 \frac{Y_i}{\bar{Y}}$$

$$\mathrm{MAV}(\hat{Y}) = min_{\pi_i | n_2} \mathbb{E}_\omega \mathbb{V}(\hat{Y}) = \frac{c_2 \bar{Y}^2}{C - n_2 c_0 - \phi(n_2)} + \frac{1}{n_2} \beta_1^2$$

$$= \frac{1}{m_1 n_2} \bar{Y}^2 + \frac{1}{n_2} \beta_1^2 \tag{9.6}$$

Remarks:

1. The optimal sampling scheme is therefore a **PPS** scheme. In practice, this can occur for the stem density $Y_i \equiv 1$ if the inclusion probability is constant (one single circle), or when Y_i is the basal area and we use the angle count technique. It is also possible to design a **PPS** when Y_i is timber volume (using critical height sampling and related techniques), but this is not implemented on a routine basis (Schreuder et al. 1993).

( ), with respect to is positive. Therefore, the mathematical optimum would be to have a very large single plot. What does this mean? The global Poisson forest is homogenous and, intuitively, looks much the same everywhere; hence, it does not make sense to waste resources by observing it in many different places. Of course, in practice, one must relativize this result. Large plots can be awkward to manage because of complicated boundary and slope adjustments. Furthermore, the assumption of a negligible boundary is likely to be violated and, finally, one needs a minimum number of sample points to get a reliable estimate of the variance. In such a case, the correct interpretation is to choose n_2 large enough to fulfill the constraints, but not much larger.

3. For a local Poisson forest, the situation is completely different and one can find an optimal n_2 by solving $\frac{\partial}{\partial n_2}\mathrm{MAV}(\hat{Y}) = 0$. If $\phi(n_2)$ is linear, one can easily write the optimal solution (Mandallaz, 1997).

4. The situation is essentially the same in cluster-sampling, where the cost constraint is of the form

$$\phi(n_2) + n_2\mathbb{E}M(x)c_{c0} + n_2\mathbb{E}M(x)c_2\sum_{i=1}^{N}\pi_i$$

This can be rewritten as

$$\phi(n_2) + n_2\mathbb{E}M(x)c_{c0} + n_2\mathbb{E}M(x)m_1c_2 = C$$

where the function $\phi(n_2)$ entails the costs of traveling between the origins of the cluster and $\mathbb{E}M(x)c_{c0}$ is the average cost of installing a cluster, **including the travel cost between cluster points**. By linearizing the travel costs according to $\phi(n_2) = \beta_1 + \beta_2 n_2$ and setting $c_{2c} = \mathbb{E}M(x)c_{c0} +$ we can write the constraint as

$$n_2c_{2c} + n_2\mathbb{E}M(x)m_1c_2 = \tilde{C} = C - \beta_1$$

5. When linearizing those travel costs it is important to subtract the intercept term β_1 from the available budget C.

Implementation of an exact **PPS** can be difficult, even with the angle count technique, particularly in dense stands and also because of hidden trees. Therefore, one tries to apply so-called discrete approximations of **PPS**, an exercise that leads to the widely used technique of concentric circles. The question is, how many circles should one take and at which thresholds (e.g. what DBH when Y_i is the timber volume)? This is the subject of the next section.

9.4 Discrete approximations of PPS

We assume that the N values Y_i of the response variable Y are partitioned into classes C_l, $l = 1, 2 \ldots K$ and that the first-stage inclusion probabilities π_i

that the optimal discrete approximation has the following form: $\pi_i = \alpha g($ where α is a positive constant and the step-wise constant function $g(\cdot)$ is given by

$$g(Y_i) = \sqrt{\mathbb{E}^*(Y^2 \mid Y \in C_l)} \quad \text{whenever } Y_i \in C_l \tag{9.7}$$

$\mathbb{E}^*$ denotes the expectation with respect to discrete distribution in the finite population of the Y_i, likewise for the probability function $\mathbb{P}^*$. We define the coefficient $\tilde{\gamma}$ as

$$\tilde{\gamma} := \frac{\sum_{i=1}^N \frac{Y_i^2}{g(Y_i)}}{\sum_{i=1}^N Y_i} = \frac{\sum_{i=1}^N g(Y_i)}{\sum_{i=1}^N Y_i} \geq 1 \tag{9.8}$$

The second equality results from

$$\sum_{i=1}^N \frac{Y_i^2}{g(Y_i)} = N\mathbb{E}^* \frac{Y^2}{g(Y)} = N \sum_{l=1}^K \mathbb{P}^*(Y \in C_l)\mathbb{E}^*(\frac{Y^2}{g(Y)} \mid Y \in C_l)$$

which, by definition of the function $g(\cdot)$ and the property of the conditional expectation, is equal to

$$N \sum_{l=1}^K \mathbb{P}^*(Y \in C_l) \frac{\mathbb{E}^*(Y^2 \mid Y \in C_l)}{\sqrt{\mathbb{E}^*(Y^2 \mid Y \in C_l)}} = N \sum_{l=1}^K \mathbb{P}^*(Y \in C_l)g(Y) = \sum_{i=1}^N g(Y_i$$

The inequality results from the fact that, for $Y_i \in C_l$, we have

$$g(Y_i) = \sqrt{\mathbb{E}^*(Y^2 \mid Y \in C_l)} \geq \sqrt{\mathbb{E}^{*2}(Y \mid Y \in C_l)} = \mathbb{E}^*(Y \mid Y \in C_l)$$

and consequently

$$\sum_{i=1}^N g(Y_i) \geq N \sum_{l=1}^K \mathbb{P}^*(Y \in C_l)\mathbb{E}^*(Y \mid Y \in C_l) = \sum_{i=1}^N Y_i$$

Now we prove that the function $g(\cdot)$ produces our optimal stepwise constant inclusion probabilities. Let us denote by $\tilde{\pi}_l$ the inclusion probabilities of any such scheme with stepwise constant inclusion probabilities, i.e. $\pi_i = \tilde{\pi}_l$ whenever $Y_i \in C_l$ and by N_l the number of Y_i in C_l (not necessarily distinct). Then

$$\sum_{i=1}^N \frac{Y_i^2}{\pi_i} \sum_{i=1}^N \pi_i = \sum_{k=1}^K \frac{1}{\tilde{\pi}_l} \sum_{Y_i \in C_l} Y_i^2 \sum_{l=1}^K N_l \tilde{\pi}_l$$

Because $\sum_{Y_i \in C_l} Y_i^2 = N_l \mathbb{E}^*(Y^2 \mid Y \in C_l)$, the right-hand side can be re-stated as

$$\left(\sum_{l=1}^K \frac{1}{\tilde{\pi}_l} N_l \mathbb{E}^*(Y^2 \mid Y \in C_l)\right)\left(\sum_{l=1}^K N_l \tilde{\pi}_l\right) \tag{9.9}$$

Let

$$a_l = \frac{\sqrt{N_l}\sqrt{\mathbb{E}^*(Y^2 \mid Y \in C_l)}}{\sqrt{\tilde{\pi}_l}} \quad \text{and} \quad b_l = \sqrt{N_l \tilde{\pi}_l}$$

$$\sum_{l \, l} \sum_{l \, l} \quad (\sum_l \quad)$$

the term 9.9 is larger than

$$\left(\sum_{l=1}^{K} N_l \sqrt{\mathbb{E}^*(Y^2 \mid Y \in C_l)}\right)^2 = \left(\sum_{l=1}^{K} \sum_{Y_i \in C_l} g(Y_i)\right)^2 = \left(\sum_{i=1}^{N} g(Y_i)\right)^2$$

By Eq. 9.8, we finally have

$$\sum_{i=1}^{N} \frac{Y_i^2}{\pi_i} \geq \frac{\left(\sum_{i=1}^{N} g(Y_i)\right)^2}{\sum_{i=1}^{N} \pi_i} = \frac{\tilde{\gamma}^2 \left(\sum_{i=1}^{N} Y_i\right)^2}{\sum_{i=1}^{N} \pi_i} = \frac{\tilde{\gamma}^2 \left(\sum_{i=1}^{N} Y_i\right)^2}{m_1} \qquad (9.10)$$

The equality and, hence, the lower bound, is achieved if and only if $b_l =$
This is equivalent to $\tilde{\pi}_l = \lambda \sqrt{\mathbb{E}^*(Y^2 \mid Y)}$. In other words, the lower bound is achieved when the stepwise constant inclusion probabilities are proportional to the $g(Y_i)$, thus completing the proof.

Under the constraint $\sum_{i=1}^{N} \pi_i = m_1$ the optimal inclusion probabilities are

$$\pi_i = m_1 \frac{g(Y_i)}{\tilde{\gamma} \sum_{i=1}^{N} Y_i}$$

$$\lambda(F)\pi_i = m_1 \frac{g(Y_i)}{\tilde{\gamma}\bar{Y}} \qquad (9.11)$$

For this choice of π_i the anticipated variances for sample sizes m_1, n_2 satisfying the cost constraint are given by

$$\mathbb{E}_\omega \mathbb{V}(\hat{Y}) = \frac{1}{m_1} \frac{1}{n_2} \tilde{\gamma}^2 \bar{Y}^2 + \frac{1}{n_2} \beta_1^2$$

$$\mathbb{E}_\omega \mathbb{V}(\hat{Y}_c) = \frac{1}{m_1} \frac{1}{n_2 \mathbb{E}M(x)} \tilde{\gamma}^2 \bar{Y}^2 + \frac{1}{n_2 \mathbb{E}M(x)} (1 + \theta_1)\beta_1^2 \qquad (9.12)$$

Remarks:

1. These results are very similar to those from Eq. 9.6 except for the coefficient $\tilde{\gamma}$.

2. The best lower bound is achieved when $\tilde{\gamma} = 1$, which is possible only when the number of classes K is equal to the number of distinct values of the i.e. with **exact PPS**.

3. The best **discrete PPS** can be found numerically for a given number of classes, usually 2 or 3 but rarely more, by checking all possible thresholds and choosing the configuration with the smallest $\tilde{\gamma}$. To do so, one must know the empirical distribution of the Y_i, at least approximately, that can be obtained from a previous or pilot inventory. In general $\tilde{\gamma}$ decreases markedly from one to two circles, then slowly thereafter.

4. The optimal thresholds are the same for simple and cluster-sampling.

5. Up to now, most if not all existing forest inventories have used inclusion

( ). We define
the coefficient γ_π, which is valid for any $\pi = (\pi_1, \pi_2, \ldots \pi_N)$, by setting

$$\gamma_\pi = \frac{\sum_{i=1}^{N} \frac{Y_i^2}{\pi_i} \sum_{i=1}^{N} \pi_i}{\left(\sum_{i=1}^{N} Y_i \right)^2} \geq 1 \qquad (9.13)$$

One can easily verify that if $\pi_i = \alpha g(Y_i)$ then $\gamma_\pi = \tilde{\gamma}^2$. Hence, for any set of inclusion probabilities, we can assess how far they are from the optimal choice.

6. One can easily derive numerically the optimal sample sizes m_1, and n_2 under the local Poisson model, and also analytically with linearized traveling costs, provided that one has some idea about β_1^2 and $(1 + \theta)\beta_1^2$.

In the next section we apply a straightforward generalization of our previous results to two-stage sampling.

9.5 Optimal one-phase two-stage sampling schemes

Let us first consider simple random sampling. Our starting point is Eq. 4.29 and 4.30. The second-stage variance $\mathbb{E}_x V(x)$ is obviously independent of the location of the trees, so that

$$\mathbb{E}_\omega \mathbb{V}_{x|\omega}(\hat{Y}^*) = \mathbb{E}_\omega \mathbb{V}_{x|\omega}(\hat{Y}) + \frac{1}{n_2 \lambda^2(F)} \left(\sum_{i=1}^{N} \frac{R_i^2}{\pi_i p_i} - \sum_{i=1}^{N} \frac{R_i^2}{\pi_i} \right)$$

By using Theorem 9.2.1 for the anticipated variance and grouping the terms with the π_i we obtain the following expression:

$$\mathbb{E}_\omega \mathbb{V}_{x|\omega}(\hat{Y}^*) = \frac{1}{n_2 \lambda^2(F)} \sum_{i=1}^{N} \frac{Y_i^2 - R_i^2}{\pi_i} + \frac{1}{n_2 \lambda^2(F)} \sum_{i=1}^{N} \frac{R_i^2}{\pi_i p_i} + \frac{1}{n_2} \beta_1^2$$

It is convenient to simplify this expression by considering the following prediction model M at the tree level:

$$Y_i \;\; = \;\; Y_i^* + R_i$$
$$\mathbb{E}_M R_i \;\; = \;\; 0, \; \mathbb{V}_M R_i = \sigma_i^2, \; \mathbb{COV}(R_i, R_j) = 0, \; \forall i \neq j \qquad (9.14)$$

In this standard linear model the external prediction Y_i^* is fixed and uncorrelated with R_i so that we have $\mathbb{E}_M(Y_i^2 - R_i^2) = (Y_i^*)^2$. **For this reason we shall formally set** $\sqrt{Y_i^2 - R_i^2} = Y_i^*$, which is correct on average. Furthermore, because the σ_i^2 and $\frac{N}{\lambda(F)}$ are bounded we also have

$$\lim_{\lambda(F) \to \infty} \mathbb{E}_M \left(\frac{1}{\lambda(F)} \sum_{i=1}^{N} Y_i - \frac{1}{\lambda(F)} \sum_{i=1}^{N} Y_i^* \right)^2 = 0$$

$$\bar{Y} = \frac{1}{\lambda(F)} \sum_{i=1}^{N} Y_i = \frac{1}{\lambda(F)} \sum_{i=1}^{N} Y_i^* = \overline{Y^*} \qquad (9.15)$$

By taking the expectation $\mathbb{E}_M \mathbb{E}_\omega \mathbb{V}_{x|\omega}(\hat{Y}^*)$ we can replace the cumbersome term $Y_i^2 - R_i^2$ with Y_i^{*2} in the anticipated variance. To simplify the notation we shall omit the expectation $\mathbb{E}_M$ in the formulae. Clearly, the same arguments can be used for cluster-sampling. Here, we obtain the following expressions for the anticipated variances from one-phase two-stage sampling schemes:

$$\mathbb{E}_\omega \mathbb{V}_{x|\omega}(\hat{Y}^*) = \frac{1}{n_2} \frac{1}{\lambda^2(F)} \sum_{i=1}^{N} \frac{Y_i^{*2}}{\pi_i} + \frac{1}{n_2} \frac{1}{\lambda^2(F)} \sum_{i=1}^{N} \frac{R_i^2}{\pi_i p_i} + \frac{1}{n_2} \beta_1^2$$

$$\mathbb{E}_\omega \mathbb{V}_{x|\omega}(\hat{Y}_c^*) = \frac{1}{n_2 \mathbb{E}M(x)} \frac{1}{\lambda^2(F)} \sum_{i=1}^{N} \frac{Y_i^{*2}}{\pi_i} + \frac{1}{n_2 \mathbb{E}M(x)} \frac{1}{\lambda^2(F)} \sum_{i=1}^{N} \frac{R_i^2}{\pi_i p}$$

$$+ \frac{1}{n_2 \mathbb{E}M(x)} (1 + \theta_1) \beta_1^2 \qquad (9.16)$$

For simple random sampling, the cost constraint reads

$$\phi(n_2) + n_2 c_0 + n_2 c_{21} \sum_{i=1}^{N} \pi_i + n_2 c_{22} \sum_{i=1}^{N} \pi_i p_i = C$$

$$\phi(n_2) + n_2 c_0 + n_2 m_1 c_{21} + n_2 m_2 c_{22} = C \qquad (9.17)$$

where c_{21} is the unit cost per first-stage tree for determining the approximate value Y_i^*, while c_{22} are the additional cost per second-stage trees to obtain the exact value Y_i. m_1, m_2 are the expected numbers of first- and second-stage trees per point. Under cluster-sampling we set likewise

$$\phi(n_2) + n_2 \mathbb{E}M(x) c_{c0} + n_2 \mathbb{E}M(x) m_1 c_{21} + n_2 \mathbb{E}M(x) m_2 c_{22} = C \qquad (9.18)$$

Let us consider simple random sampling first. From our previous results we know that the optimal first-stage inclusion probabilities, for given n_2, must be of the form $\pi_i = \alpha_1 g(Y_i^*)$ (i.e. we use the same technique as before with the predictions Y_i^* instead of Y_i, in particular for the $\hat{\gamma}$ coefficient). That is, we have first-stage inclusion probabilities that are approximately proportional to the predictions, in short a **PPP** scheme. According to the Cauchy-Schwarz inequality we also have

$$\sum_{i=1}^{N} \frac{R_i^2}{\pi_i p_i} \geq \frac{\left(\sum_{i=1}^{N} |R_i| \right)^2}{\sum_{i=1}^{N} \pi_i p_i} \qquad (9.19)$$

Hence, the lower bound can be achieved with an exact **PPS** $\pi_i p_i = \alpha$ $R_i |$. Because the second-stage inclusion probabilities are proportional to the prediction error $| R_i |$, we call this a **PPE** scheme.

To implement this in actual inventory, suppose that Y_i^* is the approximate

age error of this yield function is 20% of the (exact) tree volume. Y_i is then the timber volume based on species DBH, diameter at $7m$ and total tree height, with an error averaging less than 5%. This is why Y_i is considered to be the exact value. Yield functions are based on extensive studies, and models are available that predict $\mid R_i \mid$ as a function of Y_i^* and species. The field crew enters into a laptop species and DBH (and usually also other variables). A program then generates a random number so that this particular tree might be selected with probability p_i for further measurements, e.g. diameter at 7 and height.

The problem is how to determine, for a given n_2, the constants α_1 and in such a way as to minimize the anticipated variance while satisfying the constraint. This can be done via the Lagrange technique. Setting to zero the partial derivatives of the Lagrange function with respect to α_1, α_2 and λ (the Lagrange multiplier), we obtain a set of three equations with three unknown, which is not too difficult to solve (but is a good exercise!). To simplify the notation, we set $\overline{\mid R \mid} = \frac{1}{\lambda(F)} \sum_{i=1}^{N} \mid R_i \mid$ and define the relative prediction error as $\varepsilon = \frac{\overline{\mid R \mid}}{\bar{Y}} = \frac{\sum_{i=1}^{N} \mid R_i \mid}{\sum_{i=1}^{N} Y_i}$.

The optimal solution for a given n_2 reads

$$m_1 = \frac{C - n_2 c_0 - \phi(n_2)}{\sqrt{c_{21}}\tilde{\gamma} + \varepsilon\sqrt{c_{22}}} \frac{\tilde{\gamma}}{n_2\sqrt{c_{21}}}$$

$$m_2 = \frac{C - n_2 c_0 - \phi(n_2)}{\sqrt{c_{21}}\tilde{\gamma} + \varepsilon\sqrt{c_{22}}} \frac{\varepsilon}{n_2\sqrt{c_{22}}}$$

$$\lambda(F)\pi_i = m_1 \frac{g(Y_i^*)}{\tilde{\gamma}\bar{Y}}$$

$$\lambda(F)\pi_i p_i = m_2 \frac{\mid R_i \mid}{\overline{\mid R \mid}}$$

$$\frac{m_2}{m_1} = \frac{\varepsilon\sqrt{c_{21}}}{\tilde{\gamma}\sqrt{c_{22}}}$$

$$\mathbb{E}_\omega \mathbb{V}_{x|\omega}(\hat{Y}^*) = \frac{\bar{Y}^2(\sqrt{c_{21}}\tilde{\gamma} + \varepsilon\sqrt{c_{22}})^2}{C - n_2 c_0 - \phi(n_2)} + \frac{1}{n_2}\beta_1^2$$

$$= \frac{1}{n_2 m_1}\tilde{\gamma}^2\bar{Y}^2 + \frac{1}{n_2 m_2}\overline{\mid R \mid}^2 + \frac{1}{n_2}\beta_1^2 \qquad (9.20)$$

Remarks:

1. One can verify that the above solution satisfies the cost constraint given by Eq. 9.17.

2. The optimal scheme is a combination of **PPP** and **PPE**. In some sense, one can almost write S=P+E and PPS=PP(P+E), i.e. **PPS=PPP+ PPE**

linearizing $\phi(n_2)$ (an exercise, solution in Mandallaz (1997)). However, in practice, it is more instructive to plot the anticipated variance as a function of n_2 and then assess the feasibility of the optimal circles and the numbers of first- and second-stage trees.

4. In agreement with classical results, sample sizes are inversely proportional to the square root of the corresponding costs.

5. The term **PPP** sampling is frequently referred to as **3P-Sampling**, a procedure introduced by Grosenbaugh (1971) in a slightly different context (essentially a type of Poisson list sampling that is further described by Schreuder et al. (1993)).

The solution for cluster-sampling can be obtained from Eq. 9.20 by replacing n_2 with $n_2 \mathbb{E} M(x)$, $n_2 c_0$ with $n_2 \mathbb{E} M(x) c_{c0}$ and β_1^2 with $(1 + \theta_1)\beta_1^2$.

9.6 Optimal two-phase sampling schemes

Here, we sketch the calculations for the anticipated variance from two-phase simple random sampling. The prediction model is assumed to be a stratification with respect to the so-called **working strata** $F_{2,k}$ $k = 1, 2 \ldots P_2$, which are known, in contrast to the unknown **Poisson strata**. Furthermore, we asymptotically assume that the predictions $\hat{Y}(x)$ are equal to the true stratum mean $\bar{Y}_{2,k}$ whenever $x \in F_{2,k}$. The variance among working strata is then defined as

$$\beta_2^2 = \sum_{k=1}^{P_2} p_{2,k}(\bar{Y}_{2,k} - \bar{Y})^2 \tag{9.21}$$

where $p_{2,k} = \frac{\lambda(F_{2,k})}{\lambda(F)}$. Our starting point is Eq. 5.12, i.e.

$$\mathbb{V}(\hat{Y}_{reg}^*) = \frac{1}{n_2}\mathbb{V}_x(Y(x)) + \frac{1}{n_2}\mathbb{E}_x(V(x)) - \frac{1}{n_2}(1 - \frac{n_2}{n_1})\mathbb{V}_x(\hat{Y}(x))$$

First, $\mathbb{V}_x(\hat{Y}(x)) = \beta_2^2$. Second, only the first term is relevant for the anticipated variance, and it is given by Theorem 9.2.1. Using these terms along with the tree-level prediction model M as in Eq. 9.16, we arrive at the following result:

$$\mathbb{E}_\omega \mathbb{V}_{x|\omega}(\hat{Y}_{reg}^*) = \frac{1}{n_2\lambda^2(F)} \sum_{i=1}^{N} \frac{Y_i^{*2}}{\pi_i} + \frac{1}{n_2\lambda^2(F)} \sum_{i=1}^{N} \frac{R_i^2}{\pi_i p_i} + \frac{1}{n_2}\Lambda^2 + \frac{1}{n_1}\beta_2^2$$

where $\Lambda^2 = \beta_1^2 - \beta_2^2 \geq 0$ is the **lack-of-fit term**, which is the difference of the among Poisson strata and the among working strata variances. For an easier interpretation of this term let us assume that each working stratum is the union of Poisson strata. That is, we set $F_{2,k} = \cup_{j=1}^{P_{1k}} F_{1,kj}$, where $F_{1,kj}$ the jth Poisson-stratum of the kth working-stratum. With this notation we

$$\beta_1^2 = \sum_{k=1}^{P_1}\sum_{j=1}^{P_{1k}} p_{1,kj}(\bar{Y}_{1,kj} - \bar{Y})^2$$

Writing $\bar{Y}_{1,kj} - \bar{Y} = \bar{Y}_{1,kj} - \bar{Y}_{2,k} + \bar{Y}_{2,k} - \bar{Y}$ we then get

$$\Lambda^2 = \beta_1^2 - \beta_2^2 = \sum_{k=1}^{P_2}\sum_{j=1}^{P_{1k}} p_{1,kj}(\bar{Y}_{1,kj} - \bar{Y}_{2,k})^2$$

Hence, the lack-of-fit term is simply the remaining heterogeneity not accounted for by the working-strata.

Essentially the same decomposition technique holds for cluster-sampling such that we can now formulate the main result:

Theorem 9.6.1. *The anticipated variance under two-phase two-stage cluster random sampling is given by*

$$
\begin{aligned}
\mathbb{E}_\omega \mathbb{V}(\hat{Y}_c^*) &= \frac{1}{n_2 \mathbb{E} M(x)\lambda^2(F)} \sum_{i=1}^{N} \frac{Y_i^{*2}}{\pi_i} \\
&+ \frac{1}{n_2 \mathbb{E} M(x)\lambda^2(F)} \sum_{i=1}^{N} \frac{R_i^2}{\pi_i p_i} \\
&+ \frac{1}{n_2 \mathbb{E} M(x)}\Lambda^2 \\
&+ \frac{1}{n_1 \mathbb{E} M(x)}(1 + \theta_2)\beta_2^2
\end{aligned}
$$

where

$$
\begin{aligned}
(1 + \theta_1)\beta_1^2 &= \frac{\mathbb{V}_x\left(\sum_{k=1}^{P_1} M_{1,k}(x)(\bar{Y}_{1,k} - \bar{Y})\right)}{\mathbb{E} M(x)} \\
(1 + \theta_2)\beta_2^2 &= \frac{\mathbb{V}_x\left(\sum_{k=1}^{P_2} M_{2,k}(x)(\bar{Y}_{2,k} - \bar{Y})\right)}{\mathbb{E} M(x)} \\
\Lambda^2 &= (1 + \theta_1)\beta_1^2 - (1 + \theta_2)\beta_2^2
\end{aligned}
$$

are the inflation factors with respect to the Poisson and working strata and Λ^2 is the lack of fit. $M_{1,k}(x) = \sum_{l=1}^{M} I_{F_k}(x_l)$ is the number of points of the cluster with origin x falling into the kth Poisson stratum $F_{1,k}$. Likewise, we use $M_{2,k}(x)$ for the kth working stratum.

Remarks:

1. The first term is the anticipated variance of the predictions at the tree level as if the forest were a global Poisson forest. Our second term is due to the second-stage variance. The third derives from the lack-of-fit of the prediction model at the plot level, while the forth arises from the overall heterogeneity of the working strata.

Table 9.1 **ANOVA with working strata**

Source	Sum of Squares	Anticipated Values
Model	$\sum_{x \in s_2} M^2(x)(\hat{Y}_c(x) - \hat{Y}^*)^2$	$n_2 \mathbb{E}M(x)(1 + \theta_2)\beta_2^2$
Residual	$\sum_{x \in s_2} M^2(x)(Y_c^*(x) - \hat{Y}_c(x))^2$	$n_{2,g} \mathbb{E}M(x)(\epsilon^2 + \Lambda^2)$
Total	$\sum_{x \in s_2} M^2(x)(Y_c^*(x) - \hat{Y}^*)^2$	$n_{2,g} \mathbb{E}M(x)(\epsilon^2 + (1 + \theta_1)\beta$

2. If the working strata coincide with the Poisson strata, then the lack-of-fit term is zero. Note that when $n_1 \to \infty$ **the heterogeneity of the strata is not entirely removed unless the lack-of-fit term is indeed zero**

3. A corresponding result for simple random sampling can be obtained by setting $\mathbb{E}M(x) = 1$, $1 + \theta_1 = 1 + \theta_2 = 1$; for one-stage sampling by setting $Y_i^* = Y_i$, $R_i = 0$.

We can summarize the previous result in a classical one-way ANOVA (Table 9.1), using the working strata as our model.

The **pure error term** ε^2 is defined as

$$\epsilon^2 = \frac{1}{\lambda^2(F)} \sum_{i=1}^{N} \frac{Y_i^{*2}}{\pi_i} + \frac{1}{\lambda^2(F)} \sum_{i=1}^{N} \frac{R_i^2}{\pi_i p_i} \tag{9.22}$$

We introduce the observed coefficient of determination $R^2 = \frac{\text{Model}}{\text{Total}}$ and can write $\frac{(1-R^2)\text{Total}}{n_2 \mathbb{E}M(x)} - \epsilon^2 = \Lambda^2$. Therefore, the lack-of-fit term Λ^2 is a decreasing function of the goodness of fit of the model, as expressed by the coefficient of determination R^2, thus the expression "lack-of-fit". Note that R^2 is also a decreasing function of ϵ^2. To estimate the pure error term we rewrite it as a density

$$\epsilon^2 = \frac{1}{\lambda(F)} \sum_{i \in F} \frac{Y_i^{*2}}{\lambda(K_i)} + \frac{1}{\lambda(F)} \sum_{i \in F} \frac{R_i^2}{\lambda(K_i)p_i} \tag{9.23}$$

Hence, at a given point u we can construct the estimate

$$\epsilon^{*2}(u) = \frac{1}{\lambda(F)} \sum_{i \in s_2(u)} \frac{Y_i^{*2}}{\pi_i \lambda(K_i)} + \frac{1}{\lambda(F)} \sum_{i \in s_3(u)} \frac{R_i^2}{\pi_i \lambda(K_i)p_i^2} \tag{9.24}$$

We then obtain an estimate of ϵ^2 and of its variance by applying previous techniques with $Y^*(x + e_l) = \epsilon^{*2}(x + e_l)$. Note also that ϵ^2 can be estimated for any new sampling scheme using data from a past or pilot inventory. Simply replace K_i with $K_{i,new}$, π_i with $\pi_{i,old}$ and p_i^2 with $p_{i,old}p_{i,new}$ in Eq. 9.24. Alternatively, we can usually extract from our pilot inventory the empirical distribution of the variable $\frac{Y_i^*}{\lambda(K_i)}$ and $\frac{R_i^2}{\lambda(K_i)p_i}$ so that we can estimate the

the data at the plot level. Consequently, according to the ANOVA table, one can also estimate the lack-of-fit term Λ^2 if the cluster geometry of the pilot inventory is the same as the cluster geometry of the inventory for which we want to predict the anticipated variance (note that the inclusion probabilities appearing in the pure-error terms can differ). **The crucial point is that we do not need to know the Poisson strata to do that**; we only assume that they do exist. If the working strata are available in the form of thematic maps it is possible, using simulations and GIS, to estimate the term $(1 + \theta_{2,g})\beta$ for any cluster structure. This is impossible for the term $(1 + \theta_{1,g})\beta_{1,g}^2$ of the Poisson-strata. However, one can estimate the term β_1^2 from the pilot inventory with clusters of nominal size M: simply compute M ANOVA tables for simple random sampling and pool the resulting estimates of β_1^2, likewise for β_2^2. We might conjecture that

$$1 \leq 1 + \theta_1 \leq 1 + \theta_2 \leq \mathbb{E}M(x) + \frac{\mathbb{V}M(x)}{\mathbb{E}M(x)}$$

In short, one can estimate the lack-of-fit term if the geometry of the cluster is constant or at least work out a plausible scenario if it is not. We shall now proceed to derive the optimal two-phase two-stage sampling scheme under cluster-sampling. As special cases, these can contain one-stage and simple random sampling. To abbreviate the notation we write $\bar{M} = \mathbb{E}M(x)$. The cost constraint is now

$$
\begin{aligned}
C &= n_1 \bar{M}c_1 + \phi(n_2) + n_2 \bar{M}c_{2c} + n_2 \bar{M}\left(c_{21} \sum_{i=1}^{N} \pi_i + c_{22} \sum_{i=1}^{N} \pi_i p_i\right) \\
C &= n_1 \bar{M}c_1 + \phi(n_2) + n_2 \bar{M}c_{2c} + n_2 \bar{M}\left(c_{21} m_1 + c_{22} m_2\right) \quad (9.25)
\end{aligned}
$$

where n_1 is the number of first-phase clusters with unit cost c_1 per plot (usually the time required for the orientation and interpretation of one aerial photograph). Recall that $\bar{M}c_{2c}$ is the installation cost per cluster, including the travel cost within that cluster. For given n_1 and n_2 it is clear from the previous results that we shall choose the first-stage inclusion probabilities according to $\pi_i = \alpha_1 g(Y_i^*)$ and $\pi_i p_i = \alpha_2 \mid R_i \mid$. Using the Lagrange multiplier

$$m_1 = \frac{C - n_1 \bar{M} c_1 - n_2 \bar{M} c_{2c} - \phi(n_2)}{\sqrt{c_{21}} \tilde{\gamma} + \sqrt{c_{22}} \varepsilon} \cdot \frac{\tilde{\gamma}}{n_2 \bar{M} \sqrt{c_{21}}}$$

$$m_2 = \frac{C - n_1 \bar{M} c_1 - n_2 \bar{M} c_{2c} - \phi(n_2)}{\sqrt{c_{21}} \tilde{\gamma} + \sqrt{c_{22}} \varepsilon} \cdot \frac{\varepsilon}{n_2 \bar{M} \sqrt{c_{22}}}$$

$$\lambda(F)\pi_i = m_1 \frac{g(Y_i^*)}{\tilde{\gamma} \bar{Y}}$$

$$\lambda(F)\pi_i p_i = m_2 \frac{|R_i|}{|\bar{R}|}$$

$$\mathbb{E}_\omega \mathbb{V}(\hat{Y}_c^*) = \frac{1}{n_2 \bar{M} m_1} \tilde{\gamma}^2 \bar{Y}^2 + \frac{1}{n_2 \bar{M} m_2} \overline{|R|}^2$$

$$+ \frac{1}{n_2 \bar{M}} \Lambda^2 + \frac{1}{n_1 \bar{M}} (1 + \theta_2) \beta_2^2$$

$$= \frac{\bar{Y}^2 (\sqrt{c_{21}} \tilde{\gamma} + \sqrt{c_{22}} \varepsilon)^2}{C - n_1 \bar{M} c_1 - n_2 \bar{M} c_{2c} - \phi(n_2)}$$

$$+ \frac{1}{n_2 \bar{M}} \Lambda^2 + \frac{1}{n_1 \bar{M}} (1 + \theta_2) \beta_2^2 \qquad (9.26)$$

Remarks:

1. A corresponding result for simple random sampling can be obtained by setting $\mathbb{E} M(x) = 1$, $1 + \theta_1 = 1 + \theta_2 = 1$ and $c_{2c} = c_{20}$. For one-stage sampling set $Y_i^* = Y_i$, $R_i = 0$ and $\varepsilon = 0$, and then replace c_{21} with $c_2 = c_{21} + c_{22}$.

2. The optimal sampling scheme is, as expected, a combination of **PPP PPE**.

3. By linearizing the travel cost it is easy to derive analytically the optimal solution for n_2 and n_1 as well as the absolute lower bound for the anticipated variance (Mandallaz (2001, 2002)). It is more instructive to calculate the optimal n_1 for given n_2 and to plot the resulting anticipated variance as a function of n_2. These are useful when assessing the feasibility (e.g. the size $\lambda(F)\pi_i$ of the inclusion circles and of m_1, m_2) as well as the stability of the optimum, which is done below.

4. Applying the same technique one can also derive the optimal solution under other constraints, such as for a given m_1 or m_2, m_1 and m_2, n_1 optimal for given n_2. The relationships from Eq. 9.26 always remain valid, Mandallaz (2001, 2002).

Setting to zero the partial derivative of the anticipated variance with respect

$$n_{1,opt}(n_2) = \frac{(C - n_2 \bar{M} c_{2c} - \phi(n_2))\sqrt{1 + \theta_2 \Delta_2}}{\bar{M}\sqrt{c_1}(\sqrt{c_{21}}\tilde{\gamma} + \sqrt{c_{22}}\varepsilon + \sqrt{c_1}\sqrt{1 + \theta_2 \Delta_2})} \qquad (9.27)$$

$$m_1 = \frac{C - n_2 \bar{M} c_{2c} - \phi(n_2)}{\sqrt{c_{21}}\tilde{\gamma} + \sqrt{c_{22}}\varepsilon + \sqrt{c_1}\sqrt{1 + \theta_2 \Delta_2}} \frac{\tilde{\gamma}}{n_2 \bar{M}\sqrt{c_{21}}}$$

$$m_2 = \frac{C - n_2 \bar{M} c_{2c} - \phi(n_2)}{\sqrt{c_{21}}\tilde{\gamma} + \sqrt{c_{22}}\varepsilon + \sqrt{c_1}\sqrt{1 + \theta_2 \Delta_2}} \frac{\varepsilon}{n_2 \bar{M}\sqrt{c_{22}}}$$

$$\mathbb{E}_w \mathbb{V}(\hat{Y}_c^*) = \frac{\bar{Y}^2\left(\sqrt{c_{21}}\tilde{\gamma} + \sqrt{c_{22}}\varepsilon + \sqrt{c_1}\sqrt{1 + \theta_2 \Delta_2}\right)^2}{C - n_2 \bar{M} c_{2c} - \phi(n_2)} + \frac{1}{n_2 \bar{M}}\Lambda^2$$

where $\Delta_2 = \frac{\beta_2}{\bar{Y}}$ is the coefficient of variation among working strata. The optimal inclusion probabilities can be obtained from Eq. 9.26 with m_1 and m_2 from Eq. 9.27.

Remarks:

1. As already mentioned, the most difficult aspect is to attain realistic modeling of the travel costs between sample points or clusters (this can involve several components: e.g. travel by car from one fixed point to the next, walking time to reach the plots from the nearest fixed point, lodging expenses and so on). Generally speaking, cluster-sampling is of potential interest in areas with limited accessibility such as mountains or tropical forests. The cluster size should be chosen so that one can perform the field work within a given time unit, e.g. one day, before moving to the next cluster. Furthermore, as we know, the plots within a cluster should be as heterogenous as possible. Usually, it is worthwhile for one to exploit the occasion of walking from one cluster point to the next for conducting related investigations, e.g. transect vegetation sampling. Our previous results are useful qualitative tools for assessing the relative efficiencies of various sampling schemes. Cluster-sampling is performed in Sweden, Finland, France, Canada, the United States, and many tropical countries. Although it has sometimes been used in Switzerland at the national level (e.g. SANASILVA inventory for forest decline), it is, rightly, implemented only very rarely for inventories at the enterprize level.

2. This theory assumes that we have well-defined working strata. Although they may not necessarily be contiguous in the plane, these strata should be large enough to ensure negligible boundary effects. Consequently, we cannot have a very large number of them. If the adopted prediction model is rather complex, and especially if it contains continuous auxiliary variables, one can define the working strata according to the value of those resulting predictions.

3. From a geostatistical perspective, optimizing a sampling scheme is more difficult. Roughly speaking, ignoring the travel costs and assuming that

sible. However, the variogram is hardly ever known and its estimation at small lags is most important. Almost inevitably this necessitates the use of clusters. However, recall that, for global estimation, geostatistical and design-based estimation give very similar results. Optimization is primarily a global problem.

4. Empirical studies previously performed have tended to indicate that the graph of the anticipated variance as a function of the number of terrestrial plots (see the last relationship in Eq. 9.27) is rather flat near its minimum, so that the resulting error is fairly robust with respect to the various parameters in those schemes. In the next chapter, we shall learn that this is indeed the case for the Swiss National Forest Inventory.

5. One can generalize the results to allow for a varying terrestrial sampling density and plot/cluster structure across different sub-domains (regions) F_g of the entire domain F, thereby minimizing the error of the overall estimate for F. This can be relevant for extensive national inventories in which costs and forest structure differ markedly among regions. Such details are provided in Mandallaz (2001, 2002). For the Swiss National Forest Inventory, this more sophisticated optimization is, somewhat surprisingly, not worth the increased complexity, as will be demonstrated in the next chapter.

Extensive simulations with small- and medium-sized real forests have shown that the anticipated variance is an excellent linear predictor of the empirical variance. Simulations with two-phase one-stage systematic cluster-sampling have already been performed with 32 different schemes (1, 2 or 3 circles plus angle count factors with various DBH thresholds) to estimate timber volume per ha (here $\frac{300m^3}{ha}$). There, the regression line of the mean estimated variance (ranging from ca 300 to 600 $(\frac{m^3}{ha})^2$) has an intercept equal to 17 and a slope of 0.91, whereas the R^2 value is excellent, namely .99. Corresponding values for the regression line of the empirical variance (which under systematic sampling can be obtained only via simulations) were $-136, 0.95$ and .96. This illustrates the fact that the estimated variance (calculated under the assumption of random cluster-sampling) usually overestimates the true variance under systematic sampling, in this case by an average of 25% (Mandallaz, 2001). Hence, for a given budget, the ranking of sampling schemes according to their anticipated variances will identify the right candidates. If several schemes display roughly equal anticipated variances, it may be wise to choose the simplest.

So far, we have considered the optimization of a sampling scheme with respect to a single response variable. However, in a multivariate case the situation is clearly more difficult. The present theory is meaningful only if the working strata are the same for the various response variables $Y^{(p)}$ (N.B. the Poisson strata, even if unknown, should not depend on the choice of variable). The various parameters presented in the ANOVA Table 9.1 will be determined

imize a weighted sum of the corresponding anticipated variances. However, in practice, one will have only a few primary variables of interest, so that the choice of the weights is somewhat subjective. From a pragmatic point of view, it might be simpler to derive the optimal schemes for these few variables and find a compromise. The experience gained so far seems to indicate that the lower bound is a very slowly varying function near the optimum, so that there should be enough freedom for suboptimal solutions in the low-dimensional multivariate case.

In the next chapter, optimization procedures will be illustrated using data from the Swiss National Forest Inventory (**SNFI**:http://www.lfi.ch), with our focus there being only on evaluating timber.

9.7 Exercises

Problem 9.1. *Table 9.2 below presents the empirical distribution of DBH values (in cm) for the first Swiss National Forest Inventory discussed later in Chapter 10. The frequency f for each diameter class $[i, i+1)$ is given as a percentage. The histogram for this data set is presented in Figure 10.1. Write a computer program to find a) the optimum DBH thresholds for two and three concentric circles, and b) calculate the corresponding $\tilde{\gamma}$ and $g()$ functions defined in Eq. 9.8 and 9.7 (the easiest way to program this is by using matrix algebra software such as Mathlab, Mapple, Mathematica or SAS/IML). Apply your program to this SNFI data set. The predicted volume V in m^3 is a function of the DBH in cm given by the relationship*

$$V = exp[-12.4 + 3.26log(DBH) - 0.007(log(DBH))^4]$$

Check your results against the solutions presented in Appendix C. If your program is correct, then calculate optimal circles with your own data sets. Are your inventories close to the optimum?

N.B.: *The volume in the last diameter class $\geq$ 98cm- is calculated as if the DBH were always equal that value. However, this approximation is irrelevant for practical purposes. In addition, one may prefer to smooth the data first in order to ensure monotonous decreasing in the higher diameter classes.*

Problem 9.2. *Derive the optimal solution given in Eq. 9.26 and check that it satisfies the cost constraint. Then, linearize the travel costs between clusters by setting $\phi(n_2) = \beta_1 + \beta_2 n_2$ (see Chapter 10 for a practical example). Rewrite the cost constraint as*

$$n_1 \bar{M} c_1 + n_2 \tilde{c}_2 + n_2 \bar{M}(c_{21}m_1 + c_{22}m_2) = \tilde{C}$$

where $\tilde{c}_2 = \beta_2 + \bar{M}c_{2c}$ and $\tilde{C} = C - \beta_1$.

Derive the optimum sample sizes n_1 and n_2 by minimizing the last expression in Eq. 9.26 after having replaced $C - n_1 \bar{M}c_1 - n_2 \bar{M}c_{2c} - \phi(n_2)$ with $\tilde{C}$

Table 9.2 **Distribution of diameter classes in SNFI**

DBH	f	DBH(cm)	f	DBH(cm)	f
12	6.596	41	1.248	70	0.074
13	5.827	42	1.216	71	0.068
14	5.287	43	1.117	72	0.055
15	4.818	44	1.033	73	0.048
16	4.397	45	0.948	74	0.041
17	4.238	46	0.877	75	0.039
18	3.845	47	0.801	76	0.040
19	3.442	48	0.759	77	0.029
20	3.239	49	0.681	78	0.029
21	3.094	50	0.666	79	0.020
22	2.921	51	0.574	80	0.018
23	2.851	52	0.517	81	0.027
24	2.626	53	0.467	82	0.016
25	2.593	54	0.426	83	0.014
26	2.536	55	0.391	84	0.009
27	2.491	56	0.351	85	0.015
28	2.328	57	0.336	86	0.005
29	2.311	58	0.295	87	0.003
30	2.214	59	0.320	88	0.009
31	2.115	60	0.257	89	0.005
32	2.024	61	0.125	90	0.005
33	1.967	62	0.140	91	0.005
34	1.854	63	0.120	92	0.004
35	1.895	64	0.150	93	0.004
36	1.714	65	0.123	94	0.003
37	1.524	66	0.097	95	0.004
38	1.506	67	0.096	96	0.003
39	1.462	68	0.106	97	0.002
40	1.339	69	0.088	≥ 98	0.027

$n_1 \bar{M} c_1 - \tilde{c}_2 n_2$. Find a compact expression for the absolute lower bound of the anticipated variance at the optimum values for n_1 and n_2. How could you use that lower bound in practice?

Problem 9.3. In Problem 9.2 we obtained the optimal solutions when the anticipated variance was minimized under the cost constraint. Now, derive the solution for a dual problem, i.e. when costs are minimized under the constraint of a given anticipated variance W. First, for given sample sizes n_1 and show that the **PPP** and **PPS** rules also hold in this case and determine the expected cost of this optimal solution for a general travel cost function $\phi(n$ Second, linearize the function $\phi(n_2)$ as in Problem 9.2 and find the optimum

with those from Problem 9.2.

CHAPTER 10

The Swiss National Forest Inventory

In this chapter we illustrate the optimization technique, using data based on the first Swiss National Forest Inventory (**SNFI1**;1983-1985) and on the second Swiss National Forest Inventory (**SNFI2**;1993-1995). **SNFI1** involved a one-phase two-stage simple sampling scheme comprising plots with two concentric circles: $200m^2$ and $500m^2$. DBH thresholds were set at $12cm$ $36cm$. In all, $n_2 = 10'974$ plots were available from a $1km \times 1km$ grid with $m_1 \approx 11.7$. Its second-stage procedure samples with essentially equal probability $p_i \approx 0.33$ for $DBH_i < 60cm$ and $p_i = 1$ for $DBH_i \geq 60cm$. This resulted in $m_2 \approx 4$. **SNFI2** consisted of a two-phase two-stage sampling scheme with the same concentric plots, but with only half the original plots from the sub-grid $\sqrt{2}km \times \sqrt{2}km$. Its first phase relied on aerial photographs and was based on a $0.5km \times 0.5km$ grid, resulting in $n_1 = 51'296$, $n_2 = 6'412$, $m_1 \approx$ The second-stage procedure was changed to implement **PPE** with $m_2 \approx$ Switzerland (CH) is divided into $D = 5$ domains: Jura (JU), Swiss Plateau (SP), Pre-Alps (PA), Alps (AL) and Southern Alps (SA). Here, those regions will be identified by the index g. The working strata were based on semi-automatic interpretation of aerial photographs to determine the average tree height (with 25 points per plot). This led to four working strata within each domain: $0m - 10m$, $10m - 20m$, $20m - 30m$ and $> 30m$. The corresponding R^2's fell into the range $0.12 - 0.24$, which is not as good as the usual stand-map stratification with $R^2 \approx 0.4$. However, it required only the aerial photographs available from the Swiss Topographic Survey. The post-stratification procedure analyzed the four working strata individually within each domain, which explains why the $R^2 = 0.27$ for (CH) is higher than the R^2's in each domain. Cost parameters, as well as the p_g, $\overline{Y}_g$ were based on the first inventory, while the estimation of lack of fit and pure error had to be based on the second. Some of the results presented below relied on a more general optimization procedure with terrestrial sampling density varying in the $D = 5$ regions in such a way as to minimize the resulting anticipated variance for Switzerland as a whole. There, the general ideas remained the same but, of course, the formulae were more complicated (Mandallaz, 2001, 2002). However, one can still intuitively understand the results without having to study the technical details involved in those calculations. The results from $D = 1$ (Switzerland as a single region) were obtained via the technique given in Section 9.6.

177

$(\quad_{,g})$ for this particular case. Here, t_g is the mean time required to go, usually by car, from either the lodging facility to the topographic fixed point (marked on the aerial photograph and easily accessible) nearest to the sample plot or from one fixed point to the nearest fixed point. Likewise, c_g, the mean installation time, is the time required to access the sample plot from the nearest fixed point, usually by walking, plus the time required to locate and secure exactly the center of the permanent plot. The total installation time increases approximately linearly with the number of sample points. Let $n_{0g} = 10'974p_g$ the number of terrestrial plots available from **SNFI1** in each region. With a square root law for the travel time from fixed point to fixed point we can write $\phi(n_{2,g}) \approx n_{2,g}c_g + t_g\sqrt{n_{2,g}}\sqrt{n_{0,g}}$. **By making this choice, we insure that the total travel time is the one observed when** $n_{2,g} = n$ Then one linearize $\phi(n_{2,g})$, according to $t_g\sqrt{n_{2,g}}\sqrt{n_{0,g}} = \alpha_g + s_g n_{2,g}$, to obtain $\phi(n_{2,g}) \approx n_{2,g}c_g + s_g n_{2,g} + \alpha_g = n_{2,g}c_{2,g} + \alpha_g$ with $c_{2g} = c_g + s_g$. The R^2's for the six linear regressions are all above 0.97 in their respective ranges. Taking into account the inherent difficulty in assessing travel costs that fit is more than acceptable. Because we must subtract the intercept terms from the overall budget, we set $\tilde{C} = C - \sum_{g=1}^D \alpha_g$. Table 10.1 (based partially on research by Lanz (2000)) presents the parameters needed to calculate the values. Table 10.2 (where $\sqrt{\gamma_\pi}$ is given for easier comparison with the optimal $\tilde{\gamma}$) provides a summary of the parameters required for the optimization. The cost for the orientation and interpretation of an aerial photograph is constant $c_1 = 10$ minutes. Note that all the terrestrial costs are stated in total time units for a crew of two persons (i.e. the effective time spent is only half of the values given).

Table 10.3 shows the proportion of the variance due to pure error, working strata and the lack of fit, as well as the coefficients of variation $\frac{\beta_{2,g}}{\bar{Y}_g}$. All the inflation factors $1 + \theta_{1,g}$, $1 + \theta_{2,g}$ are equal to 1 in this simple random sampling. Pure error terms for the **SNFI** inclusion circles were obtained in each region according to their particular empirical DBH distribution. The lack-of-fit terms were then estimated by reconstructing the various ANOVA tables (according to Table 9.1) from the empirical errors obtained in each region, using the second phase only and the corresponding R^2's.

To illustrate the results from discrete **PPP** the empirical distribution of DBH (as obtained from **SNFI1** for CH) is displayed in Fig. 10.1. Fig. 10.2 presents $\tilde{\gamma}$ as a function of the DBH threshold for the two concentric circles (for variable Y_i^*, which is timber volume based only on the DBH). The minimum is $\tilde{\gamma} = 1.18$ at $DBH = 32cm$. Threshold values for an optimal scheme with three concentric circles are $DBH = 24cm$ and $DBH = 41cm$ with $\tilde{\gamma} = 1$ (see Problem 9.1). In comparison, the scheme with one circle only gives $\tilde{\gamma}$ 1.61. Hence, the efficiency gain from one to two circles is substantial, whereas the gain from two to three circles is marginal if one accounts the increased complexity into account. The SNFI concentric circles were determined on the

Table 10.1 *Installation and travel costs*

	c_g (hrs)	t_g (hrs)	$n_{2,g}$ range	α_g (hrs)	s_g (hrs)	c_{2g} (hrs)
JU	3.25	0.80	400-10,000	1,038	0.27	3.52
SP	3.01	0.74	400-10,000	1,037	0.27	3.28
PA	4.31	1.04	400-10,000	1,386	0.36	4.67
AL	5.20	1.20	600-16,000	2,537	0.41	5.61
SA	6.96	1.44	200- 6,000	1,160	0.52	7.48
CH	4.46	1.04	2,000-20,000	4,950	0.55	5.01

Legend: c_g: installation time, t_g: travel time, α_g: intercept for travel cost, s_g: slope for travel cost, c_{2g}: linearized installation and travel cost per point.

Table 10.2 *Parameters of SNFI*

	p_g (%)	$\overline{N}_g$	c_{2g} (hrs)	c_{21g} (min)	c_{22g} (min)	$\sqrt{\gamma_\pi}$	ε_g (%)	$\overline{Y}_g$ (m^3/ha)	λ_g (%)	R (%)
JU	18	468	3.52	1.9	4.8	1.28	15	328	47	15
SP	21	454	3.28	1.9	4.8	1.26	14	403	49	24
PA	19	508	4.67	2.0	4.8	1.26	16	419	53	20
AL	30	445	5.61	2.1	5.1	1.31	19	292	70	21
SA	12	425	7.48	2.1	5.1	1.48	22	178	72	12
CH	100	460	5.01	2.0	5.0	1.27	16	332	57	27

Legend: p_g: relative surface area, $\overline{N}_g$: stem density, c_{2g}: linearized installation and travel cost per point, c_{21g}: unit cost per first-stage tree, c_{22g}: unit cost per second-stage tree, $\sqrt{\gamma_\pi}$: from Equation (9.13), ε_g: relative prediction error at tree level, $\overline{Y}_g$: timber volume per ha, $\lambda_g = \frac{\Lambda_g}{\overline{Y}}$: relative lack of fit, R_g^2: coefficient of determination.

basis of empirical studies performed at the end of the 1970s (in the district of Nidwald). Although they do not correspond to the optimal $g()$ function, they did serve their purpose very well indeed.

Table 10.4 gives the key parameters for the optimal two concentric circles.

Table 10.5 gives the parameters for the optimal sampling scheme with two concentric circles (for an overall budget of $C = 44'307 hrs$ or $\tilde{C} = 37'149 hrs$ equal to the variable costs obtained with **SNFI2**). In contrast, Table 10.6 presents the optimal sampling scheme for Switzerland consisting of the D domain only, while Table 10.7 adds the constraint $m_1 = 11.7$, which is the

Table 10.3 *Components of variance in SNFI*

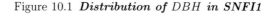

	$\dfrac{\epsilon_g^2}{\mathbb{V}_g Y(u)}$ (%)	$\dfrac{\mathbb{V}_g \widehat{Y}(u)}{\mathbb{V}_g Y(u)} = R_g^2$ (%)	$\dfrac{\Lambda_g^2}{\mathbb{V}_g Y(u)}$ (%)	$\dfrac{\beta_{2,g}}{\overline{Y}_g}$ (%)
JU	35	15	50	26
SP	28	24	48	35
PA	26	20	54	32
AL	20	21	59	42
SA	29	12	59	32
CH	24	27	49	42

Legend: $\dfrac{\epsilon_g^2}{\mathbb{V}_g Y(u)}$: pure error, $\dfrac{\mathbb{V}_g \widehat{Y}(u)}{\mathbb{V}_g Y(u)} = R_g^2$: working strata, $\dfrac{\Lambda_g^2}{\mathbb{V}_g Y(u)}$: lack of fit, $\dfrac{\beta_{2,g}}{\overline{Y}_g}$: coefficient of variation among working strata.

Figure 10.1 *Distribution of DBH in SNFI1*

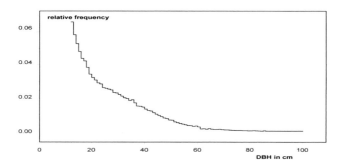

Legend: f relative frequency in %, *DBH* diameter class in *cm*.

average number of first-stage trees per plot observed in **SNFI1**. The overall budget is, of course, the same for all schemes.

Table 10.8 displays the relative empirical and anticipated errors for **SNI2** and for the various optimal sampling schemes.

Recall that the anticipated errors are based on **SNI1** data except for the lack-of-fit terms, which had to be based on **SNI2**. Therefore, the perfect agreement between the first two columns in Table 10.8 is tautological, it simply demonstrates that, up to very small rounding errors, the calculations are consistent. Fig. 10.3 displays the relative anticipated error for an optimal scheme with one domain as a function of the number n_2 of terrestrial plots when the number

Figure 10.2 *Gamma values for two concentric circles according to DBH threshold in cm*

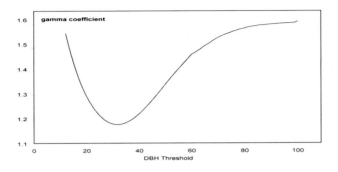

Table 10.4 *Optimal concentric circles*

	threshold (cm)	$\tilde{\gamma}$	$g_{1g}(Y_i^*)$ (m^3)	$g_{2g}(Y_i^*)$ (m^3)
JU	31	1.18	0.33	2.00
SP	31	1.17	0.33	2.07
PA	32	1.17	0.36	2.18
AL	32	1.19	0.35	2.29
SA	31	1.25	0.28	2.44
CH	32	1.18	0.36	2.18

Legend: $g_{1g}(Y_i^*)$ and $g_{2g}(Y_i^*)$: values of the step functions $g_g()$ below and above the threshold.

n_1 of aerial photographs is optimal for a given n_2. Likewise, Fig. 10.4 shows the relative error for an optimal scheme with one domain and $m_1 = 11.7$, that is when the number of first-stage trees is equal to the observed **SNI1** value.

Table 10.5 *Optimal sampling scheme with five domains*

	small circle (m^2)	large circle (m^2)	m_{1g}	m_{2g}	$n_{1,g}$	$n_{2,g}$	$\frac{n_{2,g}}{n_{1,g}}$ (%)	error (%)
JU	226	1,368	26.5	2.1	4,154	999	24	1.73
SP	170	1,067	24.3	1.8	4,846	1,547	32	1.49
PA	192	1,162	26.1	2.3	4,385	1,319	30	1.69
AL	217	1,418	21.5	2.2	6,924	1,749	25	1.88
SA	319	2,783	25.4	2.9	2,769	380	14	4.01
CH					23,079	5,995		0.85

Legend: m_{1g}: number of first-stage trees per point, m_{2g}: number of second-stage trees per point, $n_{1,g}$: number of first-phase points, $n_{2,g}$: number of second-phase points.

Table 10.6 *Optimal sampling scheme with one domain*

	small circle (m^2)	large circle (m^2)	m_1	m_2	n_1	n_2	$\frac{n_2}{n_1}$ (%)	error (%)
CH	207	1,393	25.6	2.2	23,668	5,859	25	0.86

Legend: m_1: number of first-stage trees per point, m_2: number of second-stage trees per point, n_1: number of first-phase points, n_2: number of second-phase points.

Table 10.7 *Optimal sampling scheme with one domain and $m_1 = 11.7$*

	small circle (m^2)	large circle (m^2)	m_1	m_2	n_1	n_2	$\frac{n_{2,g}}{n_{1,g}}$ (%)	error (%)
CH	95	636	11.7	1.9	22,878	6,393	28	0.89

Legend: m_1: number of first-stage trees per point, m_2: number of second-stage trees per point, n_1: number of first-phase points, n_2: number of second-phase points.

Table 10.8 *Empirical and anticipated relative errors in %*

	SNI2 e.e.	SNI2 a.e.	$D = 5$ a.e.	$D = 1$ a.e.	$D = 1$ $m_1 = 11.7$ a.e.
JU	1.81	1.81	1.73	1.70	1.80
SP	1.72	1.71	1.49	1.64	1.73
PA	1.88	1.87	1.69	1.81	1.84
AL	1.88	1.88	1.87	1.86	1.86
SA	3.18	3.18	4.01	3.06	3.15
CH	0.89	0.89	0.85	0.86	0.89

Legend: e.e. empirical relative error, a.e. anticipated relative error.

Figure 10.3 *Relative anticipated error with one domain*

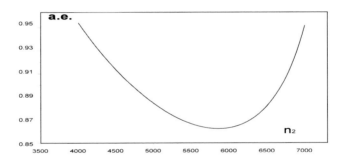

Legend: a.e. relative anticipated error in %, n_2 number of terrestrial plots.

Remarks:

1. The optimal scheme with $D = 5$ domains yields surface areas for the large circles that are too great for field work because of complex boundary adjustments and slope corrections, especially in the Alps. With respect to **SNI2** the relative error is decreased by 3.5%, or, equivalently, the cost is reduced by 7%, which does not justify this increased complexity.

2. Essentially the same can be said for the simplified optimal scheme with $D = 1$.

3. The optimal scheme with one domain and $m_1 = 11.7$ is very close to **SNI2** except for the number of aerial photographs and the surface area of the

Figure 10.4 *Relative anticipated error with one domain and* $m_1 = 11.7$

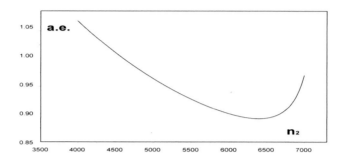

Legend: a.e. relative anticipated error in %, n_2 number of terrestrial plots

small circle, both of which are halved. Errors are the same. In this sense, the **SNI** is remarkably "optimal."

4. In practical terms the optimum is rather "flat" with respect to the resulting error, which remains within narrow bounds over a large range of n_2. In contrast, the characteristics of the design vary much more.

5. The only feasible possibility for substantially increasing the efficiency of **SNI** is to reduce the lack-of-fit term by improving the prediction model based on the aerial photographs, thus increasing the R^2 from 0.20 to 0.30 0.40.

CHAPTER 11

Estimating change and growth

A forest is a dynamic system whose characteristics vary over time. Williams (1998), Kangas and Maltamo (Eds, 2006) and Köhl et al. (2006) have provided general introductions to the subject of estimating periodic changes and growth in the forest. Here, we restrict our attention to spatial means at time points $t_1 < t_2$. To simplify the notation we shall consider only one response variable in order to drop the upper index. A particular forest F with area $\lambda(F$ assumed to be the same at both time points. Of course, this area is itself a key issue for monitoring over time, but the temporal change in F is assessed by different means (e.g. remote sensing or GIS).

To clearly define the problem we must use notation that is a bit cumbersome and somewhat redundant. However, the advantage here is one of absolute accuracy, which is unfortunately not always the case in the literature. Therefore, we will replace the real forest over F with a statistically equivalent idealized version, in the following sense. A tree is said to exist at time t_k if its height is greater than a value H_0, not necessarily the breast height of $1.3m$ (e.g. 10 this threshold is largely irrelevant for the actual calculations). Furthermore, we shall assume that trees that die or are harvested in the time interval $[t_1$ are not removed but stay in the position they were when alive. Dead or harvested trees with height above H_0 therefore exist and we ignore trees which died or were harvested prior to time t_1, no matter whether they are still in the forest (possibly on the ground) or not. The population $\mathcal{P}$ then consist of all trees existing at times t_1 or t_2, **identified by a unique label** $i = 1, 2, \ldots N$ The following variables are used:

1. $E_i(t_k) = 1$ if the ith tree exists at time t_1, otherwise $E_i(t_k) = 0$.

2. $A_i(t_k) = 1$ if the ith tree is alive at time t_k, otherwise $A_i(t_k) = 0$. A tree with $A_i(t_1) = 1$ and $A_i(t_2) = 0$, i.e. that died or is harvested after t_1 before t_2, is called a **depletion tree**.

3. $Y_i(t_k)$ is the response variable of interest (e.g. volume) of the ith tree at time t_k, provided that $E_i(t_k) = 1$. Note that $Y_i(t_k)$ could be set to zero if the tree is below an arbitrary threshold (e.g. merchantable size).

4. $T_i(t_k) = 1$ if the DBH of the ith tree is above ($\geq$) our **inventory threshold** and $T_i(t_k) = 0$ otherwise, provided that $E_i(t_k) = 1$. This indicator could obviously be defined with a variable other than DBH.

instance

- $E_i(t_1) = E_i(t_2) = 0$ is impossible by definition.
- if $E_i(t_1) = 1$ then $E_i(t_2) = 1$.
- if $E_i(t_1) = 1$ and $A_i(t_2) = 1$ then $A_i(t_1) = 1$.
- if $E_i(t_1) = 1$ and $T_i(t_1) = 1$ then $T_i(t_2) = 1$.
- $E_i(t_1)E_i(t_2)A_i(t_1)A_i(t_2)T_i(t_1)T_i(t_2) = E_i(t_1)A_i(t_2)T_i(t_1)$.

Such relationships are elementary logical consequences of our definitions and will be used tacitly in many algebraic manipulations. We shall consider the following quantities:

$$\bar{Y}(t_1) = \frac{1}{\lambda(F)} \sum_{i=1}^{N} E_i(t_1)A_i(t_1)T_i(t_1)Y_i(t_1)$$

$$\bar{Y}(t_2) = \frac{1}{\lambda(F)} \sum_{i=1}^{N} E_i(t_2)A_i(t_2)T_i(t_2)Y_i(t_2)$$

$$\Delta_{1,2} = \bar{Y}(t_2) - \bar{Y}(t_1) \tag{11.1}$$

Note that $\bar{Y}(t_k)$ entails only trees which exist, are alive and above the threshold at t_k. Quantity $\Delta_{1,2}$ is called the **net change** over the time interval $[t_1, t$ Obviously, one can consider three or more time points $t_1 < t_2 < t_3$ and extend those previous definitions.

Suppose that $\bar{Y}(t_k)$ is estimated by any of the inventory techniques presented so far, thereby yielding $\hat{\bar{Y}}(t_k)$ and $\hat{\mathbb{V}}(\hat{\bar{Y}}(t_k))$. Moreover, presume that the two inventories are totally independent of each other, in particular when we have two one-phase terrestrial inventories with **temporary** plots, such that the sample plots from each inventory are used only once and are not revisited. Then, one can estimate unbiasedly the net change by

$$\hat{\Delta}_{1,2} = \hat{\bar{Y}}(t_2) - \hat{\bar{Y}}(t_1) \tag{11.2}$$

and its variance by

$$\hat{\mathbb{V}}(\hat{\Delta}_{1,2}) = \hat{\mathbb{V}}(\hat{\bar{Y}}(t_1)) + \hat{\mathbb{V}}(\hat{\bar{Y}}(t_2)) \tag{11.3}$$

Hence, the problem allows for a simple solution. However, the drawbacks of estimating change with independent inventories are that 1) the resulting variance is usually large (the correlation between forest characteristics in the same point in space at two time points cannot be used to reduce that variance), and 2) it does not permit a detailed analysis of changes at the plot or tree level. For these reasons, inventories are frequently used in which parts or all of the terrestrial plots are revisited. First, we shall consider the case where the same one-phase one-stage simple random sampling design is used in both occasions and all the plots are revisited (after eventually taking into account changes

). Thus, we would have an inventory with **plots**. This entails some technical difficulties and we shall need the following concepts:

1. **Survivor trees**

$$\mathbb{S} = \{i \in \mathcal{P} \mid E_i(t_1) = E_i(t_2) = 1, T_i(t_1) = T_i(t_2) = 1, \; A_i(t_1) = A_i(t_2) = 1$$

Trees that exist, are alive and above the threshold at both time points.

2. **In-growth trees**

$$\begin{aligned} \mathbb{I} \;=\; & \{i \in \mathcal{P} \mid E_i(t_1) = 1, T_i(t_1) = 0, A_i(t_2) = 1, T_i(t_2) = 1\} \\ & \cup \{i \in \mathcal{P} \mid E_i(t_1) = 0, E_i(t_2) = A_i(t_2) = T_i(t_2) = 1\} \end{aligned}$$

Trees that either exist, are alive and below the threshold at t_1, but which are alive and above the threshold at t_2, or those that do not exist at t_1 grow fast enough to exist, be alive and above the threshold at t_2.

3. **Depletion trees**

$$\mathbb{D} = \{i \in \mathcal{P} \mid E_i(t_1) = 1, A_i(t_1) = 1, A_i(t_2) = 0, T_i(t_1) = 1\}$$

Trees that exist, are alive and above the threshold at t_1 but which have since died or have been cut or harvested prior to t_2 (i.e. they are not alive at t_2).

The above categories are not exhaustive. For instance they do not include trees with $E_i(t_1) = T_i(t_1) = 0$, $E_i(t_2) = T_i(t_2) = 1$, $A_i(t_2) = 0$. However, they are the only relevant ones for the quantities defined in Eq. 11.1. We now define the following components of **in-growth, survivor growth depletion**, as originally proposed by Beers (1962).

$$I_{1,2} \;=\; \frac{1}{\lambda(F)} \sum_{i=1}^{N} (1 - E_i(t_1)) E_i(t_2) A_i(t_2) T_i(t_2) Y_i(t_2) \qquad (11.4)$$

$$+ \frac{1}{\lambda(F)} \sum_{i=1}^{N} E_i(t_1) E_i(t_2) A_i(t_1) A_i(t_2) T_i(t_2) (1 - T_i(t_1)) Y_i(t_2$$

$$= \frac{1}{\lambda(F)} \sum_{i=1}^{N} E_i(t_2) A_i(t_2) T_i(t_2) Y_i(t_2) T E_i(t_1)$$

$$\Delta S_{1,2} \;=\; \frac{1}{\lambda(F)} \sum_{i=1}^{N} E_i(t_1) A_i(t_2) T_i(t_1) \Big(Y_i(t_2) - Y_i(t_1) \Big)$$

$$D_{1,2} \;=\; \frac{1}{\lambda(F)} \sum_{i=1}^{N} E_i(t_1) A_i(t_1) (1 - A_i(t_2)) T_i(t_1) Y_i(t_1)$$

where $TE_i(t_1) = E_i(t_1)(1 - T_i(t_1)) + (1 - E_i(t_1))$. It is important to realize that the depletion component as defined above does not take into account the growth of trees from time t_1 until their time of death or harvesting $\tau_i \in (t_1, t$

Tedious but elementary algebraic manipulations (along with those aforementioned relationships) lead to the following fundamental result, which is also intuitively evident

$$\bar{Y}(t_2) = \bar{Y}(t_1) + I_{1,2} + \Delta S_{1,2} - D_{1,2} \tag{11.5}$$

At this point we encounter the first disturbing complication : by construction, we have **time additivity** with respect to the true net changes, that is

$$\Delta_{1,3} = (\bar{Y}(t_3) - \bar{Y}(t_2)) + (\bar{Y}(t_2) - \bar{Y}(t_1)) = \Delta_{1,2} + \Delta_{2,3} \tag{11.6}$$

this is no longer true for the corresponding components defined in Eq. 11.4. For instance, with all $E_i(t_k) = 1$ and $A_i(t_k) = 1$, we see that for trees with $T_i(t_1) = 0$ and $T_i(t_2) = T_i(t_3) = 1$ their contribution to $I_{1,3}$ is $Y_i(t_3)$, whereas the contribution to $I_{1,2} + I_{2,3}$ is $Y_i(t_2)$. The only way to bypass this problem is to consider the forest in continuous time. This, of course, assumes that one can determine the $Y_i(t)$ at the time the tree died or was harvested, or at the time $T_i(t) = 1$ for any $t \in [t_1, t_3]$, which may be a difficult task in practice. Such a model was originally developed by Eriksson (1995), and a simplified presentation has been provided by Gregoire and Valentine (2007). This increased complexity is usually not worth the effort in standard inventories performed at 10 or 20 years interval, essentially because the growth components are primarily used as rough guidelines for forest management.

We now consider the problem of estimating the various components defined in Eq. 11.4. Usually this is done for inventories performed with either one fixed inclusion circle or via the angle count technique (Schreuder et al., 1993). We shall take a slightly more general approach here: the one-phase one-stage sampling scheme is the same at both time points t_1 and t_2, but otherwise the inclusion probabilities at the tree level are arbitrary. This includes schemes with one or an arbitrary number of circles (in most instances 2 or 3) as well as horizontal point sampling (angle count). The sets of trees $\mathbb{S}$, $\mathbb{I}$, $\mathbb{D}$ previously defined do not depend on whether trees are included in the sample or not. However, for any tree in one of these sets, it can be sampled at time t_k or not.

Let us now introduce the inclusion indicator variables $I_i(x, t_k) = 1$, if the tree is contained in the sample at point x and at time t_k, $I_i(x, t_k) = 0$ otherwise. We have the corresponding inclusion probabilities $\pi_i(t_k) = \mathbb{E}I_i(x, t$ Note that, by construction, we have the following trivial but important relationships

- if $I_i(x, t_1) = 1$ then $I_i(x, t_2) = 1$
- $I_i(x, t_1)I_i(x, t_2) = I_i(x, t_1)$

Trees with $A_i(t_2) = 1$ sampled on at least one occasion can be classified into the following relevant categories:

1. **Sample survivor trees**

$$i \in \mathbb{S}, \ E_i(t_1) = 1, \ T_i(t_1) = T_i(t_2) = 1, \ I_i(x, t_1) = I_i(x, t_2) = 1$$

2. Sample in-growth trees

$$i \in \mathbb{I}, \ E_i(t_1) = 1, \ T_i(t_1) = 0, \ I_i(x, t_1) = 1, \ T_i(t_2) = 1, \ I_i(x, t_2) = 1$$

Trees below the threshold but sampled on the first occasion, then above the threshold and sampled again on the second occasion, or

$$i \in \mathbb{I}, \ E_i(t_1) = 0, \ E_i(t_2) = 1, \ T_i(t_2) = 1, \ I_i(x, t_2) = 1$$

trees that did not exist on the first occasion, but which now exist, are alive, above the threshold and are sampled on the second occasion.

3. Sample on-growth trees

$$i \in \mathbb{I}, \ E_i(t_1) = 1, \ T_i(t_1) = 0, \ I_i(x, t_1) = 0, \ T_i(t_2) = 1, \ I_i(x, t_2) = 1$$

Trees below the threshold and not sampled on the first occasion, above the threshold and sampled on the second.

4. Sample non-growth trees

$$E_i(t_1) = 1, \ T_i(t_1) = 1, \ I_i(x, t_1) = 0, \ T_i(t_2) = 1, \ I_i(x, t_2) = 1$$

Trees above the threshold but not sampled on the first occasion, but still above the threshold and sampled on the second occasion.

This widespread terminology is injudicious, particularly for non-growth trees, which are a problem precisely because they do grow! Note that such non-growth trees cannot occur if the sampling scheme is based on a single inclusion circle, whereas they can, of course, occur when angle count or concentric circles are applied.

We also define the following estimates at the plot level (local densities):

$$
\begin{aligned}
Y(x, t_k) &= \frac{1}{\lambda(F)} \sum_{i=1}^{N} E_i(t_k) A_i(t_k) T_i(t_k) Y_i(t_k) \frac{I_i(x, t_k)}{\pi_i(t_k)} \\
\Delta_{1,2}(x) &= Y(x, t_2) - Y(x, t_1) \tag{11.7}
\end{aligned}
$$

which by construction are obviously unbiased estimates of $\bar{Y}(t_k)$ and Δ For a sample s_2 of n_2 points $x \in s_2$ that are uniformly and independently distributed in F one can use the point estimates and estimated variance as for ordinary local densities, i.e. via Theorem 4.2.1.

We now proceed to estimate growth components. Let us first consider the in-growth component. If one considers the sample in-growth trees as defined above we are led to the estimate

$$
\begin{aligned}
I_2(x) &= \frac{1}{\lambda(F)} \sum_{i=1}^{N} E_i(t_1) A_i(t_2)(1 - T_i(t_1)) T_i(t_2) Y_i(t_2) \frac{I_i(x, t_1) I_i(x, t_2)}{\pi_i(t_2)} \\
&\quad + \frac{1}{\lambda(F)} \sum_{i=1}^{N} (1 - E_i(t_1)) E_i(t_2) A_i(t_2) T_i(t_2) Y_i(t_2) \frac{I_i(x, t_2)}{\pi_i(t_2)} \tag{11.8}
\end{aligned}
$$

( ).

Because $I_i(x, t_1)I_i(x, t_2) = I_i(x, t_1)$ we see that $\mathbb{E}_x I_2(x) \neq I$. To obtain an unbiased estimate of ingrowth we must also include those sample on-growth trees. That is, we calculate

$$O(x) = \frac{1}{\lambda(F)} \sum_{i=1}^{N} E_i(t_1) A_i(t_2)(1 - T_i(t_1))(1 - I_i(x, t_1))T_i(t_2)Y_i(t_2)\frac{I_i(x, t_2}{\pi_i(t_2)}$$

(11.9)

We then define the ingrowth estimate $I_{1,2}^{(1)}(x)$ as

$$
\begin{aligned}
I_{1,2}^{(1)}(x) &= I_2(x) + O(x) &(11.10)\\[2mm]
&= \frac{1}{\lambda(F)} \sum_{i=1}^{N}(1 - E_i(t_1))E_i(t_2)A_i(t_2)T_i(t_2)Y_i(t_2)\frac{I_i(x, t_2)}{\pi_i(t_2)}\\[2mm]
&+ \frac{1}{\lambda(F)} \sum_{i=1}^{N} E_i(t_1)A_i(t_2)(1 - T_i(t_1))T_i(t_2)Y_i(t_2)\frac{I_i(x, t_2)}{\pi_i(t_2)}
\end{aligned}
$$

By elementary algebra and Eq. 11.4 we obtain $\mathbb{E}_x I_{1,2}^{(1)}(x) = I_{1,2}$ Hence, $\mathbb{E}_x I_{1,}^{(1)}$ is an unbiased estimate of the ingrowth component $I_{1,2}$.

The estimate from Eq. 11.10 is sometimes called the "revised Purdue" estimator in the literature (Schreuder et al., 1993; Gregoire, 1995, 1993).

$I_{1,2}^{(1)}(x)$ is not as simple as it looks for use in field work. The first component is due to trees that did not exist at t_1, but which are then above the threshold and sampled from point x at t_2. Such "very fast growing" trees are usually rare and can be identified, as long as a map is available of trees existing in the neighborhood of x at t_1. Furthermore, the second component may require us to know $T_i(t_1)$ and therefore $Y_i(t_1)$ for trees not sampled at t_1. In other words, it might be necessary to predict backwards, as a function of $Y_i(t_2)$, the value $Y_i(t_1)$ for trees sampled at t_2 but not at t_1. This will induce a further error that is difficult to assess and generally ignored. Alternatively, one can use an increment borer, which is expensive, however, and also not error free.

We now consider the following estimator of the survivor growth $\Delta S_{1,2}$ suggested by Martin (1982):

$$\Delta S_{1,2}^{(1)}(x) = \frac{1}{\lambda(F)} \sum_{i=1}^{N} E_i(t_1)A_i(t_2)T_i(t_1)(Y_i(t_2) - Y_i(t_1))\frac{I_i(x, t_1)}{\pi_i(t_1)}$$

(11.11)

which, by construction, is clearly unbiased, i.e. $\mathbb{E}_x \Delta S_{1,2}^{(1)}(x) = \Delta S_{1,2}$ and unproblematic for field work.

Finally, we have the obvious unbiased estimate of depletion

$$D_{1,2}^{(1)}(x) = \frac{1}{\lambda(F)} \sum_{i=1}^{N} E_i(t_1)A_i(t_1)(1 - A_i(t_2))T_i(t_1)Y_i(t_1)\frac{I_i(x, t_1)}{\pi_i(t_1)}$$

(11.12)

$$e_{1,2}^{(1)}(x) = \Delta_{1,2}(x) - \left(I_{1,2}^{(1)}(x) + \Delta S_{1,2}^{(1)}(x) - D_{1,2}^{(1)}(x) \right)$$

From the previous equations one has additivity in the mean, that is

$$\mathbb{E}_x e_{1,2}^{(1)}(x) = 0$$

However, simple but tedious calculations lead to the surprising result

$$e_{1,2}^{(1)}(x) = \frac{1}{\lambda(F)} \sum_{i=1}^{N} E_i(t_1) A_i(t_2) T_i(t_1) Y_i(t_2) \left(\frac{I_i(x,t_2)}{\pi_i(t_2)} - \frac{I_i(x,t_1)}{\pi_i(t_1)} \right) \quad (11.13)$$

In other words, we have **spatial additivity in the mean** and generally **non-additivity at a given point** x. This is a troubling feature indeed, even if one should not overemphasize its importance, as is done for non-additivity with respect to time intervals. Note that non-growth trees contribute to $e_{1,2}^{(1)}$ which is equal zero under sampling schemes based on one circle.

To achieve additivity at all points x, while retaining unbiasedness, one can for instance add the zero mean error term $e_{1,2}^{(1)}(x)$ to either $I_{1,2}^{(1)}(x)$ or $\Delta S_{1,2}^{(1)}$. The first possibility, as proposed by Martin (1982) leads to

$$I_{1,2}^{(2)}(x) = I_{1,2}^{(1)}(x) + e_{1,2}^{(1)}(x) \quad (11.14)$$

The triplet $[I_{1,2}^{(2)}(x), \Delta S_{1,2}^{(1)}(x), D_{1,2}^{(1)}(x)]$ is compatible (additive) in the sense that

$$\Delta_{1,2}(x) = I_{1,2}^{(2)}(x) + \Delta S_{1,2}^{(1)}(x) - D_{1,2}^{(1)}(x)$$

However, $I_{1,2}^{(2)}(x)$ also has problematic features. Because it includes the term $e_{1,2}^{(1)}(x)$, we see that trees not in $\mathbb{I}$ can contribute to it. This is particularly true for non-growth trees, and this term can sometimes take negative values.

The second possibility, suggested by Deusen et al. (1986) leads to

$$\Delta S_{1,2}^{(2)}(x) = \Delta S_{1,2}^{(1)}(x) + e_{1,2}^{(1)}(x) \quad (11.15)$$

$$= \frac{1}{\lambda(F)} \sum_{i=1}^{N} E_i(t_1) A_i(t_2) T_i(t_1) \left(Y_i(t_2) \frac{I_i(x,t_2)}{\pi_i(t_2)} - Y_i(t_1) \frac{I_i(x,t}{\pi_i(t_1} \right)$$

The triplet $[I_{1,2}^{(1)}(x), \Delta S_{1,2}^{(2)}(x), D_{1,2}^{(1)}(x)]$ is compatible and, therefore,

$$\Delta_{1,2}(x) = I_{1,2}^{(1)}(x) + \Delta S_{1,2}^{(2)}(x) - D_{1,2}^{(1)}(x)$$

$\Delta S_{1,2}^{(2)}(x)$ is an unbiased estimate of $\Delta S_{1,2}$. Although such a computation does not require back-prediction for trees sampled at t_2 but not at t_1, it also can take negative values.

For completeness, let us assess a third possibility, as put forth by Roesch et al.

(1989). This rests upon a slightly modified version of the error term $_{1,2}$ and of $\Delta S_{1,2}^{(1)}(x)$. Set

$$e_{1,2}^{(2)}(x) = \frac{1}{\lambda(F)} \sum_{i=1}^{N} E_i(t_1) A_i(t_2) T_i(t_1) Y_i(t_1) \left(\frac{I_i(x,t_2)}{\pi_i(t_2)} - \frac{I_i(x,t_1)}{\pi_i(t_1)} \right)$$

which has a zero mean. Note that non-growth trees can contribute to it and that it is always zero under sampling schemes based on one circle. We define the new unbiased estimators as

$$I_{1,2}^{(3)}(x) = I_{1,2}^{(1)}(x) + e_{1,2}^{(2)}(x) \tag{11.16}$$

$$\Delta S_{1,2}^{(3)}(x) = \frac{1}{\lambda(F)} \sum_{i=1}^{N} E_i(t_1) A_i(t_2) T_i(t_1)(Y_i(t_2) - Y_i(t_1)) \frac{I_i(x,t_2)}{\pi_i(t_2)}$$

Elementary but tedious algebra shows that

$$\Delta_{1,2}(x) = I_{1,2}^{(3)}(x) + \Delta S_{1,2}^{(3)}(x) - D_{1,2}^{(1)}(x)$$

so that we have additivity. Note that both $I_{1,2}^{(3)}(x)$ and $\Delta S_{1,2}^{(3)}(x)$ can require back-predictions.

One can also ensure additivity by changing the estimates $Y(x,t_k)$ (Roesch et al., 1989).

Generally speaking, additivity is not a great concern for regional or national inventories, where changes in F are far more important. Simplicity, unbiasedness and efficiency of the estimates are more crucial there. From a logical perspective, the triplet $[I_{1,2}^{(1)}(x), \Delta S_{1,2}^{(1)}(x), D_{1,2}^{(1)}(x)]$, though non additive, seems to be more convincing.

Since all the previous estimates were defined as local densities the generalization to one-phase cluster-sampling is immediate. We calculate the mean values over the clusters and weight them by the number of points in those clusters, in exactly the same manner as described in Theorem 4.3.1.

The generalization to two-stage sampling is also straightforward, at least in principle. According to Eq. 4.25 we replace $I_i(t_k)Y_i(t_k)$ with

$$I_i(x,t_k)Y_i^*(t_k) + \frac{I_i(x,t_k)J_i(x,t_k)R_i(t_k)}{p_i(t_k)}$$

It is easier to implement second-stage sampling at t_2 independent of the second-stage sampling at t_1. Back-prediction may be required not only for the $Y_i^*(t_k)$ but also for the residual $R_i(t_k)$. A simpler alternative is to calculate the net change and its various components with the $Y_i^*(t_k)$ alone, in order to express them as a proportion of the net change and then, finally, to use those proportions on the net change as calculated from the generalized densities. Two-stage sampling and the associated adjustment with the residuals is particularly interesting because of the bias reduction in local estimations

Table 11.1 *Components of change*

variable	$\hat{\bar{Y}}(t_1)$	$\hat{\bar{Y}}(t_2)$	$\hat{\Delta}_{1,2}$	$\hat{\Delta}S^{(1)}$	$\hat{I}_{1,2}^{(1)}$	$\hat{D}_{1,2}$
estimate m^3/ha	337	362	25	76	15	67
error in %	1	1	7	1.4	3	3

Legend: t_1: SNFI1, t_2: SNFI2, $t_2 - t_1 = 10$ years, joined surface area: $\lambda(F$ $0.9810^6 ha$, number of plots: 4980, estimates based on two-stage sampling.

(global models can be locally inadequate). Table 11.1 presents data from the first and second Swiss National Forest Inventories that illustrate this theory.

We see that the theoretical equation $25 = \Delta_{1,2} = \Delta S_{1,2} + I_{1,2} - D_{1,2} = 76 + 15$ $67 = 24$ is not quite satisfied, even though the discrepancy is not statistically significant. This is due to the inherent difficulty in assessing ingrowth and depletion, for which backward predictions are often necessary. On the whole, the agreement is quite good.

There is some concern that permanent plots may become no longer representative over time, particularly if those plots and their trees are somehow made visible, which is often done to facilitate fieldwork but which is wrong from a statistical point of view. Here, the representativeness of **SNFI2** was assessed by comparing the approximately initial 6000 plots with a set of 600 new plots. There was no evidence for significant differences.

An alternative to establishing permanent plots only is the so-called **sampling with partial replacement**, a procedure, in which a fraction of the original plots is revisited on the second occasion, during which new plots are also added. Using forward and backward predictions it is possible to derive optimal estimators. Details of those calculations and advanced versions of the procedures are provided by Ware and Cunia (1962), Newton et al. (1974), Deusen (1989) and de Vries (1986). The formulae are complicated but manageable for two occasions, very elaborate for three and essentially intractable beyond. This is probably why that technique was very popular at its beginning but is far less so now.

Extensive inventories are also difficult to implement from an aspect of logistics and organization. For this reason, there is a growing interest in re-distributing the work of a periodic inventory (e.g. every 10 years) over annual regional surveys, (Scott and Köhl, 1994).

Problem 11.1. *Table 11.2 below presents all the candidate trees at point and times t_1 and t_2 when the sampling technique is an angle count with basal area factor $k = 4\frac{m^2}{ha}$. Tree number 10 has been harvested within the interval $[t_1, t_2]$. Classify the other trees into the categories "sample survivor", "sample in-growth", "sample on-growth", and "sample non-growth". Then, determine the change $\Delta_{1,2}(x)$ and the various growth component triplets*

$$[I_{1,2}^{(k)}(x), \ \Delta S_{1,2}^{(l)}(x), \ D_{1,2}^{(1)}(x)]$$

defined in Chapter 11 when the response variable $Y_i(t_k)$ is the basal area in m^2 and when all trees are interior trees, that is $K_i \subset F$. In particular, assess the additivity of those growth components. The inventory threshold for DBH will be 12cm.

Table 11.2 **Candidate trees at point** x

Tree nr.	$d_i(x)$	$DBH_i(t_1)$	$DBH_i(t_2)$
1	10	60	70
2	8	40	51
3	3	20	28
4	2	10	14
5	1	8	12
6	3	10	15
7	2.5	8	12
8	5.5	20	24
9	10	36	48
10	8	40	–

Legend: $d_i(x)$, distance from the ith tree to point x in m, $DBH_i(t_k)$, DBH in cm at time t_k.

CHAPTER 12

Transect-Sampling

12.1 Generalities

Transect-sampling, or more precisely line intersect-sampling, is a relatively simple and widely used technique in ecological work and forest inventories. To get its flavor let us consider two standard forestry examples. Suppose that one needs a quick estimate of the volume of windthrown wood or of logging residue on a clear-felled area. The idea is to walk along a straight line, the transect, and to determine the volume of all the trees or logs on the forest floor that intersect this straight line. The procedure certainly selects a sample of the trees on the ground. Depending on the choice of transect and sampling scheme, our problem is to identify the inclusion probabilities and then derive an estimate for the entire population of downed logs. In the second example let us assume that one wants to estimate the percentage of the forest area where regeneration is occurring. We are confident that, by traversing a straight line, we will monitor the length of the transect that intersects with the regeneration patches. Intuitively speaking, one may argue that the proportion of the transect length that crosses those patches will somehow reflect the proportion of the surface area covered with new vegetation.

The literature on transect sampling is immense, in large part dealing with the estimation of animal abundance (e.g. Seber, 1986). This topic will not be investigated here: the main difficulty being that animals move or hide while trees do not. Instead, we shall focus on some key aspects in a design-based framework: i.e. we shall assume that the objects (e.g. the logs) are fixed and that the transect has been chosen according to a well-defined random mechanism. Historically, the model-dependent approach has been widely used in stochastic or integral geometry , with the main field of applications being stereology (the science of inference for 2-D or 3-D objects based on linear or planar probe, especially in medical research). In more applied literature these two approaches are often confused, so caution is required. Here, we shall concentrate on the most important issues, partially following the modern stereological approach (see Barabesi and Fattorini (1997) for applications to ecology but Baddeley and Jensen (2005) for a general mathematical background). Key references for "elementary" geometric probability are provided by the first chapters of Santaló (1976). The reader should also consult de Vries (1986) and Gregoire and Valentine (2007) for further details and references pertaining to the forest

units (logs, or more generally woody debris), on the forest floor with attribute Y_i (e.g. volume) so that we can estimate the quantity $\bar{Y} = \frac{1}{\lambda(F)} \sum_{i=1}^{N}$ To define the inclusion rule we assign to each unit the segment of a straight line of length l_i, subsequently calling this a "needle" because the probabilistic aspect is related to the famous needle problem of Buffon (to be detailed in Section 12.5). For a straight log the needle is taken to be its central axis. For woody debris of arbitrary shape the needle is considered the line segment that joins the two extreme ends of the debris. A particular unit is included in the sample only if its needle intersects the transect (note that in general the debris may intersect the transect but not its needle). As usual we shall assume that we are working in the orthogonal projection of a forest onto the horizontal plane (thus transect and needle lengths are always in that plane). For the second problem we consider "particles" P_i (e.g. regeneration patches) of horizontal surface area $\lambda(P_i)$, with the goal of estimating the proportion $\bar{\lambda} = \frac{1}{\lambda(F)} \sum_{i=1}^{N} \lambda(P_i)$ of the total surface area that is occupied by those particles. If the transect intersects the particle P_i then the length $l(T \cap P_i)$ of the intersection is recorded. Intuitively, we can expect to estimate $\bar{\lambda}$ according to the proportion of the length of the intersections to the transect length.

We shall first examine transects that cross the entire forest area F, so that the length of the transect is a random variable, whose distribution depends on the sampling procedure used to select it at random. As we shall see the notion of a random line is not as obvious as it may seem and there are many non-equivalent approaches. This point is frequently overlooked in the applied literature.

12.2 IUR transect-sampling

We first consider the so-called isotropic uniform random (**IUR**) sampling in a circle. The random transect $T(x, \theta)$ is identified by the angle θ of the line L through the origin perpendicular to it and by the distance x between the origin and the transect (see Fig. 12.1). Random variable θ is uniformly distributed on the interval $[0, \pi]$ and random variable x has the uniform distribution on the interval $[-R, R]$, where R is the radius of the circle. Here, θ and x independent. To simulate an **IUR** random line with respect to a forest F arbitrary shape (it can in particular comprises "holes"), it suffices to include the forest in a large circle and simulate **IUR** lines with respect to the circle, then retain only the lines intersecting the forest (rejection principle). For any set $Y \subset F$ we write $T \uparrow Y$ (T hits Y) if $T(x, \theta) \cap Y \neq \emptyset$. Let us denote by $D(F)$ the set of all x and θ values generating a transect that hits the forest, i.e. $D(F) = \{(x, \theta) \mid T(x, \theta) \uparrow F\}$.

The **IUR** lines are those with a density $f_{IU}(x, \theta)$ constant over $(x, \theta) \in D($ and zero for $(x, \theta) \notin D(F)$. This is perfectly analogous with the definition

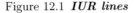

Figure 12.1 *IUR lines*

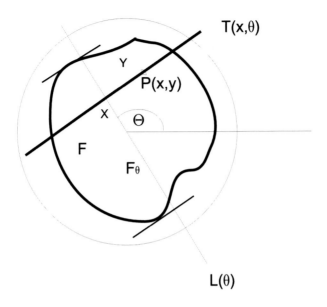

of uniformly distributed random points. Let us denote by F_θ the orthogonal projection of the forest on line $L(\theta)$ (Fig. 12.1) and by $l(\cdot)$ the length function for any one-dimensional object.

Then the density can be factored into

$$f_{IU}(x \mid \theta) = \frac{I_{F_\theta}(x)}{l(F_\theta)}$$

$$f_{IU}(\theta) = \frac{l(F_\theta)}{\pi \bar{H}(F)} I_{[0,\pi]}(\theta)$$

$$f_{IU}(x,\theta) = \frac{I_{D(F)}(x,\theta)}{\pi \bar{H}(F)} \tag{12.1}$$

where

$$\bar{H}(F) = \frac{1}{\pi} \int_0^\pi l(F_\theta) d\theta$$

is the mean linear projection of F (i.e. the mean caliper diameter). A result from integral geometry states that if F is convex then $\bar{H}(F) = \frac{1}{\pi} B(F)$, where $B(F)$ is the perimeter length of F. It is important to realize that for arbitrary F the random variables x and θ are generally not independent (although they are for the circle).

Let us now consider an arbitrary subset $Y \subset F$. We calculate the probability that the transect crosses with Y given the fact that it intersects F. First, for

$$\mathbb{P}_{IU}(T(x,\theta) \uparrow Y \mid \theta, T \uparrow F, Y \subset F) = \frac{l(Y_\theta)}{l(F_\theta)}$$

and consequently

$$\mathbb{P}_{IU}(T \uparrow Y \mid T \uparrow F) = \int_0^\pi \frac{l(Y_\theta)}{l(F_\theta)} f_{IU}(\theta) d\theta = \int_0^\pi \frac{l(Y_\theta)}{l(F_\theta)} \frac{l(F_\theta)}{\pi \bar{H}(F)} d\theta = \frac{\bar{H}(Y)}{\bar{H}(F)}$$

Using Eq. 12.1, we finally obtain for the conditional join density

$$
\begin{aligned}
f_{IU}((x,\theta) \mid T \uparrow Y, Y \subset F) &= \frac{f_{IU}(x,\theta) I_{D(Y)}(x,\theta)}{\mathbb{P}_{IU}(T \uparrow Y \mid T \uparrow F, Y \subset F)} \\
&= \frac{I_{D(F)}(x,\theta)}{\pi \bar{H}(F)} \frac{I_{D(Y)}(x,\theta)}{\frac{\bar{H}(Y)}{\bar{H}(F)}} = \frac{I_{D(Y)}(x,\theta)}{\pi \bar{H}(Y)}
\end{aligned}
$$

Hence, we have the following relationship

$$
\begin{aligned}
\mathbb{P}_{IU}(T \uparrow Y \mid T \uparrow F) &= \frac{\bar{H}(Y)}{\bar{H}(F)} \\
f_{IU}((x,\theta) \mid T \uparrow Y, Y \subset F) &= \frac{I_{D(Y)}(x,\theta)}{\pi \bar{H}(Y)}
\end{aligned}
\tag{12.2}
$$

In words, the **IUR lines through F that hit an arbitrary $Y \subset F$ are IUR lines through Y**, in perfect accord with the fact that as long as points uniformly distributed in F are in $Y \subset F$, they are uniformly distributed in We can now calculate the expected value of the transect length. For a fixed orientation one has

$$\mathbb{E}_{IU}l(T(x,\theta) \cap F \mid \theta) = \int_{x \in F_\theta} l(T(x,\theta) \cap F) \frac{dx}{l(F_\theta)} = \frac{\lambda(F)}{l(F_\theta)} =: \bar{L}_{IU}(\theta)$$

because the second equality is nothing else but the Cavalieri's principle for plane surface areas. For the overall expected value one similarly arrives at

$$
\begin{aligned}
\bar{L}_{IU} := \mathbb{E}_{IU}l(T(x,\theta) \cap F) &= \frac{1}{\pi \bar{H}(F)} \int_0^\pi d\theta \int_{F_\theta} l(T(x,\theta)) dx \\
&= \int_0^\pi \frac{1}{\pi \bar{H}(F)} \lambda(F) d\theta = \frac{\lambda(F)}{\bar{H}(F)}
\end{aligned}
$$

For instance, the mean transect length with respect to a circle of radius R

$$\bar{L}_{IU,circle} = \frac{\pi R^2}{2R} = \frac{\pi}{2} R$$

Now we can proceed to derive an estimator for the "needle case." We write $T \uparrow l_i$ if the needle l_i (with length $l(l_i) = | l_i |$) intersects with the transect. The projection of the needle onto line $L(\theta)$ perpendicular to the transect is denoted by $l_i(\theta)$ with length $l(l_i(\theta)) = l(l_i)cos(\theta_i - \theta) =: | l_i(\theta) |$, where θ_i is the angle between the needle and the direction $\theta = 0$. The conditional inclusion

probability is obviously given by $(\ )= \quad (\ (\quad)\quad |\ )= \ l(F_\theta)$
that the $\pi_i(\theta)$ are maximal when the transect is perpendicular to the needles.
Moreover, the unconditional inclusion probabilities can be obtained from the
previous arguments with $Y = l_i$, so that we have $\pi_i = \mathbb{P}_{IU}(T \uparrow l_i) = \frac{\bar{H}}{H}$
To calculate the mean caliper of the needle we can assume without loss of
generality that $\theta_i = 0$ and we obtain

$$\bar{H}(l_i) = \frac{1}{\pi} \int_0^\pi |l_i \| \cos(\theta) | \, d\theta = \frac{2}{\pi} | l_i |$$

We summarize these results as:

Theorem 12.2.1. *The inclusion probabilities and expected transect lengths
under IUR sampling are given by*

$$\pi_i(\theta) = \frac{| l_i(\theta) |}{l(F_\theta)}$$

$$\pi_i = \frac{2 | l_i |}{\pi \bar{H}(F)}$$

$$\bar{L}_{IU}(\theta) = \frac{\lambda(F)}{l(F_\theta)}$$

$$\bar{L}_{IU} = \frac{\lambda(F)}{\bar{H}(F)}$$

Let us introduce the indicator variable $I_i(T) = 1$ if $T \uparrow l_i$ and 0 otherwise.
Then we have from Theorem 12.2.1 this result:

Theorem 12.2.2. *The following estimators are, with respect to the orientation of the transect, conditionally and unconditionally unbiased estimators of
the Horvitz-Thompson type for the spatial mean* $\bar{Y} = \frac{1}{\lambda(F)} \sum_{i=1}^N Y_i$

$$\hat{\bar{Y}}_\theta = \frac{1}{\bar{L}_{IU}(\theta)} \sum_{i=1}^N \frac{I_i(T)Y_i}{| l_i(\theta) |}$$

$$\hat{\bar{Y}} = \frac{\pi}{2\bar{L}_{IU}} \sum_{i=1}^N \frac{I_i(T)Y_i}{| l_i |}$$

Remarks:

1. A special case deserving special consideration would be when Y_i is the volume of a log. Although this can be difficult to determine accurately, it can
be easily approximated by $\hat{Y}_i = \frac{\pi}{4} d_i^2(u) l_i = g_i(u) l_i$, where u is the distance
from one end of the log to the point of intersection with the transect,
is the log (needle) length, $d_i(u)$ is the diameter of the log at the intersection and $g_i(u)$ is the cross-cut area. For a given unit Y_i and $\hat{Y}_i$ may differ
substantially from one another. However, when considering u as a random
variable, it is intuitively clear that u is uniformly distributed on the interval $[0, l_i]$, so that $\mathbb{E}_u \hat{Y}_i = Y_i$. Thus, on average, Theorem 12.2.2 will yield

unbiased estimates of the volume. One can consult de Vries (1986) for a rigorous argument.

2. The main drawback of the **IUR** sampling scheme is that one needs the mean transect lengths. If a map of F is available in a GIS, this can be obtained either through direct computations or by simulating a large number of virtual transects (of which a small number will be used for field work). In practice, however, transects across the entire forest are feasible only for relatively small areas. Furthermore, one may intuitively prefer a few long transects rather than many small ones. In the next section we shall learn how one can sample transects with a probability proportional to their lengths.

The above estimates can be easily adapted for solving the particle problems. Indeed, by Cavalieri's principle, one has

$$\mathbb{E}_{IU}\left(l(T(x,\theta) \cap P_i) \mid \theta\right) = \frac{1}{l(F_\theta)} \int_{x \in F_\theta} l(T(x,\theta) \cap P_i) dx = \frac{\lambda(P_i)}{l(F_\theta)}$$

Because **IUR** lines through F hitting P_i are **IUR** lines through P_i we also have by Eq. 12.2 and Theorem 12.2.2:

$$\mathbb{E}_{IU} l(T \cap P_i) = \mathbb{E}_{IU}\left(l(T \cap P_i) \mid T \uparrow P_i\right)\mathbb{P}(T \uparrow P_i) = \frac{\lambda(P_i)}{\bar{H}(P_i)} \frac{\bar{H}(P_i)}{\bar{H}(F)} = \frac{\lambda(P_i}{\bar{H}(F}$$

Therefore, by Theorem 12.2.1, we have

Theorem 12.2.3. *The following estimators are, with respect to the orientation of the transect, conditionally and unconditionally unbiased estimators of the ratio $\bar\lambda = \frac{1}{\lambda(F)}\sum_{i=1}^{N}\lambda(P_i)$*

$$\hat{\bar\lambda}_\theta = \frac{1}{\bar{L}_{IU}(\theta)}\sum_{i=1}^{N} I_i(T)l(T \cap P_i)$$

$$\hat{\bar\lambda} = \frac{1}{\bar{L}_{IU}}\sum_{i=1}^{N} I_i(T)l(T \cap P_i)$$

where $I_i(T) = 1$ if the transect T intersects the particle P_i and $I_i(T) = 0$ otherwise.

Note that the estimates in 12.2.3 are not of the Horvitz-Thompson type. In practice, one will draw n i.i.d transects T_k, $k = 1, 2 \ldots n$, thereby yielding point estimates $\hat{\bar{Y}}_k = \frac{\pi}{2L}\sum_{I=1}^{N}\frac{I_i(T_k)Y_i}{|l_i|}$ and then obtain the overall estimate and its variance in the usual way by calculating

$$\hat{\bar{Y}} = \frac{1}{n}\sum_{k=1}^{n}\hat{\bar{Y}}_k$$

$$\hat{\mathbb{V}}(\hat{\bar{Y}}) = \frac{1}{n(n-1)}\sum_{k=1}^{n}(\hat{\bar{Y}}_k - \hat{\bar{Y}})^2 \tag{12.3}$$

This will be performed similarly for θ, θ

We now briefly consider another well-known stereological problem. A set C smooth curves is embedded in F. We seek to determine the total length $l(C)$, or the length per unit area $\nu = \frac{l(C)}{\lambda(F)}$. Our estimation is based on $N(T(x,\theta) \cap$ the total number of intersections made with C by the random **IUR** transect through F. The curves can be organized in any way inside F, although ν useless parameter unless distribution is fairly even. Let dC be an infinitesimal linear element of curve. Then, obviously $N(T \cap dC)$ is 1 if $T \uparrow dC$ otherwise 0. Furthermore, $N(T \cap C) = \int_C N(T \cap dC)$ and $\mathbb{E}_{IU}(T \cap C) = \int_C \mathbb{E}_{IU} N(T \cap dC)$ $\int_C \mathbb{P}_{IU}(T \uparrow dC)$. We have seen that, under IUR sampling, the probability of hitting a needle is $\frac{2|l_i|}{\pi \bar{H}(F)}$. Hence, $\mathbb{E}_{IU} N(T \cap dC) = \mathbb{P}_{IU}(T \uparrow dC) = \frac{2l(dC)}{\pi \bar{H}(F)}$ that by integration over the total curve C, we get

$$\mathbb{E}_{IU} N(T \cap C) = \frac{2}{\pi} \frac{l(C)}{\bar{H}(F)} = \frac{2}{\pi} \frac{\bar{L}_{IU} l(C)}{\lambda(F)} \tag{12.4}$$

If one replaces the mean transect length by the observed length one could consider the estimate $\hat{\nu} = \frac{\pi}{2} \frac{N(T \cap C)}{l(T \cap F)}$, which is usually biased. Nonetheless, this difficulty can vanish when the sampling scheme selects the transects with a probability proportional to their lengths, i.e. **PPL**. This is precisely the subject of the next section.

12.3 PPL transect-sampling

A second way to select a random line through the planar domain F is by **passing it through a point chosen uniformly over it, in a direction determined uniformly and independently**. Let the line have coordinates (x, y, θ), where (x, θ) are the polar coordinates of the foot of the perpendicular from the line to the origin of coordinates (as in **IUR** sampling); y will then be the distance along the line from the foot of the perpendicular to the uniformly chosen interior point $P(x, y)$ (see Fig. 12.1). For the corresponding densities $f_L(\cdot)$ (N.B. index L stands for the L in **PPL** to distinguish this from the **IUR** densities with index IU), we will have the following:

$$f_L(x, y \mid \theta) = \frac{I_F(x, y)}{\lambda(F)}$$

$$f_L(\theta) = \frac{1}{\pi} I_{[0,\pi]}(\theta)$$

$$f_L(x, y, \theta) = \frac{I_F(x, y) I_{[0,\pi]}(\theta)}{\pi \lambda(F)} \tag{12.5}$$

$(\quad)$ is gained by integrating over

$$f_L(x,\theta) = \frac{1}{\pi\lambda(F)}l(T(x,\theta)\cap F)$$

$$= \frac{l(T(x,\theta)\cap F)}{\frac{\lambda(F)}{\bar{H}(F)}}\frac{1}{\pi\bar{H}(F)}$$

$$= \frac{l(T(x,\theta)\cap F)}{\mathbb{E}_{IU}l(T(x,\theta)\cap F)}f_{IU}(x,\theta)$$

We arrive at the important relationship of

$$f_L(x,\theta) = \frac{l(T\cap F)}{\bar{L}_{IU}}f_{IU}(x,\theta) \tag{12.6}$$

This means that the **PPL** scheme selects transects obtained via the **IUR** scheme with a probability proportional to their lengths. This is not intuitively obvious from the onset. Using Eq. 12.1 and Theorem 12.2.1 we can obtain the conditional density

$$f_L(x\mid\theta) = \frac{l(T(x,\theta)\cap F)}{\mathbb{E}_{IU}l(T(x,\theta)\cap F)}\frac{I_{D(F)}(x,\theta)}{\bar{H}(F)} = \frac{l(T(x,\theta)\cap F)I_{D(F)}(x,\theta)}{\lambda(F)}$$

$$\tag{12.7}$$

As an exercise, the reader can check that the mean transect length under PPL for a circle of radius R is

$$\bar{L}_{L,circle} = \frac{16}{3\pi}R$$

Therefore, the ratio $\frac{\bar{L}_{L,circle}}{\bar{L}_{IU,circle}}$ is equal to $\frac{32}{3\pi^2}\approx 1.08$ for the circle, which is surprisingly close to 1. In general, by combining Eq. 12.6 and Theorem 12.2.1, we have for the ratio of the mean transect lengths

$$\frac{\mathbb{E}_L l(T\cap F)}{\mathbb{E}_{IU}l(T\cap F)} = 1 + \frac{\mathbb{V}_{IU}l(T\cap F)}{\bar{L}_{IU}^2} \geq 1$$

The conditional inclusion probabilities for needle intersection are

$$\pi_i(\theta) = \frac{1}{\lambda(F)}\int_{l_i(\theta)}l(T(x,\theta)\cap F)dx$$

which, typically, cannot be calculated. However, an obvious approximation would be to use the observed transect length instead of its mean value over $l_i(\theta)$ and then set

$$\hat{\pi}_i(\theta) = \frac{\mid l_i(\theta)\mid l(T\cap F)}{\lambda(F)}$$

likewise, by using $\mathbb{E}_\theta\mid l_i(\theta)\mid = \frac{2}{\pi}\mid l_i\mid$ we can estimate the unconditional inclusion probabilities by

$$\hat{\pi}_i = \frac{2}{\pi}\frac{l(T\cap F)\mid l_i\mid}{\lambda(F)}$$

$$\hat{\bar{Y}}_L(\theta) = \frac{1}{\lambda(F)}\sum_{i=1}^{N}\frac{I_i(T)Y_i}{\hat{\pi}_i(\theta)} = \frac{1}{l(T\cap F)}\sum_{i=1}^{N}\frac{I_i(T)Y_i}{|\,l_i(\theta)\,|}$$

$$\hat{\bar{Y}}_L = \frac{1}{\lambda(F)}\sum_{i=1}^{N}\frac{I_i(T)Y_i}{\hat{\pi}_i} = \frac{\pi}{2l(T\cap F)}\sum_{i=1}^{N}\frac{I_i(T)Y_i}{|\,l_i\,|}$$

Using Eq. 12.7, we obtain

$$\frac{1}{|\,l_i(\theta)\,|}\mathbb{E}_{x|\theta}\left(\frac{I_i(T)}{l(T\cap F)}\right) = \frac{1}{|\,l_i(\theta)\,|}\frac{1}{\lambda(F)}\int_{l_i(\theta)}\frac{l(T\cap F)}{l(T\cap F)}dx = \frac{1}{\lambda(F)}$$

Likewise, we get

$$\mathbb{E}_{(x,\theta)}\frac{I_i(T)}{l(T\cap F)} = \mathbb{E}_\theta\mathbb{E}_{(x|\theta)}\frac{I_i(T)}{l(T\cap F)} = \mathbb{E}_\theta\frac{1}{\lambda(F)}\int_{l_i(\theta)}\frac{l(T\cap F)}{l(T\cap F)}dx$$

$$= \frac{1}{\lambda(F)}\mathbb{E}_\theta\,|\,l_i(\theta)\,| = \frac{1}{\lambda(F)}\frac{2}{\pi}\,|\,l_i\,|$$

Therefore, the two pseudo Horvitz-Thompson estimators are design-unbiased. The gist of our argument is that the observed transect length $l(T\cap F)$ in the estimator and in the density cancel out under **PPL** sampling. We can then state the result as

Theorem 12.3.1. *The following conditional and unconditional estimators are, under PPL transect sampling, unbiased estimates of the spatial mean* $\bar{Y} = \frac{1}{\lambda(F)}\sum_{i=1}^{N}Y_i$

$$\hat{\bar{Y}}_L(\theta) = \frac{1}{l(T\cap F)}\sum_{i=1}^{N}\frac{I_i(T)Y_i}{|\,l_i(\theta)\,|}$$

$$\hat{\bar{Y}}_L = \frac{\pi}{2l(T\cap F)}\sum_{i=1}^{N}\frac{I_i(T)Y_i}{|\,l_i\,|}$$

For the particle problem we obtain via Eq. 12.7

$$\mathbb{E}_L\left(\frac{l(T\cap P_i)}{l(T\cap F)}I_i(T)\,|\,\theta\right) = \frac{1}{\lambda(F)}\int_{P_{\theta,i}}\frac{l(T\cap F)}{l(T\cap F)}l(T\cap P_i)dx = \frac{\lambda(P_i)}{\lambda(F)}$$

This, of course, is also unconditional (with $P_{\theta,i}$ denoting as usual the orthogonal projection of P_i on the direction $L(\theta)$ perpendicular to the transect). Hence, we have the result:

Theorem 12.3.2. *The estimator*

$$\hat{\lambda}_L = \frac{1}{l(T\cap F)}\sum_{i=1}^{N}I_i(T)l(T\cap P_i)$$

is, under PPL transect sampling, an unbiased estimate of the proportion

$\frac{1}{\lambda(F)} \sum_{i=1} (\quad)$, both conditionally (for a given orientation θ) or unconditionally (θ random).

In some applications the particles are circles C_i of diameters D_i with associated response variable Y_i. The intersection probabilities are

$$\pi_i(\theta) = \mathbb{P}(T \uparrow C_i \mid \theta) = \frac{1}{\lambda(F)} \int_{C_{i,\theta}} l(T \cap F) dx$$

This can be approximated by $\frac{D_i l(T \cap F)}{\lambda(F)}$ independent of θ. Hence, one can consider the conditional and unconditional estimator to be

$$\hat{\bar{Y}} = \frac{1}{l(T \cap F)} \sum_{i=1}^{N} \frac{I_i(T) Y_i}{D_i} \tag{12.8}$$

Using the same arguments as in the proofs for Theorems 12.3.1 and 12.3.2 we learn that the estimator given by Eq. 12.8 is design-unbiased. Let us consider two applications:

1. **Animal abundance estimation**
   An observer moves along a transect and is aware of the presence of an animal, such as a bird, only if it is flushed. One might assume a model in which each animal is the center of an imaginary "flushing" circle of diameter $D_i = 2R_i$. Radius R_i is estimated by the observer at the moment of flushing. In this case, the estimator from Eq. 12.8 with $Y_i \equiv 1$ is known as the Hayne estimator, a tool much used in wildlife counts.

2. **The Strand's estimator** (developed by inventorist L. Strand in 1958)
   While moving along a transect and sighting to **one side** of it with a Relascope (critical angle α, counting factor $k = 10^4 sin^2(\frac{\alpha}{2})$) **in a direction perpendicular to that transect**, the observer checks which trees are in the sample. The angle count technique assigns to each tree of diameter D limit circle of diameter $\frac{D_i}{sin(\frac{\alpha}{2})}$. A tree is "in" if the transect intersects with its limit circle. Because we are sampling trees on just one side, the inclusion probabilities are only one-half of the value $\frac{l(T \cap F) D_i}{\lambda(F)}$ obtained above. Note also that they are proportional to the diameters of the trees and no longer proportional to the square of the diameters, i.e. the basal area, as in classical angle count sampling. The estimator from Eq. 12.8 can then be written as

$$\hat{\bar{Y}} = \frac{2 sin(\frac{\alpha}{2})}{l(T \cap F)} \sum_{i=1}^{N} \frac{I_i(T) Y_i}{D_i}$$

If D_i is in cm $l(T \cap F)$ in m and $\lambda(F)$ in ha, this is equivalent to

$$\hat{\bar{Y}} = \frac{10^4 2\sqrt{k}}{l(T \cap F)} \sum_{i=1}^{N} \frac{I_i(T) Y_i}{D_i}$$

The mean and variance formulae for combining n i.i.d **PPL** transects are the same as under **IUR** sampling.

estimates by replacing the expected transect lengths with the observed ones, a very nice result indeed.

Let us return to the the problem of estimating $\nu = \frac{l(C)}{\lambda(F)}$. We have seen that under **IUR** sampling the estimator $\hat{\nu} = \frac{\pi}{2} \frac{N(T \cap C)}{l(T \cap F)}$ is generally biased. With **PPL** sampling we gain by Eq. 12.6 and 12.4

$$\mathbb{E}_L \frac{N(T \cap C)}{l(T \cap F)} = \int_{D(F)} \frac{N(T \cap C)}{l(T \cap F)} \frac{l(T \cap F)}{\bar{L}_{IU}} f_{IU}(x, \theta) = \frac{\mathbb{E}_{IU} N(T \cap C)}{\bar{L}_{IU}} = \frac{2}{\pi}$$

hence, we arrive at the result previously claimed

$$\mathbb{E}_L \hat{\nu} = \mathbb{E}_L \frac{\pi}{2} \frac{N(T \cap C)}{l(T \cap F)} = \nu = \frac{l(C)}{\lambda(F)} \tag{12.9}$$

The above relationship has been used to estimate the total length of road networks, rivers and even root systems within a certain area, simply by counting the intersections with transects.

For completeness let us mention a third very intuitive possibility (after **IUR** and **PPL**) for generating random lines through F. Here, we draw two uniformly independent points in F and take the line joining them as a random through F. Surprisingly, this leads to a different sampling scheme, where **IUR** transects are selected with a probability proportional to the third power of the transect length (Coleman, 1979). However, this scheme does not seem to have any useful practical implications. In the next section, we briefly outline the procedures with transect of fixed length.

12.4 Transects with fixed length

The following results were first presented, in a more general setup, by Kaiser (1983). Fortunately, they can be obtained almost without changes from the **PPL** theory of the previous section. We consider a straight line of fixed length L. Its midpoint is uniformly distributed in the forest F and its orientation chosen independently and according to an arbitrary distribution function f (except for the fixed length this is essentially the **PPL** scheme). Usually f will be degenerate (i.e. all the transects will have the same orientation) or will be uniformly distributed on the interval $[0, \pi]$. In the first instance, a unit is sampled if its needle intersects the transect. In the second instance, the particle P_i is sampled if the transect intersects **entirely** with P_i. To handle partial intersections, one chooses an end of the transect at random (with probability 0.5) and samples partially intersected particle on this end, ignoring partial intersections on the other end. Up to boundary effects at the forest edge, the

$$\pi_i(\theta) = \frac{Ll(P_{i,\theta})}{\lambda(F)}$$

$$\pi_i = \frac{L\mathbb{E}_\theta l(P_{i,\theta})}{\lambda(F)} \qquad (12.10)$$

If the transect intersects the boundary, the portion outside of that forest area F should be reflected and moved sideways so that no particle or needle can be intersected more than once by the same transect. This procedure yields unbiased estimates for convex F. In general, the bias is expected to be small if L is much smaller than $\bar{H}(F)$. **Exactly the same estimates given in Section 12.3 can be used after the observed transect length $l(T \cap$ is replaced with the fixed transect length L.**

It is somewhat surprising that the mathematics is simpler for transects of random length. The approach presented for these three transect sampling schemes allows for a unified treatment. **Their formulae are the same as long as one utilizes adequate transect lengths, i.e. those that are expected, observed and nominal transect lengths.**

Kaiser (1983) also considered slightly more general estimates that depend on further random variables (under the design) than those presented here. Transects that consist of many segments can be useful in such application as those for a closed polygon, where a triangle brings the field crew back to their starting point. The major difficulty is to define a sampling protocol that deals with multiple inclusions and boundary effects (Affleck et al., 2005). For completeness and recreation, we present in the next section the famous Buffon's needle problem and its relationship to transect sampling.

12.5 Buffon's needle problem

The original problem was formulated in 1777 by Georges Louis Leclerc, Comte de Buffon (universal scientist and the founder of geometric probability). His goal was to calculate the probability that a randomly thrown needle of length L would intersect one out of (infinitely) many parallel lines at a distance D each other ($L \leq D$). To be more precise, our probabilistic model here is the following: the position of the needle is identified by the distance x from its middle point M to the parallel line below M and the angle θ (see Fig. 12.2). and θ are viewed as independent and uniformly distributed random variables on $[0, d]$ and $[o, \pi]$ respectively. We have an intersection if either $\frac{L}{2}sin(\theta) > x$ or $\frac{L}{2}sin(\theta) > D - x$. For given θ this happens with probability

$$p(\theta) = \frac{1}{D}\left(\int_0^{\frac{L}{2}sin(\theta)} dx + \int_{D-\frac{L}{2}sin(\theta)}^{D} dx \right) = \frac{L}{D}sin(\theta)$$

$$p = \frac{1}{\pi} \int_0^\pi p(\theta)d\theta$$

which is easily found to be

$$p = \frac{2}{\pi} \frac{L}{D} \tag{12.11}$$

Thus one can estimate π by throwing a large number of needles on the floor ($L = D$ being optimal for the resulting variance).

Figure 12.2 **Buffon's needle problem**

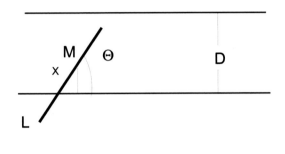

Figure 12.3 **Buffon's needle and transect**

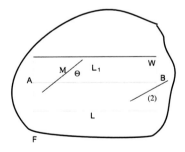

Let us now examine the model-dependent version of the transect problem. We consider in Fig. 12.3 a transect AB of length L and a needle L_1, whose length is denoted by $| L_1 |$. A rectangle of area WL is contained in the forest area F. The needle midpoint is randomly thrown in F in the following sense: 1) its midpoint M is uniformly distributed in F (the probability that it lies in the rectangle is therefore $p_1 = \frac{WL}{\lambda(F)}$), 2) when M is in the rectangle, the position of the needle is indicated by the orthogonal distance m of M to the transect and the acute angle θ between the needle's direction and the transect.

are uniformly and independently distributed on their respective intervals. We assume that needle length is small in comparison with the transect length, so that boundary effects of an intersection of type (2) occur with a negligible probability. The density of (m, θ) is constant equal to $\frac{4}{W\pi}$. We shall call this the **purely random needle model** NM. The needle intersects the transect if and only if $m \leq \frac{|L_1|}{2} sin(\theta)$. Our probability of intersection, given that M in the rectangle, is then

$$p_2 = \frac{4}{W\pi} \int_0^{\frac{\pi}{2}} \frac{|L_1|}{2} sin(\theta)d\theta = \frac{2|L_1|}{\pi W}$$

The unconditional probability of intersection p is equal to $p_1 p_2 = \frac{2L|L}{\pi\lambda(F)}$ Hence, we have proven the following result:

Theorem 12.5.1. *Under the assumption of negligible boundary effects the probability that a randomly thrown needle l_i of length $|l_i|$ will intersect a straight line transect of length L is given by*

$$p_i = \frac{2L|l_i|}{\pi\lambda(F)}$$

One then obtains exactly the same intersection probability for the dual problem of fixed needles and a random transect (the end point of the transect is uniformly distributed in F and its orientation on $[0, 2\pi]$). This result is obvious for reason of symmetry, de Vries (1986). However, the above formula is wrong if the orientation of the needles is clustered in one direction, as, for example, with windthrown wood.

Under the model-dependent approach with purely random needles we can estimate the spatial mean $\bar{Y} = \frac{1}{\lambda(F)} \sum_{i=1}^N Y_i$ by the Horvitz-Thompson estimator

$$\hat{\bar{Y}} = \frac{1}{\lambda(F)} \sum_{i=1}^N \frac{I_i(T)Y_i}{p_i} = \frac{\pi}{2L} \sum_{i=1}^N \frac{Y_i}{|l_i|} \tag{12.12}$$

in which the intersection probabilities p_i can be viewed as inclusion probabilities for a given transect, with $I_i(T)$ being as usual 1 or 0 depending on whether or not unit i is included in the sample. The model-dependent Horvitz-Thompson estimator from Eq. 12.12 has exactly the same formal structure as the design-based version given in Theorem 12.3.1 (after replacing the observed length $l(T \cap F)$ with the fixed length L).

Obviously $\hat{\bar{Y}}$ is model-unbiased. Its model-dependent variance (under the purely random needle model) is given by

$$\mathbb{V}_{NM}(\hat{\bar{Y}}) = \frac{\pi}{2L} \frac{1}{\lambda(F)} \sum_{i=1}^N \frac{Y_i^2}{l_i} - \frac{1}{\lambda(F)^2} \sum_{i=1}^N Y_i^2 \tag{12.13}$$

$$(\ ) = \quad (1 \quad) \text{ and also because the}$$

are not correlated under that model.

One can apply the same argument to obtain an unbiased model-dependent estimate of estimator from Eq. 12.13 on the basis of a single transect, namely

$$\hat{\mathbb{V}}_{NM}(\hat{\bar{Y}}) = \left(\frac{\pi}{2L}\right)^2 \sum_{i=1}^{N} I_i(T)\frac{Y_i^2}{l_i^2} - \frac{1}{\lambda(F)}\frac{\pi}{2L}\sum_{i=1}^{N} I_i(T)\frac{Y_i^2}{l_i} \qquad (12.14)$$

Of course, the model-dependent approach cannot be trusted blindly. Again, de Vries (1986) provides further examples.

12.6 Exercises

Problem 12.1. *Write a computer program to simulate Buffon's needle problem, thereby estimating π. What is the optimum choice for needle length as compared to the distance D between the parallel lines? Perform a real experiment by throwing a needle, at least 100 times, onto a sufficiently large sheet of paper printed with parallel lines. Do your estimates for π, based on these simulated and real experiments, differ significantly from the true value of $\pi \approx 3.14159$?*

Problem 12.2. *We consider the following two-stage procedure with transect sampling: forest area F is embedded into N cells (usually of simple shape such as a square) F_i, with $F \subset \cup_{i=1}^{N} F_i$. A sample s_1 of such cells is drawn according to a sampling scheme with inclusion probabilities π_i and π_{ij}. We assume that the forested areas $\lambda(F \cap F_i) = \lambda_i$ are known for $i \in s_1$, as well as $\lambda(F)$. For each cell $i \in s_1$ a small number of transects are performed, so that one obtains a point estimate $\hat{\bar{Y}}_i$ and an estimated variance $\hat{V}_i$ according to Eq. 12.3. The choices are arbitrary for transect procedures (**IUR** vs. **PPL** and point estimates for the quantity of interest $\bar{Y}_i = \frac{1}{\lambda_i}\sum_{j \in F_i} Y_j$. The Y_j be, for example, the volume of the jth log in the ith cell or the surface area $\lambda(P_j)$ of the jth particle. The only requirements are that the same transect be utilized for all cells and that all the transects be independent of each other. That is, one must generate a random point and a random direction for each transect in each cell $i \in s_1$.*

Consider the estimate

$$\hat{\bar{Y}} = \frac{1}{\lambda(F)} \sum_{i \in s_1} \lambda_i \frac{\hat{\bar{Y}}_i}{\pi_i}$$

Using Eq. B.3 and B.4 show that $\hat{\bar{Y}}$ is unbiased for $\frac{1}{\lambda(F)}\sum_{j \in F} Y_j$ and derive its theoretical variance. Taking a sample copy of the theoretical variance yields an intuitive estimate $\hat{V}(\hat{\bar{Y}})$ of the variance. Calculate the expected value $\mathbb{E}_{1,2}(\hat{V}(\hat{\bar{Y}}))$ and suggest a new unbiased estimate $\hat{\mathbb{V}}(\hat{\bar{Y}})$ for the variance. Tailor your results to simple random sampling of n out N cells.

APPENDIX A

Simulations

A.1 Preliminaries

To illustrate the various sampling procedures software can be used to simulate various one-stage schemes, including angle and concentric circles techniques, random and systematic simple or cluster sampling, as well as one-phase and two-phase. The variables that can be analyzed are: number of stem, basal area and timber volume density over many user defined sub-populations (e.g. species, state of health), and ratios of such variables. These data are from the full census of a small area of the Zürichberg forest described in chapter 8. It comprises 17ha and about 4,900 trees. The projection in the horizontal plane of the forest, of its stand map and of the trees is stored in a grid file.The software offers the possibility of performing any necessary boundary adjustments at the forest edge (with an accuracy of 1% to 5% for the $\lambda(K_i \cap F)$). Because such a data set is very expensive to collect, and neither the diameter at 7 nor tree heights have been recorded here, so that two-stage sampling cannot be simulated. However, in that type of sampling, there is perfect alignment between theory and practice as far as second-stage variance is concerned. To simplify the layout the figures have been grouped in Section A.5.

A.2 Simple random sampling

The first example depicts the results obtained for the following sampling scheme:

- Simple random sampling with one circle of $200m^2$.
- 17 plots with an accuracy of 5% for the boundary adjustments. This density, 1 sample point per ha, is the usual choice for an inventory at the enterprise level in Switzerland.
- 200 runs for the simulations.
- The response variable is timber value in m^3 per ha.

Fig. A.1 and A.2 are histograms of the point estimate and estimated variance.

- The mean of the 200 point estimate does not differ significantly from the true value and the distribution of those point estimates is approximately normal.

- The number of trees sampled per point varies from 3 to 9 with an average of 5.5.

- The empirical mean of the 200 point estimates is taken as the true variance. The mean of the 200 estimated variances does not differ significantly from the true variance. Distribution of the 200 estimated variance differs from a normal distribution and resembles, as expected, a chi square distribution.

- The empirical coverage probability of a 95% confidence interval, calculated under the assumption of normally distributed point estimates, is 92.5%.

- This example shows perfect agreement, on average, between theory and practice. However, it also reveals that some simulation runs can be rather far from the truth.

For further illustration we consider the same sampling scheme with a systematic grid of $100m \times 100m$, i.e. 1 sample point per ha. The numbers of points n_2 is now random, varying from 13 to 22, with an average of 17.3. Fig. A.3 and A.4 display the results.

Remarks:

- Results under random and systematic sampling are very similar.

- The estimate of the variance assuming simple random sampling (the most commonly practiced procedure), slightly overestimates the true empirical variance (which is only available under simulations). In this case the over-estimation is small, being 10% for the variance and 5% for the error.

- If one does not adjust for boundary effects at the forest edge the point estimate underestimate the true value by roughly 7% on average. By comparison, the sampling error is ca 15% when $n_2 = 17$ points. However, the sampling error can be decreased by taking larger sample sizes, **while the systematic error that is induced by not adjusting for boundary effects will remain constant and, eventually, be larger than the sampling error**.

Finally, Fig. A.5 present an histogram of the estimated proportion of broadleaf species using exactly the same systematic sampling scheme. Again, the agreement between point estimates and true values is excellent and the variance estimates under the assumption of random sampling are conservative. This empirical finding seems to be the rule. Thus, it is appropriate to apply the usual variance formulae in systematic sampling as well.

The sampling scheme is as follows:

- A cluster consists of nominally 4 points that are the corners of a $40m \times 40$ square. The number of points per cluster ranges from 2 and 4, averaging 2.9.
- Our systematic square grid with fixed orientation and random start has a mesh of $140m \times 140m$. The number of non-void clusters is 10 to 15 (average of 12.8).
- This scheme uses the **SNFI** method, i.e. small circle of $200m^2$ and large circle of $500m^2$ with a DBH threshold of $36cm$. The number of trees per plot varies between 7 and 11 with an average of 8.9.
- The response variable is basal area in m^2 per ha.

Fig. A.6, A.7 and A.8 display the main features of this scheme.

Remarks:

- Bias due to cluster sampling is negligible.
- Assuming random cluster sampling, the mean estimated variance overestimates the empirical mean of the systematic cluster sampling by roughly 40%. This overestimation is larger than in the first example, primarily because the average overall number of points is 39 instead of 17. Of course, the regular pattern of systematic cluster points surveyed is more pronounced in a small area surveyed than in a large one.
- The intra-cluster correlation is significantly different from 0, with an average of 0.12. In agreement with common experience the basal area is less dependent than stem density on the particular stand structure.

A.4 Two-phase simple systematic sampling

The sampling scheme is as follows:

- First-phase sample points are on a $50m \times 50m$ grid with random start and orientation.
- The second-phase sample points are on a sub-grid $100m \times 100m$. This scheme uses a single circle of $200m^2$.
- The response variable is stem density.
- Three estimation techniques used here are sample mean, a design-based regression estimator for stratification according to the development stage (with $n_1 \approx 4n_2 \approx 68$ and $n_1 = \infty$) and the model-dependent estimator ($n_1 = \infty$).

to 100.

Fig. A.9, A.10, A.11 and A.12 present some of our results.

Remarks:

- For comparison, the sample mean based on the small sample has an empirical mean of 281.8, an empirical variance of 1764 and a mean estimated variance of 3445. Hence, this overestimation is rather large.

- The regression estimate is unbiased, its empirical variance of 1497 is only slightly smaller than the empirical variance of the sample mean. However, its mean estimated variance is very close to its empirical variance, both of which are roughly half the mean estimated variance of the sample mean. Because the empirical variance usually is not available in practice, one would conclude that, as expected, the regression estimate is more accurate than the sample mean.

- The design-based regression estimate for $n_1 = \infty$ is also unbiased, with an empirical mean of 279.6, an empirical variance of 1386 and a mean estimated variance of 1035 (so that we have a slight underestimation, which is rather rare). On the other hand, the mean estimated variance is, as expected, smaller than when we use $n_1 \approx 68$.

- Our model-dependent estimate is also unbiased. Its mean estimated variance somewhat underestimates its empirical variance and is also smaller than the corresponding design-based variance ($n_1 = \infty$).

- The model-assisted estimate (g-weights) is very close to the design-based estimate.

Conclusions:

These simulations demonstrate the excellent agreement between theory and practice with respect to the point estimates. They also reveal that the variance estimate for systematic sampling, assuming random sampling, is conservative, i.e. it underestimates the true variance. This effect can sometimes be rather substantial because of the small size of the forest. It is likely to be much smaller, if not negligible, for national forest inventories. The empirical coverage probabilities are close to their nominal values of 95% (ranging from 88% to 98%). The distribution of the point estimates is generally "bell-shape", with an occasional, significant, departure from the normal distribution (longer tails). This effect is partially counter-balanced by the conservative variance estimate so that the coverage probabilities are acceptable. In short, one sees that we are close to the truth on average, but not all the time!

Figure A.1 *Histogram of the estimated volume density*

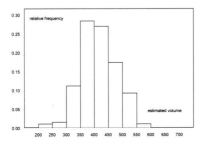

Simple random sampling

Statistics: true mean= 419.7 $\frac{m^3}{ha}$, empirical mean=415 $\frac{m^3}{ha}$

empirical variance=3749 $\left(\frac{m^3}{ha}\right)^2$

Figure A.2 *Histogram of the estimated variance*

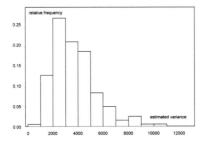

Simple random sampling

Statistics: true variance =3749 $\left(\frac{m^3}{ha}\right)^2$, empirical mean=3709 $\left(\frac{m^3}{ha}\right)^2$

Figure A.3 *Histogram of the estimated volume density*

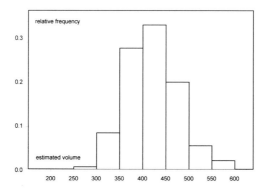

Systematic random sampling

Statistics: true mean= 419.7 $\frac{m^3}{ha}$, empirical mean=422 $\frac{m^3}{ha}$
empirical variance=3339 $\left(\frac{m^3}{ha}\right)^2$

Figure A.4 *Histogram of the estimated variance*

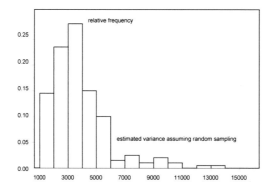

Systematic random sampling

Statistics: true variance =3339 $\left(\frac{m^3}{ha}\right)^2$, empirical mean=3788 $\left(\frac{m^3}{ha}\right)^2$

Figure A.5 *Histogram of the proportion of broadleaf species*

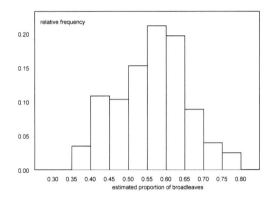

Systematic random sampling

Statistics: true value=0.53, empirical mean=0.56, empirical variance 0.011
mean estimated variance 0.017

Figure A.6 *Histogram of the estimated basal area*

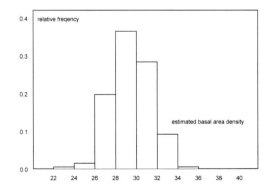

Systematic random cluster sampling

Statistics: true value=29.68 $\frac{m^2}{ha}$, empirical mean=29.43 $\frac{m^2}{ha}$
empirical variance 3.5 $\left(\frac{m^2}{ha}\right)^2$

Figure A.7 *Histogram of the estimated variance*

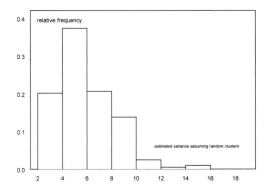

Systematic random cluster sampling

Statistics: true value$=3.5\left(\frac{m^2}{ha}\right)^2$, empirical mean$=5.5\left(\frac{m^2}{ha}\right)^2$

Figure A.8 *Histogram of the intra cluster correlation*

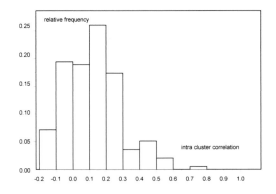

Systematic random cluster sampling

Statistics: empirical mean$=0.12$

Figure A.9 *Histogram of the regression estimate: stem density*

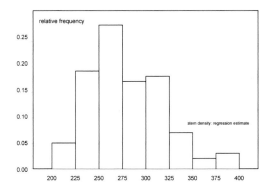

Systematic two-phase sampling

Statistics: true mean =280.2, empirical mean=278.4
empirical variance=1497.

Figure A.10 *Histogram of the estimated variance*

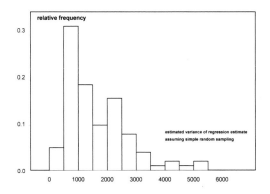

Systematic two-phase sampling

Statistics: true variance =1497, empirical mean=1639

Figure A.11 *Histogram of the model-dependent estimate*

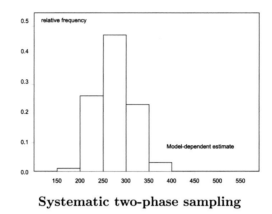

Systematic two-phase sampling

Statistics: true mean =280.2, empirical mean=278.3
empirical variance=1479.

Figure A.12 *Histogram of the estimated model-dependent variance*

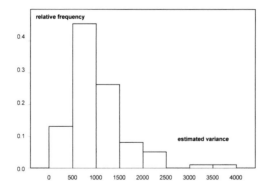

Systematic two-phase sampling

Statistics: true variance =1479, empirical mean=1040

Conditional expectations and variances

In many problems, such as with post-stratification or multi-stage/multi-phase sampling, direct calculation for expectation and even more so for variances are difficult. An elegant way to bypass this is through repeated use of conditioning arguments, which are briefly discussed here.

Let us consider two discrete random variables X and Y with a joint probability function $f_{X,Y}(x,y) = \mathbb{P}(X = x, Y = y)$ and marginal probability functions of $\mathbb{P}(X = x) = f_X(x) = \sum_y f_{X,Y}(x,y)$ and $\mathbb{P}(Y = y) = f_Y(y) = \sum_x f_{X,Y}(x,y)$ Then, the conditional probability function of X given Y is stated as

$$\mathbb{P}(X = x \mid Y = y) = f_{X|Y}(x \mid y) = \frac{f_{X,Y}(x,y)}{f_Y(y)}$$

Consider an arbitrary function $h(X,Y)$ of the two random variables X Y. To calculate its expectation we note that

$$\mathbb{E}_{X,Y} h(X,Y) = \sum_{x,y} h(x,y) f_{X,Y}(x,y) = \sum_y f_Y(y) \left(\sum_x h(x,y) \frac{f_{X,Y}(x,y)}{f_Y(y)} \right)$$

which can be rewritten as

$$\mathbb{E}_{X,Y} h(X,Y) = \sum_y f_Y(y) \left(\sum_x h(x,y) f_{X|Y}(x \mid y) \right)$$

The term in brackets on the right-hand side is the conditional expectation of $h(X,Y)$ given that $Y = y$. This is denoted by $\mathbb{E}_X(h(X,Y) \mid Y = y$ $\mathbb{E}_{X|Y} h(X,Y)$. It is a function of the realization y of Y and does no longer depend on x. The second sum over y is the expectation of this function with respect to the marginal distribution of Y. Hence, we have the following fundamental decomposition rule for expectation.

$$\mathbb{E}_{X,Y} h(X,Y) = \mathbb{E}_Y \left(\mathbb{E}_{X|Y} h(X,Y) \right) \tag{B.1}$$

Recall the definition of the variance for an arbitrary random variable Z: $\mathbb{V}Z$ $\mathbb{E}(Z - \mathbb{E}Z)^2 = \mathbb{E}Z^2 - \mathbb{E}^2 Z$, where $\mathbb{E}^2(Z) = (\mathbb{E}(Z))^2$. Because of B.1 we have the identity

$$\mathbb{E}_{X,Y} h^2(X,Y) - \mathbb{E}_{X,Y}^2 h(X,Y) \quad = \quad \mathbb{E}_Y \left[\mathbb{E}_{X|Y} h^2(X,Y) - \mathbb{E}_{X|Y}^2 h(X,Y) \right]$$

$$+ \quad \mathbb{E}_Y \left[\mathbb{E}_{X|Y}^2 h(X,Y) - \left(\mathbb{E}_Y \mathbb{E}_{X|Y} h(X,Y) \right. \right.$$

( ) while the first term on the right-hand side is the expected value under Y of the conditional variance of $h(X,Y)$ given Y. This can be rewritten as $\mathbb{E}_Y \mathbb{E}_{X|Y}(h(X,Y)$ $\mathbb{E}_{X|Y} h(X,Y))^2$. The second term is the variance, under Y, of the conditional expectation of $h(X,Y)$ given Y. Hence, we have the important decomposition

$$\mathbb{V}_{X,Y} h(X,Y) = \mathbb{E}_Y \mathbb{V}_{X|Y} h(X,Y) + \mathbb{V}_Y \mathbb{E}_{X|Y} h(X,Y) \qquad \text{(B.2)}$$

This holds also for continuous random variables (with probability density functions replacing probability functions) as well as for multivariate random variables. In sampling theory Y is often a set of observed sample sizes or of indicator variables.

The above result permits a nice geometrical interpretation. A random variable $Z = h(X,Y)$ is viewed as a vector in an abstract space, with the squared length $|Z|^2$ given by its variance. The conditional expectation $\mathbb{E}_{X|Y} Z$ can be viewed as the orthogonal projection $Z^\perp$ of Z on the subspace generated by Y (mathematically the set of Y-measurable functions). Then B.2 reads as $|Z|^2 = |Z - Z^\perp|^2 + |Z^\perp|^2$, which is nothing else than the celebrated theorem of Pythagoras in a new setting. The random variable Z is decomposed in orthogonal components.

To illustrate the above general result consider two-stage sampling and an estimator $\hat{Y}$. The index 1 refers to the first stage and the index 2 to the second stage. To calculate the overall expected value and variance, we simply apply the above results to obtain

$$\mathbb{E}_{1,2}(\hat{Y}) = \mathbb{E}_1(\mathbb{E}_{2|1}(\hat{Y})) \qquad \text{(B.3)}$$

$$\mathbb{V}_{1,2}(\hat{Y}) = \mathbb{E}_1(\mathbb{V}_{2|1}(\hat{Y})) + \mathbb{V}_1(\mathbb{E}_{2|1}(\hat{Y})) \qquad \text{(B.4)}$$

That is, one calculates expectation and variance of the second stage given the first-stage sample and then the expectation and variance with respect to the first stage. This will also obviously be true for two-phase sampling with exactly the same notation. The decomposition rule can be iterated. Let us calculate the variance for a three-stage sampling procedure of an estimator Then one can write

$$\mathbb{V}_{1,2,3}(\hat{Y}) = \mathbb{E}_1(\mathbb{V}_{(2,3|1)}(\hat{Y})) + \mathbb{V}_1(\mathbb{E}_{(2,3|1)}(\hat{Y}))$$

and decompose again, first

$$\mathbb{V}_{(2,3|1)}(\hat{Y}) = \mathbb{E}_{(2|1)}(\mathbb{V}_{(3|1,2)}(\hat{Y})) + \mathbb{V}_{(2|1)}(\mathbb{E}_{(3|1,2)}(\hat{Y}))$$

and second

$$\mathbb{E}_{(2,3|1)}(\hat{Y}) = \mathbb{E}_{(2|1)}(\mathbb{E}_{(3|1,2)}(\hat{Y}))$$

To finally obtain the overall variance under three-stage sampling as

$$\mathbb{V}_{1,2,3}(\hat{Y}) = \mathbb{E}_1 \mathbb{E}_{(2|1)} \mathbb{V}_{(3|1,2)}(\hat{Y}) + \mathbb{E}_1 \mathbb{V}_{(2|1)} \mathbb{E}_{(3|1,2)}(\hat{Y}) + \mathbb{V}_1 \mathbb{E}_{(2|1)} \mathbb{E}_{(3|1,2)}(\hat{Y}$$
$$\text{(B.5)}$$

written as
$$\mathbb{V}(\hat{Y}) = \mathbb{E}_1(\mathbb{V}_{(>1|1)}(\hat{Y})) + \mathbb{V}_1(\mathbb{E}_{(>1|1)}(\hat{Y})) \tag{B.6}$$
where the omnibus symbol > 1 represents all stages of sampling after the first.

Of course, one could have a combination of multi-stage with multi-phase sampling and use formally the identical decomposition rule.

Solutions to selected exercises

C.1 Chapter 2

Problem 2.1

$\pi_1 = \frac{3}{4}$, $\pi_2 = \frac{3}{4}$, $\pi_3 = \frac{1}{2}$

$$\Delta = \frac{1}{16} \begin{pmatrix} 3 & -1 & -2 \\ -1 & -3 & -2 \\ -2 & -2 & 4 \end{pmatrix}$$

Problem 2.2

$$\sum_{k \in \mathcal{P}} \Delta_{kl} = \sum_{k \in \mathcal{P}} \mathbb{COV}(I_k, I_l) = \mathbb{COV}(\sum_{k \in \mathcal{P}} I_k, I_l) = \mathbb{COV}(n_s, I_l) = 0$$

if the sample size n_s is constant. Conversely, the relationship

$$0 = \sum_{k,l \in \mathcal{P}} \Delta_{kl} = \sum_{k,l \in \mathcal{P}} \mathbb{COV}(I_k, I_l) = \mathbb{V}(\sum_{k \in \mathcal{P}} I_k) = \mathbb{V}(n_s)$$

implies a constant sample size with zero variance.

Problem 2.3

1. No, because the sums of the rows or of the columns of Δ is not equal to zero, so that, according to the first part of Problem 2.2, the sample size cannot be constant.

2. $\mathbb{V}(I_k) = \frac{6}{25} = \pi_k(1 - \pi_k)$. This leads to the quadratic equation $\pi_k^2 - \pi_k + \frac{6}{25}$ 0 with solutions $\frac{3}{5}$ and $\frac{2}{5}$. Hence we have $\pi_1 = \pi_2 = \pi_3 = \frac{3}{5} > \pi_4 = \pi_5 =$ With $\boldsymbol{\pi} = (\pi_1, \pi_2, \pi_3, \pi_4, \pi_5)^t$ we have $\boldsymbol{\Pi} = \boldsymbol{\Delta} + \boldsymbol{\pi}\boldsymbol{\pi}^t$ and

$$\boldsymbol{\Pi} = \frac{1}{5} \begin{pmatrix} 3 & 3 & 3 & 0 & 0 \\ 3 & 3 & 3 & 0 & 0 \\ 3 & 3 & 3 & 0 & 0 \\ 0 & 0 & 0 & 2 & 2 \\ 0 & 0 & 0 & 2 & 2 \end{pmatrix}$$

One sees that some pairs of units cannot be selected together and that some are always selected together. We are led to the conclusion that only two

and $p(\{4,5\}) = \frac{2}{5}$.

Problem 2.4

1. $S_Y^2 = \frac{1}{N-1}\left(\sum_{i=1}^{N} Y_i^2 - \frac{1}{N}\left(\sum_{i=1}^{N} Y_i\right)^2\right)$

2. $\hat{\mathbb{V}}_{srs}(\bar{Y}_s) = (1 - \frac{n}{N})\frac{1}{n}S_Y^2$

3. Because of the independence of those random variables I_i we have $\mathbb{V}_{BE}(\hat{\bar{Y}})$
   $\frac{1}{n^2}\sum_{i=1}^{N}\mathbb{V}(I_i)Y_i^2$. With $\mathbb{V}(I_i) = \frac{n}{N}(1 - \frac{n}{N})$ we get

$$\mathbb{V}_{BE}(\hat{\bar{Y}}_s) = (1 - \frac{n}{N})\frac{1}{nN}\sum_{i=1}^{N}Y_i^2$$

4.
$$\mathbb{V}_{BE}(\tilde{Y}_s) \approx \mathbb{E}\left(\frac{\frac{1}{N}\sum_{i=1}^{N}I_iY_i}{\frac{1}{N}\sum_{i=1}^{N}I_i} - \bar{Y}\right)^2$$

which is approximately equal to

$$\frac{\mathbb{E}\left(\frac{1}{N}\sum_{i=1}^{N}(Y_i - \bar{Y})\right)^2}{\left(\frac{n}{N}\right)^2} = \left(\frac{N}{n}\right)^2\mathbb{V}\left(\frac{1}{N}\sum_{i=1}^{N}I_i(Y_i - \bar{Y})\right)$$

Because of a) the independence of the I_i and b) $\mathbb{V}(I_i) = \frac{n}{N}(1 - \frac{1}{N})$, we obtain

$$\mathbb{V}_{BE}(\tilde{Y}_s) = (1 - \frac{n}{N})\frac{1}{nN}\sum_{i=1}^{N}(Y_i - \bar{Y})^2$$

5. The Horvitz-Thompson estimate of $\sum_{i=1}^{N}Y_i^2$ is obviously $\frac{N}{n}\sum_{i\in s}Y_i^2$ and so one obtains

$$\hat{\mathbb{V}}_{BE}(\hat{\bar{Y}}_s) = (1 - \frac{n}{N})\frac{1}{n^2}\sum_{i\in s}Y_i^2$$

For $\hat{\mathbb{V}}_{BE}(\tilde{Y}_s)$ we must use 2.30 with $\hat{N} = \frac{n_s}{n}N$, so that we get

$$\hat{\mathbb{V}}_{BE}(\tilde{Y}_s) = (1 - \frac{n}{N})(\frac{1}{n_s})^2\sum_{i\in s}(Y_i - \tilde{Y}_s)^2$$

6. Under simple random sampling $\hat{\bar{Y}}_s = \tilde{Y}_s$ and the variances are also equal. With Bernoulli sampling we have

$$\frac{\mathbb{V}(\hat{\bar{Y}}_s)}{\mathbb{V}(\tilde{Y}_s)} = 1 + CV^2$$

where CV is the coefficient of variation for Y_i in the population. In addition, $\mathbb{V}_{BE}(\tilde{Y}_s) = \frac{N-1}{N}\mathbb{V}_{srs}(\hat{\bar{Y}}_s)$. Hence, under a random sample size with constant inclusion probabilities the weighted mean is practically equivalent to the ordinary sample mean under simple random sampling.

$$\sum_{i=1}^{N} i = \frac{N(N+1)}{2}$$

$$\sum_{i=1}^{N} i^2 = \frac{N(N+1)(2N+1)}{6}$$

$$S_Y^2 = \frac{N(N+1)}{12}$$

or direct calculations to get

$$\bar{Y} = 50.5, \ S_Y^2 = 841\frac{2}{3}$$

$$\mathbb{V}_{srs}(\hat{\bar{Y}}_s) = 75.75, \ \mathbb{V}_{BE}(\hat{\bar{Y}}_s) = 304.515, \ \mathbb{V}_{BE}(\tilde{Y}_s) = 74.9925$$

This clearly confirms the theoretical findings. The exact probability based on the binomial distribution is 0.9364 and the approximate probability, based on the fact that $\frac{n_s - Np}{\sqrt{Np(1-p)}}$ (where $p = \frac{n}{N}$) follows an approximately standard normal distribution, is 0.9044.

Problem 2.5

1. $\bar{Y}_{s_k} = 46 + (k-1)$. $\pi_i \equiv \frac{1}{10}$. $\pi_{ij} = \frac{1}{10}$ if i and j are in the same sample otherwise $\pi_{ij} = 0$. Therefore, one cannot estimate the variance with a single sample.

2.

$$\mathbb{E}(\hat{\bar{Y}}_{sys}) = \sum_{k=1}^{10} p(s_k)\bar{Y}_{s_k} = \bar{Y} = 50.5$$

3.

$$\mathbb{V}(\hat{\bar{Y}}_{sys}) = \sum_{k=1}^{10} p(s_k)(\hat{\bar{Y}}_{sys} - \bar{Y})^2 = \frac{1}{10}\sum_{k=1}^{10}(46 + k - 1 - 50.5)^2 = 8.25$$

which is about nine times smaller than under simple random sampling.

4. The classical variance estimate is a constant equal to 82.5 for all samples. It overestimates the correct variance by a factor of ten.

5. One has by Eq. 2.72 and Problem 2.4

$$\mathbb{V}(\hat{\bar{Y}}_{sys}) = \frac{0.9}{1}\frac{10 \cdot 10 - 1}{10^2(10-1)}841.67(1 + (10-1)\rho)$$

and therefore $\rho = -0.1$ which is very close to the minimum possible value $\rho = -\frac{1}{M-1} = -\frac{1}{9} = -0.11$ (the variance must be positive).

$\{\qquad\} + 10(\qquad 1)\qquad\qquad 10$ (i.e. the linear trend in these data coincides with the samples) one obtains $\mathbb{V}(\hat{Y}_{sys}) = 825$. This is 100 times large than from the former systematic sample and roughly 10-fold greater than under simple random sampling! The intra-cluster correlation is in this case $\rho = 0.99$ and close to the maximum. It is possible to balance the systematic samples to achieve even $\bar{Y}_{s_k} = 50.5 \ \forall k$, hence, $\mathbb{V}(\hat{Y}_{sys}) = 0$. It suffices to take

$$S_0 = \{1, 21, 41, 61, 81\}, \quad S_1 = \{20, 40, 60, 80, 100\}$$

and set $s_k = \{S_0 + (k-1)\} \cup \{S_1 - (k-1)\}$. The intra-cluster correlation is at its minimum possible value of $\rho = -0.11$. The population obviously displays a perfect linear trend with respect to the labels. Under a random permutation, this structure will be destroyed and the variance found under systematic sampling will be close to the variance under simple random sampling.

7. It suffices for us to consider each systematic sample as a cluster and to take a sample of at least two clusters under simple random sampling.

Problem 2.6

$$\hat{p}_d = \frac{1}{n} \sum_{i \in s} Y_i$$

$$\mathbb{V}(\hat{p}_d) = (1 - \frac{n}{N}) \frac{1}{n} \frac{N}{N-1} p_d(1 - p_d)$$

$$\hat{\mathbb{V}}(\hat{p}_d) = (1 - \frac{n}{N}) \frac{1}{n-1} \hat{p}_d(1 - \hat{p}_d)$$

Problem 2.8

The variance is $\mathbb{V}(\hat{Y}) = \sum_{i=1}^{N} Y_i^2 \frac{1-\pi_i}{\pi_i}$. The Lagrange function is

$$L(\pi_i, \lambda) = \sum_{i=1}^{N} Y_i^2 \frac{1 - \pi_i}{\pi_i} + \lambda(\sum_{i=1}^{N} \pi_i - n)$$

Differentiating with respect to λ yields the constraint and for a given π_{j_0} the relationship $-\frac{Y_{j_0}^2}{\pi_{j_0}^2} + \lambda = 0$. This leads to $\sqrt{\lambda} = \frac{1}{n} \sum_{i=1}^{N} Y_i$ and consequently to the **PPS** scheme

$$\pi_i = \frac{nY_i}{\sum_{i=1}^{N} Y_i}$$

Problem 2.9

The strata mean are $\bar{Y}_k = 10.5 + (k-1)20$ and the within strata variances are a constant, namely $S_{k,Y}^2 = 35$. One obtains

$$\mathbb{V}(\hat{Y}_{strat}) = \sum_{k=1}^{5} \left(\frac{20}{100}\right)^2 \frac{1 - \frac{2}{20}}{2} 35 = 3.15$$

sample size of 10.

Problem 2.10

The variances $S_{u_i}^2$ within each primary unit are always equal to 9.1666. Totals for the primary units are $T_i = 55 + (i-1)100, i = 1,2\ldots 10$ with $\sum_{i=1}^{10} T$ $5050 = 100\bar{Y}$. The variance of the totals for the primary units is $S_{U_I}^2$ $91666.6666 = 9.1666^4$. Then, the variance of the estimate for the total is given by Eq. 2.56

$$V(\hat{Y}_{II\pi}) = 10^2 \frac{1 - \frac{5}{10}}{5} 9.166610^4 + \frac{10}{5}\sum_{i=1}^{10} 10^2 \frac{1 - \frac{2}{10}}{2} 9.1666 \approx 9210^4$$

This produces a population mean of

$$V(\hat{\bar{Y}}_{II\pi}) = \frac{1}{N^2}V(\hat{Y}_{II\pi}) \approx 92$$

which is slightly larger than gained via simple random sampling. Hence, the best scheme is stratified random sampling, followed by simple random sampling, whereas systematic sampling can be either very efficient or entirely inefficient.

C.2 Chapter 3

Problem 3.1

The ordinary sample mean for all elements drawn is in this case the HT estimator of the mean. Tedious but simple calculations with Eq. 3.18 yield $\sigma_3^2 = 0.5$, $\sigma_2^2 = 10$ and $\sigma_1^2 = 916.667$, obviously by far the largest component of variance. This leads to $V(\bar{Y}_s) = 91.75$ which is, as can be intuitively expected in this simple example, practically the same as the value under two-stage sampling.

Problem 3.2

We have $\hat{R} = \frac{29.07}{131.25} = 0.22$. The estimated correlation is

$$\hat{\rho} = \frac{\hat{S}_{X,Y}}{\sqrt{\hat{S}_X^2}\sqrt{\hat{S}_Y^2}}$$

The ratio estimate of the mean is

$$\hat{\bar{Y}}_{ratio} = \bar{X}\hat{R} = 118.32 \times 0.22 = 26.21ha$$

Variance according to Theorem 3.3.1 is

$$\hat{V}(\hat{\bar{Y}}_{ratio}) = \frac{1 - \frac{n}{N}}{n}\frac{1}{n-1}\sum_{i\in s}(Y_i - \hat{R}X_i)^2$$

$$\frac{1}{n-1}\sum_{i\in s}(Y_i - \hat{R}X_i)^2 = \frac{1}{n-1}\sum_{i\in s}(Y_i - \bar{Y}_s - \hat{R}(X_i - \bar{X}_s))^2$$

which is equal to

$$\hat{S}_X^2 + \hat{R}^2\hat{S}_Y^2 - 2R\hat{S}_{X,Y} = 708 + 0.22^2 9173 - 2 \times 0.22 \times 1452.6 = 512.82$$

This gives

$$\sqrt{\hat{\mathbb{V}}(\hat{\bar{Y}}_{ratio})} = \sqrt{\frac{1 - \frac{100}{2010}}{100}512.82} = \sqrt{4.87} = 2.21$$

and provides an approximate 95% confidence interval $26.211.96 \times 2.21 = [21.88, 30.54]$. The g weight variance estimate Eq. 3.37 is

$$\left(\frac{\bar{X}}{\bar{X}_s}\right)^2 \frac{1 - \frac{100}{2010}}{100}512.82 = 4.87 \times 0.81 = 3.95$$

which yields the approximate 95% confidence interval $26.213.9 = [22.31, 30.11]$.

Problem 3.3

The condition $\hat{\mathbb{V}}(\hat{\bar{Y}}_{ratio}) < \hat{\mathbb{V}}(\bar{Y}_s)$ is, according to Problem 3.2 above, equivalent to

$$\hat{S}_X^2 + \hat{R}^2\hat{S}_Y^2 - 2R\hat{S}_{X,Y} < \hat{S}_Y^2$$

and hence to

$$\hat{\rho} > \frac{1}{2}\frac{\frac{\hat{S}_X}{\bar{X}_s}}{\frac{\hat{S}_Y}{\bar{Y}_s}}$$

or, as expressed with the coefficients of variation

$$\hat{\rho} = \frac{1}{2}\frac{\widehat{CV}_X}{\widehat{CV}_Y}$$

C.3 Chapter 4

Problem 4.1

Because the inclusion circles for the trees either overlap or are disjoined it is impossible, in a natural forest, to always draw just one tree. Mathematically, of course, one could construct a forest whose trees are sufficiently far apart to preclude any such overlapping of their inclusion circles, and then precisely define forest area F as the union of those circles.

Problem 4.2

The radius of the limit circle is given by

$$R_i(m) = \frac{DBH_i(cm)}{2\sqrt{k}} = \frac{DBH_i(cm)}{4}$$

$$(\ ) = 0, \quad (\ ) = 1, \quad (\ ) = 1,$$
$I_4(x) = 0$, $I_5(x) = 1$, $I_6(x) = 1$ and $I_7(x) = 1$. Note that we set $Y_3 = 0$ and $Y_7 = 0$ because these trees are below the threshold even if they are included with the angle count. By definition they do not belong to the population. We obtain $n(x) = 3$ and $Y(x) = 4n(x) = 12\frac{m^2}{ha}$.

Problem 4.3

1. It follows from Problem 2.7!
2. The condition
$$\mathbb{E}\big(n(x) \mid I_i(x) = 1\big) - \mathbb{E}(n(x)) = 1$$
   means that the number of trees sampled does not depend on where we are in the forest. Hence, a forest will appear globally homogenous with respect to the spatial distribution of the trees.
3. We have $\pi_i = \frac{G}{k\lambda(F)}$ so that under the above condition
$$\mathbb{V}(Y(x)) = \mathbb{V}(G(x)) = \frac{1}{\lambda(F)} \sum_{i=1}^{N} k \frac{G_i^2}{G_i} = k\bar{G}$$
   because the first term of the generalized Yates-Grundy formula vanishes. Furthermore, $G(x) = kn(x)$ and
$$\mathbb{V}(G(x)) = \mathbb{V}(kn(x)) = k^2 \mathbb{V}(n(x))$$
   Because we have proved that $\mathbb{V}(G(x)) = k\bar{G}$ and because $\bar{G} = k\mathbb{E}(n(x))$ we obtain $k^2\mathbb{V}(n(x)) = k^2\mathbb{E}(n(x))$ and finally $\mathbb{V}(n(x)) = \mathbb{E}(n(x))$.
4. A discrete random variable whose variance is equal to its expected value must have the famous Poisson distribution. Hence, for a given forest and a random point $x \in F$ the number of trees sampled follows such a distribution. Conversely, if we have a Poisson forest (in the sense that given N the number of trees in F, their locations are independently and uniformly distributed in F), then we can show that for a given x the random variable $n(x)$ follows a Poisson distribution (see Holgate, 1967).

Problem 4.4

Between any two trees the bisector determines which is the nearest neighbor to the point x. All the bisectors form a tessellation of the rectangle. Each tree lies within a single convex cell, its Voronoi polygon, with surface area v_i. The Horvitz-Thompson estimate of the spatial mean is then $\sum_{i=1}^{N} \frac{Y_i}{v_i}$. The drawback is that one has to know the locations of **all the trees** (not just those being sampled) prior to sampling in order to calculate the inclusion probabilities. Of course, one can attempt to model the surface area of a Voronoi polygon but that would then be no longer within the framework of design-based inference, and practically untractable except for Poisson forests. This idea can be generalized to Voronoi polygons of higher order to deal with the nearest k neighbors, but the geometry gets complicated. The reader can consult (Kleinn and Vilcko, 2006) to see some beautiful pictures.

The set of trees sampled from the n_2 points $x \in s_2$ is $s = \cup_{x \in s_2} s_2(x) = \{T_i \neq 0\}$ with $T_i = \sum_{x \in s_2} I_i(x)$. The number of times T_i that the i-th tree is sampled, from different points, is a binomial random variable with parameter (n_2, π_i). We have

$$Y(x) = \frac{1}{\lambda(F)} \sum_{i=1}^{N} \frac{I_i(x) Y_i}{\pi_i}$$

$$\hat{Y} = \frac{1}{n_2} \sum_{x \in s_2} Y(x) = \frac{1}{n_2 \lambda(F)} \sum_{i=1}^{N} \frac{Y_i}{\pi_i} \sum_{x \in s_2} I_i(x)$$

Consequently,

$$\hat{Y} = \frac{1}{n_2 \lambda(F)} \sum_{i=1}^{N} \frac{T_i Y_i}{\pi_i} = \hat{Y}_{pwr}$$

which establishes the equivalence. To calculate the variance we note that

$$\mathbb{E}(T_i) = n_2 \pi_i, \quad \mathbb{V}(T_i) = n_2 \pi_i (1 - \pi_i)$$

$$\mathbb{COV}(T_i, T_j) = \mathbb{E}(T_i T_j) - n_2^2 \pi_i \pi_j$$

Furthermore

$$T_i T_j = \sum_{x \in s_2} I_i(x) I_j(x) + \sum_{x \neq y \in s_2} I_i(x) I_j(y)$$

Hence, we arrive at $\mathbb{E}(T_i T_j) = n_2 \pi_{ij} + n_2(n_2 - 1)\pi_i \pi_j$ and finally

$$\mathbb{COV}(T_i, T_j) = n_2(\pi_{ij} - \pi_i \pi_j)$$

which leads to the variance formula given for the local density. The infinite population approach is therefore equivalent to a finite setup with the pwr-Hansen-Hurwitz estimator used in place of the Horvitz-Thompson estimator. As already mentioned, the infinite population approach is mathematically much simpler.

Problem 4.6

The inclusion circle $\tilde{K}_i$ for the i-th tree is the circle parallel to it and at a distance r_α from the circular cross-section of the tree: by elementary geometry its surface area is given by

$$\lambda(\tilde{K}) = G_i + L_i r_\alpha + \pi r_\alpha^2$$

where G_i is the basal area of the tree and L_i is the length of the circumference of the cross-section at $1.30m$. This a special case of the famous Steiner's formula, which remains valid for parallel sets of any convex body (Santal´ 1976). The inclusion probability for the i-th tree is therefore

$$\pi_i^\alpha = \frac{G_i + L_i r_\alpha + \pi r_\alpha^2}{\lambda(F)}$$

$$S_\alpha = \mathbb{E}(\hat{S}_\alpha) = \bar{G} + \bar{L}r_\alpha + \bar{N}\pi r_\alpha^2 \quad \text{for } \alpha = 1, 2, 3$$

where we have set $\bar{G} = \frac{1}{\lambda(F)}\sum_{i=1}^{N} G_i$, $\bar{L} = \frac{1}{\lambda(F)}\sum_{i=1}^{N} L_i$ and $\bar{N} = \frac{N}{\lambda(F)}$.

Solving the above linear system of three equations with three unknown and using $\hat{S}_\alpha$ instead of S_α we can estimate simultaneously the dendrometric quantities $\bar{G}$, the basal area density, $\bar{L}$ the density of the total perimeter length of the cross-sections (assumed to be convex and not only circular) and the stem density $\bar{N}$. Calculations for variances and covariances are slightly more complicated. The diligent reader can consult (Masuyama, 1953) for more details. It is fair to say that this interesting finding is more a mathematical curiosity than a routine procedure. However, it also demonstrates that, when applying such standard methods, it is important to use the distance from the points to the centers of the trees (and not to their outer bark) in order to eliminate a systematic positive bias (i.e. an overestimation).

Problem 4.7

We utilize Theorem 4.2.2. The constant inclusion probabilities cancel out (that is, the local densities are proportional to the observed absolute frequencies) and we immediately have

$$\hat{P} = \frac{\sum_{x \in s_2} r(x)}{\sum_{x \in s_2} n(x)} = \frac{\sum_{x \in s_2} n(x)\hat{p}(x)}{\sum_{x \in s_2} n(x)}$$

Noting that $r(x) - \hat{P}n(x) = n(x)(\hat{p}(x) - \hat{P})$ we easily obtain from Theorem 4.2.2 the estimated variance as

$$\hat{\mathbb{V}}(\hat{P}) = \frac{1}{n_2(n_2 - 1)} \sum_{x \in s_2} \left(\frac{n(x)}{\bar{n}}\right)^2 (\hat{p}(x) - \hat{P})^2$$

Hence, our estimate is the overall proportion of trees having the given characteristic, ignoring the plot structure of these data. It is also a weighted mean of the proportions within plots, the weights being the number of trees in those plots. The variance structure is formally equivalent to cluster sampling (a "plot" being now a cluster of trees) as shown in Theorem 4.3.1. **However, a word of caution**: although direct use of standard statistical software packages to calculate the weighted mean of $\hat{p}(x)$ with weights $n(x)$ does yield the correct point estimate $\hat{P}$, the variance obtained in this way is incorrect! The reason is that, in this context, the weight $n(x)$ is itself a random variable. Note also that the usual variance estimate based on a binomial distribution, i.e. $\frac{1}{n_2}\hat{P}(1 - \hat{P})$, is obviously erroneous.

Problem 4.8 From the first group we can say that $\hat{\mathbb{V}}(Y(x)) = 20 \cdot 2412 = 48412$. From the second group we can infer that

$$\hat{\mathbb{V}}(\hat{Y}_c) = 4028 = \frac{\hat{\mathbb{V}}(Y(x))}{8\bar{M}}\left(1 + \rho(\bar{M} - 1) + \rho\frac{\hat{\mathbb{V}}(M(x))}{\bar{M}}\right)$$

from which we then conclude that $\hat{\rho} = 0.44$.

Problem 5.1

ANOVA results can be summarized as follows:

Stand	Number of plots	$\bar{N}$	$\bar{G}\frac{m^2}{ha}$
4	3	655.55	28.54
5	7	181.48	26.24
6	9	152.38	20.66

Differences between stands are significant for the number of stems ($R^2 = 0.46$) but not for the basal area ($R^2 = 0.05$). The point estimates $\hat{Y}$ and their standard errors $err(\hat{Y}) = \sqrt{\hat{V}(\hat{Y})}$ for a one-phase purely terrestrial inventory are

Variable	$\hat{Y}$	$err(\hat{Y})$
Stem	245.61	65.51
Basal area	24.54	3.59

They are, as they should be, the same as the values shown in chapter 8, table 8.4.

For two-phase estimates with $n_1 = \infty$ we take the weighted means of the stand means – the weights being equal to the true proportions of their surfaces areas. Then, the variances are due to the residual terms. One obtains

Variable	$\hat{Y}$	$err(\hat{Y})$
Stem	290.43	52.70
Basal area	25.02	3.63

Two-phase point estimates are closer to the true values (given in table 8.4) than are one-phase estimates. The reduction in variance is, as expected from the ANOVA, only relevant for stem density.

C.5 Chapter 6

Problem 6.1

Using the technique presented with Problem 4.7 we obtain the estimated probabilities

$$\hat{p} = (\hat{p}_{11}, \hat{p}_{12}, \hat{p}_{21}, \hat{p}_{22})^t = (0.19898, 0.1352, 0.32143, 0.34439)^t$$

mated $(4, 4)$ variance-covariance matrix of $\hat{p}$ as

$$\hat{\Sigma}_{\hat{p}} = \begin{pmatrix} 0.0030248 & 0.0017917 & -0.001591 & -0.0024271 \\ 0.0017917 & 0.0015503 & -0.000766 & -0.0017800 \\ -0.001591 & -0.000766 & 0.0019997 & 0.0009329 \\ -0.002471 & -0.001780 & 0.0009329 & 0.0033179 \end{pmatrix}$$

For the first method, the vector $\hat{p}$ is reduced to its initial three components and the variance covariance-matrix $\hat{\Sigma}_{\hat{p}}$ to its first three rows and three columns. One then obtains $\hat{h}_{11} = 0.0250694$ and the Hessian is found to be

$$\mathbf{H} = (0.14541, -0.52041, -0.33418)$$

This leads to the scalar $\hat{\Sigma}_{\hat{h}} = 0.0002294$ and, finally, to

$$\chi^2 = 2.739$$

with one degree of freedom. Our observed significance level is then P^{obs} 0.098. The hypothesis of independence between species and state of health is therefore compatible with these data.

For the second method, we obtain for the vector the log-probabilities

$$\hat{l} = (-1.614551, -2.001, -1.134975, -1.065981)^t$$

and, by Eq. 6.58, the estimated variance-covariance matrix

$$\hat{\Sigma}_{\hat{l}} = \begin{pmatrix} 0.0763975 & 0.066601 & -0.0248820 & -0.0360590 \\ 0.0666010 & 0.084815 & -0.0176220 & -0.0382240 \\ -0.0248820 & -0.017622 & 0.0193549 & 0.0084270 \\ -0.0360590 & -0.038224 & 0.0084270 & 0.0279747 \end{pmatrix}$$

The design matrix $\mathbf{X}$ is

$$\mathbf{X} = \begin{pmatrix} 1 & 1 & 1 \\ 1 & -1 & -1 \\ -1 & 1 & -1 \\ -1 & -1 & 1 \end{pmatrix}$$

and we choose, as suggested in Eq. 6.59, $\mathbf{F} = \mathbf{X}^t$. We then arrive at

$$\hat{f} = \mathbf{F}\hat{l} = (-1.414595, 0.317455, 0.455443)^t$$

with the variance-covariance matrix

$$\hat{\Sigma}_{\hat{f}} = \begin{pmatrix} 0.5921721 & 0.0370761 & -0.043719 \\ 0.0370761 & 0.0396361 & -0.002465 \\ -0.043719 & -0.002465 & 0.077336 \end{pmatrix}$$

This leads to a weighted least squares estimate of

$$\hat{\theta} = (\hat{u}_{1(1)}, \hat{u}_{2(1)}, \hat{u}_{12(11)})^t = (-0.353649, 0.0793637, 0.1138607)^t$$

$$\hat{\Sigma}_{\hat{f}} = \begin{pmatrix} 0.0370108 & 0.0023173 & -0.002732 \\ 0.0023173 & 0.0024773 & -0.000154 \\ -0.002732 & -0.0001540 & 0.0048335 \end{pmatrix}$$

Thus, we have the scalar $\hat{\theta}_2 = \hat{u}_{12(11)} = 0.1138607$ with an estimated scalar variance of $\hat{\Sigma}_{22} = 0.048335 = \hat{\sigma}_2^2$. We then want to test the hypothesis H $\theta_2 = u_{12(11)} = 0$, which is equivalent to the independence between species and state of health. This is straightforward if one considers the test statistic $\chi^2 = \frac{\hat{\theta}_2^2}{\hat{\sigma}_2^2} = 2.68$ with an observed significance level of $P^{obs} = 0.10$. The hypothesis of independence is therefore compatible with the data. Note that both test statistics are very close to each other. Although the second method is, in this case, more difficult to implement, it can, in contrast to the first, be utilized for more complex problems.

For a classic χ^2 square we pool these 20 observations to obtain the contingency table

Species	Healthy	Damaged	Totals
Conifers	78	53	131
Broadleaf	126	135	261
Totals	204	188	392

The classic chi-square is found to be $\chi^2 = 4.43$ with $p^{obs} = 0.035$, which is significant, but .... wrong!

C.6 Chapter 7

Problem 7.1

The Punctual Kriging estimate is simply a linear interpolator between two neighboring points of x_0. Note in particular that $\hat{Y}(x_0)$ is an exact interpolator, i.e. $\hat{Y}(x_0) = Y(x_k)$ if $x_0 = x_k$, $k = 2, 3$. Adding extra observations to the left or to the right does not in any way alter this fact. Only immediate neighbors contribute to the Kriging estimate at x_0. This so-called **screen effect** is partially removed as soon as a nugget effect is introduced, although the weights λ_i will decrease with the distance between the x_i and x_0. An extreme case occurs with a pure nugget effect where the weights are a constant equal to $\frac{1}{n_2}$. Qualitatively similar findings hold for other variograms and also in the plane, where that screen effect will obviously be less pronounced (because only points on a line can have a perfect screen effect). See Wackernagel (2003) for numerical examples.

To calculate the term $\bar{\gamma}(x_j, V_0)$ we consider K points $x_{0k} \in V_0$, $k = 1, 2 \ldots K$ lying on a systematic grid and we set

$$\bar{\gamma}(x_j, V_0) := \frac{1}{K} \sum_{k=1}^{K} \gamma(x_i - x_{0k})$$

which is of course only an approximation. At each point $x_{0k} \in V_0$ we obtain the Punctual Kriging estimate $\hat{Y}(x_{0k}) = \sum_{i \in s_2} \lambda_i(x_{0k}) Y(x_i)$ with the Lagrange Multiplier $\mu(x_{0k})$ and a mean square error of

$$\sigma^2(x_{0k}) = \sum_{i \in s_2} \lambda_i(x_{0k}) \gamma(x_i - x_{0k}) + \mu(x_{0k})$$

Because of the linearity of the Kriging equations one can see that the weights $\tilde{\lambda}_i(V_0) = \frac{1}{K} \sum_{k=1}^{K} \lambda_i(x_{0k})$ and the Lagrange Multiplier $\tilde{\mu}(V_0) = \frac{1}{K} \sum_{k=1}^{K} \mu($ are solutions of the Block Kriging equations (up to the afore-mentioned approximation). Hence, the Block Kriging estimate can be obtained by taking the mean of the Punctual Kriging estimates, i.e.,

$$\hat{Y}(V_0) = \frac{1}{K} \sum_{k=1}^{K} \hat{Y}(x_{0k})$$

On the other hand, if the $\tilde{\lambda}_i(V_0)$ and the approximation

$$\bar{\gamma}(x_j, V_0) = \frac{1}{K} \sum_{k=1}^{K} \gamma(x_j - x_{0k})$$

are used in Eq. 7.16 we get for the mean square error

$$\sigma^2(V_0) = \sum_{i \in s_2} \left(\frac{1}{K} \sum_{k=1}^{K} \lambda_i(x_{0k}) \right) \left(\frac{1}{K} \sum_{k=1}^{K} \gamma(x_i - x_{0k}) \right) + \tilde{\mu}(V_0) - \bar{\gamma}(V_0, V_0)$$

As a very coarse approximation the first two terms can be replaced by the mean of the mean square errors, i.e. by $\frac{1}{K} \sum_{k=1}^{N} \sigma^2(x_{0k})$. Thus, for point estimation Block Kriging can be approximated by taking the mean of punctual Kriging estimates. The mean of the Punctual Kriging variances is likely to be an overestimation of the true Block Kriging variance.

C.7 Chapter 9

Problem 9.1

The following program is written with the linear algebra software SAS/IML. First, the data are read into a matrix X, which is then extended to a matrix *newx* with three more columns containing the cumulative frequencies, the volume and the square of the volume. It calculates the minimum $\tilde{\gamma}$ for two

value of the functions $g()$ at the optimum. The program stores in the matrix *gam* the results as a function of that threshold in order to obtain plots such as those in Fig. 10.2.

```
n = nrow(X);p = ncol(X) + 5; cumx = 0;
    do i = 1 to nrow(x);
        newx[i, 1] = x[i, 1];newx[i, 2] = x[i, 2]/100;
        newx[i, 3] = cumx + newx[i, 2];cumx = newx[i, 3];
        newx[i, 4] = exp(-12.4 + 3.826 * log(x[i, 1]) - 0.007 * (log(x[i, 1])) * *
        newx[i, 5] = newx[i, 4] * *2;
    end;
gam = J(nrow, 5, 1);
A0 = 1;AN = n;F0 = 0;FN = 1;
do A1 = 1 to n;
    F1 = newx[A1, 3]; DBH1 = newx[A1, 1];y2 = 0;meany = 0;
    do k1 = A0 to A1;
        y2 = y2 + (newx[k1, 2]/(F1 - F0)) * newx[k1, 5];
        meany = meany + newx[k1, 2] * newx[k1, 4];
    end;
    g1y = sqrt(y2); termy2 = (F1 - F0) * sqrt(y2);
    if A1 < AN then
    do;
    y2 = 0;
        do k1 = A1 + 1 to AN;
            y2 = y2 + (newx[k1, 2]/(FN - F1)) * newx[k1, 5];
            meany = meany + newx[k1, 2] * newx[k1, 4];
        end;
    g2y = sqrt(y2);termy2 = termy2 + (FN - F1) * sqrt(y2);
    end;
    gamma1 = termy2/meany;
    gam[A1, 1] = DBH1;gam[A1, 2] = gamma1;
    gam[A1, 3] = g1y; gam[A1, 4] = g2y;
end;
minim = 1000;
do A1 = 1 to AN;
    if gam[A1, 2] < minim then
    do;
        threshold = gam[A1, 1];
        gamma = gam[A1, 2];
        g1 = gam[A1, 3]; g2 = gam[A1, 4];
        minim = gam[A1, 2];
    end;
end;
print threshold, gamma, g1, g2;
```

same matrix $newx$ as before. To simplify the structure, it first calculates the mean volume $meany$ in a separate step.

```
meany=0;
do I = 1 to n;
meany=meany+newx[i,2]*newx[i,4];
end;
minim = 1000; A0 = 1; AN = n; F0 = 0; FN = 1;
do A2 = 1 to AN;
    F2 = newx[A2, 3]; DBH2 = newx[A2, 1];
        if A2 < AN then
        do;
            y2 = 0;
            do k1 = A2 + 1 to AN;
                y2 = y2 + (newx[k1, 2]/(FN − F2)) * newx[k1, 5];
            end;
            g3y = sqrt(y2);
            end;
        if A2 = AN then
        do;
            g3y = newx[k1, 4];
        end;
    do A1 = 1 to A2;
        F1 = newx[A1, 3]; DBH1 = newx[A1, 1]; y2 = 0;
            do k1 = A0 to A1;
                y2 = y2 + (newx[k1, 2]/(F1 − F0)) * newx[k1, 5];
            end;
        g1y = sqrt(y2);
        if A1 < A2 then
        do;
        y2 = 0;
            do k1 = A1 + 1 to A2;
                y2 = y2 + (newx[k1, 2]/(F2 − F1)) * newx[k1, 5];
            end;
        g2y = sqrt(y2);
        end;
        if A1 = A2 then
            do;
                g2y=newx[A1,4];
            end;
        gamma = ((F1 − F0) * g1y + (F2 − F1) * g2y + (1 − F2) * g3y)/meany
    if gamma < minim then
        do;
            threshold1 = dbh1; threshold2 = dbh2;
            mingamma = gamma;
```

$$minim = gamma;$$
```
    end;
  end; (of A1 loop)
end; (of A2 loop)
print, " optimal circles", ;
print threshold1 threshold2 mingamma g1 g2 g3;
```

Therefore, the numerical results from the data provided in this exercise are:

2 circles
$DBH1 = 32cm$, $\tilde{\gamma}_{opt} = 1.18$, $g_1 = 0.36m^3$, $g_2 = 2.18m^3$.
3 circles
$DBH1 = 24cm$, $DBH2 = 41cm$, $\tilde{\gamma}_{opt} = 1.08$, $g_1 = 0.18m^3$, $g_2 = 0.96m$
$g_3 = 2.85m^3$.

Problem 9.2 The equations

$$\frac{\partial \mathbb{E}_\omega \mathbb{V}(\hat{Y}_c^*)}{\partial n_2} = \frac{\partial \mathbb{E}_\omega \mathbb{V}(\hat{Y}_c^*)}{\partial n_1} = 0$$

lead to

$$n_1 = \frac{\tilde{C}\sqrt{1 + \theta_2}\beta_2}{\bar{M}\sqrt{\tilde{c}_1}\psi}, \quad n_2 = \frac{\tilde{C}\Lambda}{\sqrt{\bar{M}}\sqrt{\tilde{c}_2}\psi}$$

where

$$\psi = \sqrt{c_{21}}\tilde{\gamma}\bar{Y} + \sqrt{c_{22}}\,|\,R\,| + \frac{1}{\bar{M}}\sqrt{\tilde{c}_2}\Lambda + \sqrt{c_1}\sqrt{1 + \theta_2}\beta_2$$

The absolute lower bound MAV for the anticipated variance is obtained by inserting the above n_1 and n_2 into the last expression from Eq. 9.26. One then arrives at

$$\text{MAV}(\hat{Y}_c^*) = \frac{\psi^2}{\tilde{C}}$$

Given the budget and the parameters that occur in factor ψ one can easily determine if the objective for the error can be achieved in the best possible case. Conversely, given the requirement for the error, one can identify which budget would be necessary.

Problem 9.3

We must minimize the expected costs

$$n_1\bar{M}c_1 + \phi(n_2) + n_2\bar{M}c_{2c} + n_2\bar{M}\left(c_{21}\sum_{i=1}^{N}\pi_i + \sum_{i=1}^{N}\pi_i p_i\right)$$

under the constraint given in Theorem 9.6.1

$$W = \mathbb{E}_\omega \mathbb{V}(\hat{Y}_c^*) = \frac{1}{n_2 \mathbb{E}M(x)\lambda^2(F)} \sum_{i=1}^{N} \frac{Y_i^{*2}}{\pi_i} + \frac{1}{n_2 \mathbb{E}M(x)\lambda^2(F)} \sum_{i=1}^{N} \frac{R_i^2}{\pi_i p}$$

$$+ \frac{1}{n_2 \mathbb{E}M(x)} \Lambda^2 + \frac{1}{n_1 \mathbb{E}M(x)}(1+\theta_2)\beta_2^2$$

By the inequality Eq. 9.10 we have

$$\sum_{i=1}^{N} \pi_i \geq \frac{\left(\sum_{i=1}^{N} g(Y_i^*)\right)^2}{\sum_{i=1}^{N} \frac{Y_i^{*2}}{\pi_i}}$$

the lower bound being achieved, for a given partition of the Y_i^* into classes and the step-wise constant π_i, for the $g(\cdot)$ function given in Eq. 9.7. Likewise, by Eq. 9.19, we have

$$\sum_{i=1}^{N} \pi_i p_i \geq \frac{\left(\sum_{i=1}^{N} |R_i|\right)^2}{\sum_{i=1}^{N} \frac{R_i^2}{\pi_i p_i}}$$

The lower bound is achieved when $\pi_i p_i$ is proportional to $|R_i|$. Hence, for a given n_2 and n_1 we must search for the optimal solution in the class π $\alpha_1 g(Y_i^*)$ and $\pi_i p_i = \alpha_2 |R_i|$. Using Eq. 9.8, we obtain the Lagrange function

$$L(\alpha_1, \alpha_2, \lambda) = n_1 \bar{M} c_1 + \phi(n_2) + n_2 \bar{M} c_{2c} +$$

$$n_2 \bar{M}\left(c_{21}\alpha_1\tilde{\gamma}\sum_{i=1}^{N} Y_i^* + c_{22}\alpha_2 \sum_{I=1}^{N} |R_i| + \right.$$

$$\lambda\left(\frac{1}{\lambda^2(F)n_2\bar{M}\alpha_1}\tilde{\gamma}\sum_{i=1}^{N} Y_i^* + \frac{1}{\lambda^2(F)n_2\bar{M}\alpha_2}\sum_{i=1}^{N} |R_i|\right)$$

$$\lambda\left(+\frac{\Lambda^2}{n_2\bar{M}} + \frac{(1+\theta_2)\beta_2^2}{n1\bar{M}} - W\right)$$

The equations

$$\frac{\partial L}{\partial \alpha_1} = \frac{\partial L}{\partial \alpha_2} = \frac{\partial L}{\partial \lambda}$$

are straightforward to solve, such that one obtains the relationships

$$\lambda(F)\pi_i = m_1 \frac{1}{\tilde{\gamma}\bar{Y}} g(Y_i^*), \quad \lambda(F)\pi_i p_i = m_2 \frac{|R_i|}{|R|}$$

where m_1 and m_2 are respective first-stage- and second-stage sample sizes

$$m_1 = \frac{\tilde{\gamma}\bar{Y}\sqrt{c_{21}} + \overline{|R|}\sqrt{c_{22}}}{\tau(n_1, n_2)} \frac{\tilde{\gamma}\bar{Y}}{n_2\bar{M}\sqrt{c_{21}}}$$

$$m_2 = \frac{\sqrt{}}{\tau(n_1,n_2)} - \frac{||}{n_2 \bar{M}\sqrt{c_{22}}}$$

where

$$\tau(n_1,n_2) = W - \frac{\Lambda^2}{n_2 \bar{M}} - \frac{(1+\theta_2)\beta_2^2}{n_1 \bar{M}}$$

(we use the approximation given by Eq. 9.15 $\overline{Y^*} \approx \bar{Y}$).

The total cost for this optimum choice is then

$$C_{min}(n_1,n_2) = n_1 \bar{M} c_1 + \phi(n2) + n_2 \bar{M} c_{2c} + \frac{\left(\tilde{\gamma}\bar{Y}\sqrt{c_{21}} + \overline{|R|}\sqrt{c_{22}}\right)^2}{\tau(n_1,n_2)}$$

Linearizing the travel cost as in Problem 9.2, we have to minimize, with respect to n_1 and n_2, the cost

$$\tilde{C}(n_1,n_2) = n_1 \bar{M} c_1 + n_2 \tilde{c}_2 + \frac{\left(\tilde{\gamma}\bar{Y}\sqrt{c_{21}} + \overline{|R|}\sqrt{c_{22}}\right)^2}{\tau(n_1,n_2)}$$

To solve the equations

$$\frac{\partial \tilde{C}(n_1,n_2)}{\partial n_1} = \frac{\partial \tilde{C}(n_1,n_2)}{\partial n_2} = 0$$

divide them by one another to obtain the relationship

$$\frac{n_2}{n_1} = \frac{\sqrt{\bar{M}c_1}\Lambda}{\sqrt{\tilde{c}_2}\sqrt{1+\theta_2}\beta_2}$$

which is also valid for Problem 9.2. Substituting into the remaining equations, we arrive after some algebra at

$$n_1 = \frac{\sqrt{1+\theta_2}\beta_2 \psi}{\bar{M}\sqrt{c_1}W}$$

where, as in Problem 9.2,

$$\psi = \sqrt{c_{21}}\tilde{\gamma}\bar{Y} + \sqrt{c_{22}}\overline{|R|} + \frac{1}{\bar{M}}\sqrt{\tilde{c}_2}\Lambda + \sqrt{c_1}\sqrt{1+\theta_2}\beta_2$$

Inserting those optimum values for n_1 and n_2 into the cost function, we can finally obtain the lower bound

$$\tilde{C}_{min} = \frac{\psi^2}{W}$$

The similarity between solutions of dual Problems 9.2 and 9.3 is striking. In particular, under linearized travel costs, the best achievable lower bounds for the anticipated variance and the expected costs are linked by a very nice relationship of

$$\mathbb{MAV}(\hat{Y}_c^*)\tilde{C}_{min} = \psi^2$$

Although this is no longer exactly true under non-linear travel costs, it can be useful as a guideline to assess what is feasible in the best possible case.

Problem 11.1

Recall that the radius R_i (in m) of the inclusion circle is given by $R_i = \frac{DBH}{2}$ (DBH_i in cm). Hence, the surface area (in ha) of the inclusion zone is

$$\lambda(F)\pi_i(t_k) = \lambda(K_i(t_k) \cap F) = \lambda(K_i(t_k)) = \frac{\pi DBH_i^2 10^{-4}}{4k}$$

while the basal area in square meter is $G_i = \frac{\pi DBH_i^2 10^{-4}}{4}$. Consequently, when the response variable is basal area (i.e. $Y_i(t_k) = G_i(t_k)$) all quantities of the form $\frac{1}{\lambda(F)} \frac{Y_i(t_k) I_i(x, t_k)}{\pi_i(t_k)}$ are 0 or 1 times k, depending on the remaining indicator variables. Furthermore, in this simple case, it is useful to rewrite the term

$$\frac{1}{\lambda(F)}(Y_i(t_2) - Y_i(t_1)) \frac{I_i(x, t_1)}{\pi_i(t_1)}$$

as

$$\left(\frac{Y_i(t_2)}{Y_i(t_1)} - 1\right) \frac{1}{\lambda(F)} \frac{Y_i(t_1) I_i(x, t_1)}{\pi_i(t_1)}$$

which (for basal area) is either 0 or $(\frac{DBH_i(t_2)}{DBH_i(t_1)})^2 - 1$ times k, depending on the remaining indicator variables. Likewise, note that

$$\frac{1}{\lambda(F)} Y_i(t_1) \frac{I_i(x, t_2)}{\pi_i(t_2)} = \frac{Y_i(t_1)}{Y_i(t_2)} \frac{1}{\lambda(F)} \frac{Y_i(t_2) I_i(x, t_2)}{\pi_i(t_2)}$$

Trees 1, 2 and 3 are sample survivors, trees 4 and 5 are sample ingrowth. Trees 6 and 7 are sample on-growth. Trees 8 and 9 are sample non-growth and Tree number 10 is a depletion tree.

After some arithmetical work, this leads to the following results (rounded to two decimals places):

1. $Y(x, t_1) = 16.00 \frac{m^2}{ha}$, $Y(x, t_2) = 36.00 \frac{m^2}{ha}$, $\Delta_{1,2}(x) = 20.00 \frac{m^2}{ha}$

2. $I_{1,2}^{(1)}(x) = 16.00 \frac{m^2}{ha}$, $\Delta_{1,2}^{(1)}(x) = 7.79 \frac{m^2}{ha}$, $D_{1,2}^{(1)}(x) = 4.00 \frac{m^2}{ha}$
   The triplet

$$[I_{1,2}^{(1)}(x), \ \Delta S_{1,2}^{(1)}(x), \ D_{1,2}^{(1)}(x)] = [16.00, 7.79, 4.00]$$

is non-additive because

$$I_{1,2}^{(1)}(x) + \Delta_{1,2}^{(1)} S(x) - D_{1,2}^{(1)}(x) = 16.00 + 7.79 - 4.00 = 19.79 \neq 20.00 = \Delta_{1,2}$$

Therefore, the error term is

$$e_{1,2}^{(1)}(x) = 20.00 - 19.79 = 0.21 \frac{m^2}{ha}$$

3. For the second triplet we obtain $I_{1,2}(\ ) = I_{1,2}(\ ) + I_{1,2}(\ ) = 16$ $16.21\frac{m^2}{ha}$ so that the triplet

$$[I_{1,2}^{(2)}(x), \ \Delta S_{1,2}^{(1)}(x), \ D_{1,2}^{(1)}(x)] = [16.21, 7.79, 4.00]$$

is additive.

4. Setting $\Delta S_{1,2}^{(2)}(x) = \Delta S_{1,2}^{(1)}(x) + e_{1,2}^{(1)}(x) = 7.79 + 0.21 = 8.00\frac{m^2}{ha}$ we obtain the additive triplet

$$[I_{1,2}^{(1)}(x), \ \Delta S_{1,2}^{(2)}(x), \ D_{1,2}^{(1)}(x)] = [16.00, 8.00, 4.00]$$

5. The proposal by Roesch et al. (1989) requires a bit more work. One obtains $e_{1,2}^{(2)}(x) = 0.47\frac{m^2}{ha}$ and $I_{1,2}^{(3)}(x) = I_{1,2}^{(1)}(x) + e_{1,2}^{(2)}(x) = 16.00 + 0.47 = 16.47\frac{m}{ha}$ Finally, according to Eq. 11.16, one arrives at $\Delta S_{1,2}^{(3)}(x) = 7.53\frac{m^2}{ha}$ and at the following additive triplet

$$[I_{1,2}^{(3)}(x), \ \Delta S_{1,2}^{(3)}(x), \ D_{1,2}^{(1)}(x)] = [16.47, 7.53, 4.00]$$

C.9 Chapter 12

Problem 12.1

The probability of intersection is $p = \frac{2L}{\pi D}$. The number r of intersections in n throws is binomial with an expected value np and a variance $np(1 - $ The maximum likelihood estimate of p is $\hat{p} = \frac{r}{n}$ with an estimated variance $\hat{\sigma}^2 = \frac{\hat{p}(1-\hat{p})}{n}$. The obvious point estimate of π is $\hat{\pi} = \frac{1}{\hat{p}}\frac{2L}{D}$. Using Eq. 6.57 we get the approximation

$$\hat{V}(\hat{\pi}) = \frac{1}{\hat{p}^4}\frac{4L^2}{D^2}\hat{\sigma}^2 = \frac{\pi^2}{n}\frac{1-\hat{p}}{\hat{p}}$$

Therefore, the theoretical variance is minimized when p is maximized. Under the condition $L \leq D$ this means that one must choose $L = D$. For confidence intervals it is better to calculate the (asymptotic) interval for p according to $[p_{lower}, p_{upper}] = [\hat{p} - 1.96\hat{\sigma}, \hat{p} + 1.96\hat{\sigma}]$ and to inverse this interval, i.e. $[\pi_{lower}, \pi_{upper}] = \frac{2L}{D}[p_{upper}^{-1}, p_{lower}^{-1}]$.

The interested reader can consult numerous Web sites for further information and simulation programs (keywords "Buffon's needle problem").

Problem 12.2

The point estimate is

$$\hat{\bar{Y}} = \frac{1}{\lambda(F)}\sum_{i \in s_1}\lambda_i\frac{\hat{\bar{Y}}_i}{\pi_i}$$

Using Eq. B.3 we get

$$\mathbb{E}_{1,2}(\hat{\bar{Y}}) = \mathbb{E}_1(\mathbb{E}_{2|1}(\hat{\bar{Y}})) = \mathbb{E}_1\left(\frac{1}{\lambda(F)}\sum_{i \in s_1}\lambda_i\frac{\bar{Y}_i}{\pi_i}\right)$$

$$\frac{1}{\lambda(F)} \sum_{i=1}^{N} \lambda_i \bar{Y}_i = \bar{Y}$$

so that we have unbiasedness. For the variance we use Eq. B.4

$$\mathbb{V}_{1,2}(\hat{\bar{Y}}) = \mathbb{E}_1(\mathbb{V}_{2|1}(\hat{\bar{Y}})) + \mathbb{V}_1(\mathbb{E}_{2|1}(\hat{\bar{Y}}))$$

which is equal to

$$\frac{1}{\lambda^2(F)} \mathbb{E}_1 \left(\sum_{i \in s1} \frac{\lambda_i^2 V_i}{\pi_i^2} \right) + \frac{1}{\lambda^2(F)} \mathbb{V}_1 \left(\sum_{i \in s_1} \frac{\lambda_i \bar{Y}_i}{\pi_i} \right)$$

Hence, we obtain

$$\mathbb{V}(\hat{\bar{Y}}) = \frac{1}{\lambda^2(F)} \sum_{i=1}^{N} \frac{\lambda_i^2 \bar{Y}_i^2 (1 - \pi_i)}{\pi_i}$$

$$+ \frac{1}{\lambda^2(F)} \sum_{i \neq j=1}^{N} \frac{\pi_{ij} - \pi_i \pi_j}{\pi_i \pi_j} \lambda_i \bar{Y}_i \lambda_j \bar{Y}_j$$

$$+ \frac{1}{\lambda^2(F)} \sum_{i=1}^{N} \frac{\lambda_i^2 V_i}{\pi_i}$$

The sample copy is

$$\hat{V} = \frac{1}{\lambda^2(F)} \sum_{i \in s_1} \frac{\lambda_i^2 \hat{\bar{Y}}_i^2 (1 - \pi_i)}{\pi_i^2}$$

$$+ \frac{1}{\lambda^2(F)} \sum_{i \neq j \in s_1} \frac{\pi_{ij} - \pi_i \pi_j}{\pi_i \pi_j \pi_{ij}} \lambda_i \hat{\bar{Y}}_i \lambda_j \hat{\bar{Y}}_j$$

$$+ \frac{1}{\lambda^2(F)} \sum_{i \in s_1} \frac{\lambda_i^2 \hat{V}_i}{\pi_i^2}$$

Furthermore, $\mathbb{E}_{2|1}(\hat{\bar{Y}}_i^2) = \bar{Y}_i^2 + V_i$ and, because the $\hat{\bar{Y}}_i$ are independent (which implies $\mathbb{E}_{2|1} \hat{\bar{Y}}_i \hat{\bar{Y}}_j = \bar{Y}_i \bar{Y}_j$) we obtain after some algebra

$$\mathbb{E}_{1,2} \hat{V} = \mathbb{V}_{1,2}(\hat{\bar{Y}}) + \frac{1}{\lambda^2(F)} \sum_{I=1}^{N} \frac{\lambda_i^2 V_i (1 - \pi_i)}{\pi_i} := \mathbb{V}_{1,2}(\hat{\bar{Y}}) + \delta$$

Hence, the sample-copy estimate of variance is biased. However, the bias can be estimated by

$$\hat{\delta} = \frac{1}{\lambda^2(F)} \sum_{i \in s_1} \frac{\lambda_i^2 \hat{V}_i (1 - \pi_i)}{\pi_i^2}$$

so that $\widehat{\mathbb{V}}(\hat{\bar{Y}}) = \hat{V} - \hat{\delta}$ is an unbiased estimate of the variance. Finally, one

$$\widehat{\mathbb{V}}(\hat{\bar{Y}}) \quad = \quad \frac{1}{\lambda^2(F)} \sum_{i \in s_1} \frac{\lambda_i^2 \hat{Y}_i^2 (1 - \pi_i)}{\pi_i^2}$$

$$+ \frac{1}{\lambda^2(F)} \sum_{i \neq j \in s_1} \frac{\pi_{ij} - \pi_i \pi_j}{\pi_i \pi_j \pi_{ij}} \lambda_i \hat{Y}_i \lambda_j \hat{Y}_j$$

$$+ \frac{1}{\lambda^2(F)} \sum_{i \in s_1} \frac{\lambda_i^2 \hat{V}_i}{\pi_i}$$

For simple random sampling, it is convenient to define $p_i = \frac{\lambda_i}{\lambda(F)}$ and $Z_i = p_i \hat{Y}_i$. Then, one has $\hat{\bar{Y}} = N \frac{1}{n} \sum_{i \in s_1} Z_i = N \bar{Z}_{s_1}$ and, by Eq. 2.16, the following result:

$$\widehat{\mathbb{V}}(\hat{\bar{Y}}) = N^2 \left(1 - \frac{n}{N}\right) \frac{1}{n(n-1)} \sum_{i \in s_1} (Z_i - \bar{Z}_{s_1})^2 + \frac{N}{n} \sum_{i \in s_1} p_i^2 \hat{V}_i$$

Bibliography

Affleck, D., Gregoire, T. and Valentine, H. (2005) Design-unbiased estimation in line intersect sampling using segmented transects. *Environmental and Ecological Statistics*, **12**, 139–154.

Ardilly, P. and Tillé, Y. (2006) *Sampling Methods: Exercises and Solutions* Springer.

Baddeley, A. and Jensen, E. V. (2005) *Stereology for Statisticians*. Chapman and Hall / CRC.

Barabesi, L. and Fattorini, L. (1997) Line intercept sampling with finite population: a stereological approach. *Metron LV*, 23–37.

Basu, D. (1991) *An essay on the logical foundations of survey sampling*. Foundations of statistical inference, 203-242, V.P. Godambe and D.A. Sprott, Eds., Toronto:Holt, Rinehart and Winston.

Beers, T. W. (1962) Components of forest growth. *Journal of Forestry*, 245–248.

Cassel, C., Särndal, C. and Wretman, J. (1977) *Foundations of Inference in Survey Sampling*. Wiley, New York.

Chaudhuri, A. and Stenger, H. (1992) *Survey Sampling Theory and Methods* Marcel Dekker, Inc., New York.

Chaudhuri, A. and Vos, J. (1988) *Unified Theory and Strategies of Survey Sampling*. North-Holland, Amsterdam.

Christensen, R. (1987) *Plane Answers to Complex Questions. The Theory of Linear Models*. Springer Verlag, New York.

Christensen, R. (1990) *Linear Models for Multivariate, Time Series and Spatial Data*. Springer Verlag, New York.

Cochran, W. (1977) *Sampling Techniques*. Wiley, New York.

Coleman, R. (1979) *An Introduction to Mathematical Stereology*. Institute of Mathematics, University of Aarhus.

Cressie, N. (1991) *Statistics for Spatial Data*. John Wiley and sons, Inc., New York.

de Vries, P. (1986) *Sampling Theory for Forest Inventory*. Springer Verlag, Berlin.

Deusen, P. C. V. (1989) Multiple-occasion partial replacement sampling for growth components. *Forest Science*, **35**, 388–400.

Deusen, P. C. V., Dell, T. and Thomas, C. (1986) Volume growth estimation from permanent horizontal point. *Forest Science*, **32**, 415–422.

Eriksson, M. (1995) Compatible and time-additive change components estimators for horizontal-point-sampled data. *Forest Science*, **41**, 796–822.

Fienberg, S. (1980) *The Analysis of Cross-classified Categorical Data*. The MIT Press, Cambridge, Massachusetts.

Fuller, W. A. (1995) Estimation in the presence of measurement error. *International Statistical Review*, **63**, 121–147.

Godambe, V. (1955) A unified theory of sampling from finite population. *J.R. Statist. Soc. B*, **17**, 269–278.

Godambe, V. and Joshi, V. (1965) Admissibility and bayes estimation in sampling finite populations i. *Annals Mathematical Statistics*, **36**, 1707–1722.

Gourieroux, C. (1981) *Théorie des Sondages*. Economica, Paris.

Gregoire, T. and Dyer, M. (1989) Model fitting under patterned heterogeneity of variance. *Forest Science*, **35**, 105–125.

Gregoire, T. and Valentine, H. (2007) *Sampling Strategies for Natural Resources and the Environment*. Chapman and Hall, New York.

Gregoire, T. G. (1993) Estimation of forest growth from successive surveys. *Forest Ecology and Management*, **56**, 267–278.

Gregoire, T. G. (1995) Variance derivations and related aspects of growth component estimators with successive variable-radius-plot samples. *Forest Ecology and Management*, **71**, 211–216.

Grosenbaugh, L. R. (1952) Plotless timber estimates. new, fast, easy. *Journal of Forestry*, **50**, 32–37.

Grosenbaugh, L. R. (1971) Stx 1-11-71 for dendrometry of multistage 3p samples. Technical report, USDA For. Ser. Publ. FS 277.

Holgate, P. (1967) The angle count method. *Biometrika*, **54**, 615–623.

Huber, P. (1967) The behaviour of maximum likelihood estimates under non-standard conditions. *Proceedings of the 5th Berkeley Symposium on Mathematical Statistic*, **1**, 221–233.

ISATIS (1994) *Manual*. Geovariances, 77210, Avon, France.

Johnson, E. (2000) *Forest Sampling Desk Reference*. CRC Press, Boca Raton, Florida.

Kaiser, L. (1983) Unbiased estimation in line intercept sampling. *Biometrics* **39**, 965–976.

Kangas, A. and Maltamo, M. (Eds, 2006) *Forest Inventory Methodology and Applications*. Springer.

Kendall, M., Ord, J. and Stuart, A. (1983) *The Advanced Theory of Statistics, Volume 3, 4th edition*. Charles Griffin and Co Ltd, London.

Kleinn, C. and Vilcko, F. (2006) Design-unbiased estimation for point-to-tree distance sampling. *Canadian Journal of Forest Research*, **36**, 1407–1414.

Köhl, M., Magnussen, S. and Marchetti, M. (2006) *Sampling Methods, Remote Sensing and GIS Multiresource Forest Inventory*. Springer-Verlag.

Lanz, A. (2000) Optimal sample design for extensive forest inventories. Ph.D. thesis, ETH Zurich, Chair of Forest Inventory and Planning, http://e-collection.ethb.ethz.ch/.

Lehmann, E. L. (1999) *Elements of Large Sample Theory*. Springer, New York.

Mandallaz, D. (1991) A unified approach to sampling theory for forest inventory based on infinite population models. Ph.D. thesis, ETH Zurich, Chair of Forest Inventory and Planning, http://e-collection.ethb.ethz.ch/.

Mandallaz, D. (1993) Geostatistical methods for double sampling schemes: Applications to combined forest inventory. Technical report, ETH Zurich, Chair of Forest Inventory and Planning, Habilitation thesis, http://e-collection.ethz.ethz.ch/.

Mandallaz, D. (1997) The anticipated variance: a tool for the optimization of forest inventories. Technical report, ETH Zurich, Chair of Forest Inventory and Planning, http://e-collection.ethb.ethz.ch/.

Mandallaz, D. (2000) Estimation of the spatial covariance in universal kriging: Application to forest inventory. *Environmental and Ecological Statistics* 263–284.

Mandallaz, D. (2001) Optimal sampling schemes based on the anticipated variance with lack of fit. Technical report, ETH Zurich, Chair of Forest Inventory and Planning, http://e-collection.ethb.ethz.ch/.

Mandallaz, D. (2002) Optimal sampling schemes based on the anticipated variance with lack of fit. *Canadian Journal of Forest Research*, **32**, 2236–2243.

Mandallaz, D. and Lanz, A. (2001) Further results for optimal sampling schemes based on the anticipated variance. *Canadian Journal of Forest Research*, **31**, 1845–1853.

Mandallaz, D., Schlaepfer, R. and Arnould, J. (1986) Dépérissement des forˆ essai d'analyse des dépendances. *Annales des Sciences Forestières*, **43**, 441–458.

Mandallaz, D. and Ye, R. (1999) Optimal two-phase two-stage sampling schemes based on the anticipated variance. *Canadian Journal of Forest Research*, **29**, 1691–1708.

Martin, G. (1982) A method for estimating ingrowth on permanent horizontal sample points. *Forest Science*, **28**, 110–114.

Masuyama, M. (1953) A rapid method of estimating basal area in timber survey-an application of integral geometry to aerial sampling problems. *Sankhya*, **12**, 291–302.

ern, B. (1986)
tics, Berlin.

Matheron, G. (1965) *Les Variable Régionalisées et leur Estimation*. Masson, Paris.

Newton, C., Cunia, T. and Bickford, C. (1974) Multivariate estimators for sampling with partial replacement on two occasions. *Forest Science*, **20** 106–116.

Pardé, J. and Bouchon, J. (1988) *Dendrométrie*. ENGREF, Nancy.

Ramakrishnan, M. (1975a) Choice of an optimum sampling strategy. *The Annals of Statistics*, **3**, 669–679.

Ramakrishnan, M. (1975b) A generalization of the Yates-Grundy variance estimator. *Sankhya, Serie C*, 204–206.

Rao, C. R. (1967) *Linear Statistical Inference and Its Applications*. Wiley, New York.

Rao, J. and Scott, A. (1984) On chi-squared tests for multiway contingency tables with cell proportion estimated from survey data. *The Annals of Statistics*, **12**, 46–60.

Ripley, B. D. (1981) *Spatial Statistics*. Wiley.

Roesch, F., Green, E. J. and Scott, C. (1989) New compatible estimators for survivor growth and ingrowth from remeasured horizontal point samples. *Forest Science*, **35**, 281–293.

Rondeux, J. (1993) *La Mesure des Arbres et des Peuplements Forestiers*. Les Presses Agronomiques de Gembloux, Gembloux.

Royall, R. (1970) On finite population sampling theory under certain linear regression models. *Biometrika*, **57**, 377–387.

Santaló, L. A. (1976) *Integral Geometry and Geometric Probability*. Addison-Wesley.

Särndal, C., Swenson, B. and Wretman, J. (2003) *Model Assisted Survey Sampling*. Springer Series in Statistics, New York.

Schabenberger, O. and Gotway, C. (2005) *Statistical Methods for Spatial Data Analysis*. Chapman and Hall.

Schreuder, H., Gregoire, T. and Wood, G. (1993) *Sampling Methods for Multiresource Forest Inventory*. Wiley, New York.

Scott, C. T. and Köhl, M. (1994) Sampling with partial replacement and stratification. *Forest Science*, **40**, 30–46.

Seber, G. (1986) A review of estimating animal abundance. *Biometrics*, **42** 267–292.

Tillé, Y. (2001) *Théorie des Sondages: Échantillonnage et Estimation en Populations Finies*. Dunod, Paris.

Tillé, Y. (2006) *Sampling Algorithms*. Springer Verlag, New York.

Wackernagel, H. (2003)

Ware, K. D. and Cunia, T. (1962) *Continuous Forest Inventory with Partial Replacement of Samples.* Society of American Foresters, Forest Science Monograph 3, Washington D.C.

Williams, M. S. (1998) Distinguishing between change and growth in forest surveys. *Canadian Jornal Forest Research*, **28**, 1099–1106.

Zöhrer, F. (1980) *Forstinventur*. Verlag Paul Parey, Hamburg.

Index